Environmental Chemistry for a Sustainable World

Other Publications by the Editors

Books
Scientific Writing for Impact Factor Journals
https://www.novapublishers.com/catalog/product_info.php?products_id=42211

Environmental Chemistry
http://www.springer.com/978-3-540-22860-8

Organic Contaminants in Riverine and Groundwater Systems
http://www.springer.com/978-3-540-31169-0

Sustainable Agriculture
Volume 1: http://www.springer.com/978-90-481-2665-1
Volume 2: http://www.springer.com/978-94-007-0393-3

Book series
Environmental Chemistry for a Sustainable World
http://www.springer.com/series/11480

Sustainable Agriculture Reviews
http://www.springer.com/series/8380

Journal
Environmental Chemistry Letters
http://www.springer.com/10311

More information about this series at http://www.springer.com/series/11480

Sophie Fourmentin • Grégorio Crini
Eric Lichtfouse
Editors

Cyclodextrin Fundamentals, Reactivity and Analysis

Springer

Editors
Sophie Fourmentin
Unité de Chimie Environnementale
et Interactions sur le Vivant (UCEIV,
EA 4492)
Dunkerque, France

Grégorio Crini
UFR Sciences et Techniques, Laboratoire
Chrono-environnement
Université de Bourgogne Franche-Comté,
Besançon, France

Eric Lichtfouse
CEREGE INRA
Aix en Provence, France

ISSN 2213-7114 ISSN 2213-7122 (electronic)
Environmental Chemistry for a Sustainable World
ISBN 978-3-030-09418-8 ISBN 978-3-319-76159-6 (eBook)
https://doi.org/10.1007/978-3-319-76159-6

Printed on acid-free paper

This Springer imprint is published by the registered company Springer International Publishing AG part of Springer Nature.
The registered company address is: Gewerbestrasse 11, 6330 Cham, Switzerland

Preface

We know a lot about Cyclodextrins, but this is just the tip of the iceberg

(József Szejtli, 1980)

The 2016 Nobel Prize in Chemistry was awarded to Jean-Pierre Sauvage, Sir James Fraser Stoddart and Bernard Lucas Feringa for the design and synthesis of molecular machines. This prize revealed a major recognition for supramolecular chemistry involving host-guest relationships such as cyclodextrin complexes. This recognition will surely contribute to a renewed interest for cyclodextrins. Cyclodextrins have been discovered more than 100 years ago by the pioneering work of the pharmacist and chemist Antoine Villiers in France. Cyclodextrins are obtained by enzymatic degradation of starch and still fascinate researchers because they are remarkable macrocyclic molecules with major theoretical and practical impacts in chemistry, biology, biochemistry, health science and agriculture. Cyclodextrins have broken barriers between different disciplines and, nowadays, various scientists work together to find new concepts and applications.

The most characteristic feature of cyclodextrins is their ability to form inclusion complexes with various molecules through host-guest interactions, which are at the origin of many applications (Fig. 1). Actually, cyclodextrins and their chemically modified derivatives have a wide variety of practical applications in almost all sectors of industry including pharmacy, medicine, foods, cosmetics, toiletries, catalysis, chromatography, biotechnology, textile industry, supramolecular chemistry and nanotechnology. New contributions and industrial developments are further expected in biomedicine, e.g. magnetic resonance imaging and chemotherapy, fermentation processing and enzymology, sensor applications, cosmeto-textiles and agrochemistry.

This book is the first volume of two volumes on cyclodextrins published in the series *Environmental Chemistry for a Sustainable World*. After a brief description of the cyclodextrin fundamentals, the first chapter by Grégorio Crini et al. provides an overview of cyclodextrin research during the last 5 years. The second chapter by Michal Řezanka discusses the synthesis of novel cyclodextrin systems by selective

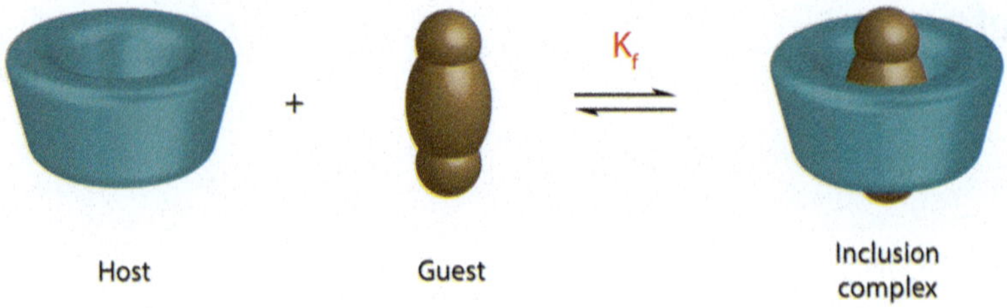

Fig. 1 Cyclodextrin hosts are able to trap a wide variety of guest chemicals. This complexing ability has led to many applications in various sectors

modifications. Then Eric Monflier et al. describe the synthesis of nanostructured porous materials based on cyclodextrins, and applications in heterogeneous catalysis and photocatalysis. The use of thermal analyses for assessing cyclodextrin inclusion complexes is reviewed in Chap. 4 by Daniel I. Hădărugă et al. Experimental methods for measuring binding constants of cyclodextrin inclusion compounds are presented by David Landy. The second volume reviews cyclodextrin applications in medicine, food, environment and liquid crystals.

The editors extend their thanks to all authors. The books would not have been possible without the assistance of several friends and colleagues deserving acknowledgement. They have helped by choosing contributors, reviewing chapters and in many other ways. Finally, we would like to thank the staff at Springer Nature for highly professional editing of the publication.

Dunkerque, France — Sophie Fourmentin
Besançon, France — Grégorio Crini
Aix-en-Provence, France — Eric Lichtfouse

Contents

Contributors

Geza N. Bandur Department of Applied Chemistry, Organic and Natural Compounds Engineering, Polytechnic University of Timişoara, Timişoara, Romania

Rudina Bleta Univ. Artois, CNRS, Centrale Lille, ENSCL, Univ. Lille, UMR 8181, Unité de Catalyse et Chimie du Solide (UCCS), Lens, France

Éva Fenyvesi CycloLab Cyclodextrin Research & Development Ltd., Budapest, Hungary

Marc Fourmentin Laboratoire de Physico-Chimie de l'Atmosphère (LPCA, EA 4493), Université du Littoral Côte d'Opale, Dunkerque, France

Nicoleta G. Hădărugă Department of Food Science, Banat's University of Agricultural Sciences and Veterinary Medicine "King Michael I of Romania" from Timişoara, Timişoara, Romania

Daniel I. Hădărugă Department of Applied Chemistry, Organic and Natural Compounds Engineering, Polytechnic University of Timişoara, Timişoara, Romania

David Landy Unité de Chimie Environnementale et Interactions sur le Vivant (UCEIV, EA 4492), Université du Littoral Côte d'Opale, Dunkerque, France

Eric Monflier Univ. Artois, CNRS, Centrale Lille, ENSCL, Univ. Lille, UMR 8181, Unité de Catalyse et Chimie du Solide (UCCS), Lens, France

Nadia Morin-Crini UFR Sciences et Techniques, Laboratoire Chrono-environnement, Université de Bourgogne Franche-Comté, Besançon, France

Anne Ponchel Univ. Artois, CNRS, Centrale Lille, ENSCL, Univ. Lille, UMR 8181, Unité de Catalyse et Chimie du Solide (UCCS), Lens, France

Michal Řezanka Institute for Nanomaterials, Advanced Technologies and Innovation, Technical University of Liberec, Liberec, Czech Republic

Giangiacomo Torri Istituto di Chimica e Biochimica G. Ronzoni, Milano, Italy

About the Editors

Sophie Fourmentin was born in Bar/S/Aube (France) in 1968. She received her Ph.D. degree in organic chemistry from the University of Lille in 1994. In 1996, she became Assistant Professor at the University of Littoral-Côte d'Opale in Dunkerque. She developed a new subject area based on "applications of cyclodextrins in the remediation of organic pollutants". In 2006, she completed her HDR (authorization to supervise research activities) entitled "Complexes Cyclodextrine/Polluants Organiques: Caractérisation, Application à la remédiation des Composés Organiques Volatils". In 2008 she was promoted Professor, and she managed the Laboratoire de Synthèse Organique et Environnement until 2010. Since January 2010, she supervises the supramolecular team in the UCEIV laboratory. Her research works are at the interface between host–guest chemistry and environmental chemistry. The aim is to take benefit of the properties of cyclodextrins to enhance conventional remediation processes (volatil organic compounds (VOC) absorption, Fenton oxidation). Her experience in the characterization of host–guest complexes in the case of complex VOC has been extended to the study of the encapsulation of flavours and essential oils, for applications in food, fragrance or pharmaceutical industry. She is the author of 91 articles (h-index: 23), a patent and coordinated two books on cyclodextrins.

Grégorio Crini, 51, is researcher at University of Bourgogne Franche-Comté (UMR Chrono-environnement), Besançon. His current interests focus on the design of novel polymer networks and the environmental aspects of polysaccharide chemistry. He published over 180 papers in international journals and books, and he is a highly cited researcher. The total citation of his publications is over 7500 according to ISI Web of Science, h-index: 32.

Eric Lichtfouse, born in 1960, is a biogeochemist at the University of Aix-Marseille, CEREGE, Aix-en-Provence, France. He got a Ph.D. in organic geochemistry at the University of Strasbourg in 1989 for the discovery of new fossil steroids in sediments and petroleum. He has invented the ^{13}C-dating method allowing to measure the dynamics of soil organic molecules, thus opening the field of molecular-level investigations of soil carbon sequestration. He is Chief Editor of the journal *Environmental Chemistry Letters*, of 3.6 impact factor. His former journal *Agronomy for Sustainable Development* has reached the second place in the category Agronomy, and has been awarded by the Essential Science Indicators. He is lecturing scientific writing and communication in universities worldwide. He has published the book *Scientific Writing for Impact Factor Journals*. This textbook describes in particular the micro-article, a new tool to identify the novelty of experimental results. Further details are available on ResearchGate, Slideshare, LinkedIn, ResearcherID, Orcid and Google Scholar citations.

Chapter 1
Fundamentals and Applications of Cyclodextrins

Grégorio Crini, Sophie Fourmentin, Éva Fenyvesi, Giangiacomo Torri, Marc Fourmentin, and Nadia Morin-Crini

Contents

This chapter is dedicated to the memory of Professor Benito Casu (*Istituto di Chimica e Biochimica G. Ronzoni*, Milan, Italy).

G. Crini (✉) · N. Morin-Crini
UFR Sciences et Techniques, Laboratoire Chrono-environnement,
Université de Bourgogne Franche-Comté, Besançon, France
e-mail: gregorio.crini@univ-fcomte.fr; nadia.crini@univ-fcomte.fr

S. Fourmentin et al. (eds.), *Cyclodextrin Fundamentals, Reactivity and Analysis*,
Environmental Chemistry for a Sustainable World,
https://doi.org/10.1007/978-3-319-76159-6_1

Abstract Cyclodextrins are natural oligosaccharides obtained from starch. They were discovered in 1891 by Villiers, and attracted major scientific and industrial interests from the late 1970s. Actually, cyclodextrins are among the most remarkable macrocyclic molecules with major theoretical and practical interest for chemistry and biology. Cyclodextrins belong to the family of cage molecules due to their structure, which is composed of a hydrophobic cavity that can encapsulate other molecules. Indeed, the most characteristic feature of cyclodextrins is their ability to form inclusion complexes with various molecules through host-guest interactions. Cyclodextrins and their derivatives have a wide variety of practical applications including pharmacy, medicine, foods, cosmetics, toiletries, catalysis, chromatography, biotechnology, nanotechnology, and textile production. Cyclodextrins are also the object of numerous fundamental studies. Between 2011 and 2015, 18,430 cyclodextrin-related publications have been published. In this chapter, after a brief description of cyclodextrin basics, we highlight selected works on cyclodextrins published over the last 5 years by various research groups.

Abbreviations

AHLs	N-acyl-L-homoserine lactones
AIP	Auto-inducing peptides
AIT	Allyl-isothiocyanate
BBB	Blood-brain barrier
CC	Cholesterol crystals
CD	Cyclodextrin
CD-ICs	Cyclodextrin inclusion complexes
CNS	Central nervous system
EFSA	European Food Safety Administration
ES	Electrospinning
GC	Gas chromatography

S. Fourmentin
Unité de Chimie Environnementale et Interactions sur le Vivant (UCEIV, EA 4492), Dunkerque, France
e-mail: sophie.fourmentin@univ-littoral.fr

É. Fenyvesi
CycloLab Cyclodextrin Research & Development Ltd., Budapest, Hungary
e-mail: fenyvesi.e@cyclolab.hu

G. Torri
Istituto di Chimica e Biochimica G. Ronzoni, Milano, Italy
e-mail: torri@ronzoni.it

M. Fourmentin
Laboratoire de Physico-Chimie de l'Atmosphère (LPCA, EA 4493), Université du Littoral Côte d'Opale, Dunkerque, France
e-mail: marc.fourmentin@univ-littoral.fr

HP-β-CD	Hydroxypropyl-β-cyclodextrin
IT	Imprinting technique
LC	Liquid chromatography
MIP	Molecularly imprinted polymers
MOF	Metal organic frameworks
NF	Nanofibers
NMR	Nuclear magnetic resonance
NPC	Niemann-Pick type C disease
PAH	Polycyclic aromatic hydrocarbons
PAN	Polyacrylonitrile
PCB	Polychlorinated biphenyls
PDT	Photodynamic therapy
QS	Quorum sensing
RAMEB	Randomly methylated-β-CD
SFC	Supercritical fluid chromatography
VOC	Volatile organic compounds

1.1 Cyclodextrins – An Introduction

1.1.1 Cyclodextrin Description

Cyclodextrins (CDs) are cyclic oligomers obtained from the enzymatic degradation of starch, one of the most essential polysaccharide in the nature. Cyclodextrins make up a family of cyclic oligosaccharides. They are composed of six or more D-glucopyranoside units linked in α(1–4), like in amylose, a component of starch. They are also called cycloamyloses. Typical native cyclodextrins contain six, seven or eight glucose units and are denoted α-, β- and γ-CDs, respectively. They are produced in a highly pure form at an industrial scale.

Cyclodextrins are hollow, truncated-cone-shaped molecules made up of several glucose units linked together covalently by oxygen atoms and held in shape via hydrogen bonding between the secondary hydroxyl groups on adjacent units at the wider rim of the cavity. Indeed, cyclodextrins are ring molecules and they are toroidal or cone-shaped. They are not cylindrical due to the lack of free rotation at the level of bonds between glucopyranose units. The primary hydroxyl groups are located on the narrow side and the secondary groups on the wider side. The molecular structures and characteristics of the three analogues are schematized in Fig. 1.1 and their main characteristics in Table 1.1. Other cyclodextrins, with less than six and more than eight units, denoted large CDs, may also be isolated at the laboratory scale. Indeed, depending on the particular transferase used during enzymatic degradation of starch, the nature of starch and the reaction conditions, different types of cyclodextrins result.

β-Cyclodextrin (β-CD) is the most studied and most frequently used, based on its cheapness, availability, and complex-forming capacities towards a large range of

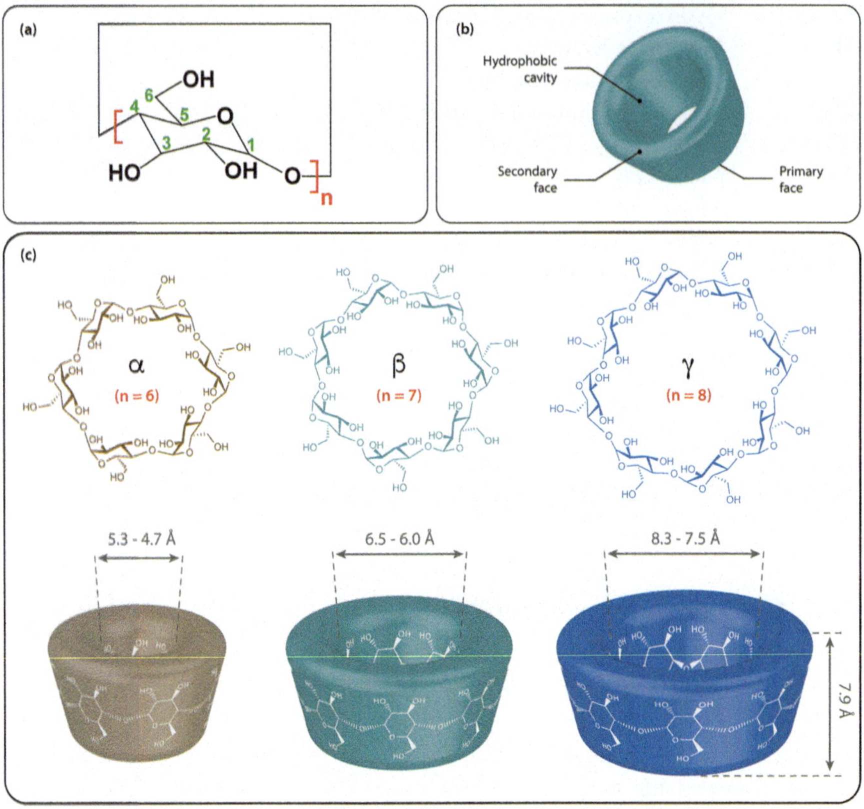

Fig. 1.1 Schematic representations of (a) the general chemical structure and (b) the tridimensional structure of cyclodextrins, and (c) chemical structure and dimensions for α-, β- and γ-cyclodextrin (n = 6, 7 and 8, respectively)

Table 1.1 The three native cyclodextrins and their main characteristics

Cyclodextrin	α	β	γ
Chemical Abstracts Service Registry Number	100016-20-3	7585-39-9	17465-86-0
Glucopyranose units	6	7	8
Formulae	$C_{36}H_{60}O_{30}$	$C_{42}H_{70}O_{35}$	$C_{48}H_{80}O_{40}$
Molecular weight (g/mol)	972.9	1135.0	1297.1
Central cavity diameter: external/internal (Å)	5.3/4.7	6.5/6.0	8.3/7.5
Height of torus (Å)	7.9	7.9	7.9
Approximate volume of cavity (Å)	174	262	427
Water solubility at 25°C (g/L)	145	18.5	232
Number of water molecules within cavity	6-8	11-12	13-17
pKa	12.3	12.2	12.1

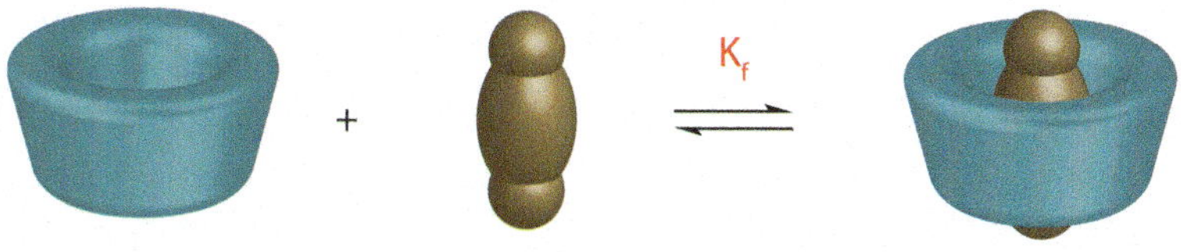

Fig. 1.2 Schematic representation of the inclusion phenomena between a cyclodextrin molecule (the host) and a solute (the guest) to form solute-cyclodextrin complexes

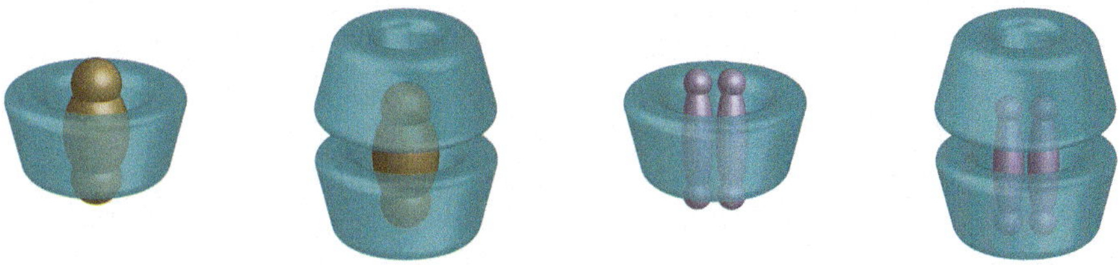

Fig. 1.3 Schematic drawing of main CD complexes stoichiometries

substances. Indeed, as others cyclodextrins, the most characteristic feature of the β-CD molecule is its ability to form inclusion compounds with various substances through host-guest interactions (Fig. 1.2). Its central cavity which is composed of seven glucose units is hydrophobic when the external part is hydrophilic because the presence of 21 hydroxyl groups. The core of its structure can trap or encapsulate other substances. This remarkable encapsulation properties can modify and/or improve the physical, chemical, and/or biological characteristics of the guest molecule (Szejtli 1988; Morin-Crini et al. 2015).

Formation of an inclusion complex is the result of an association/dissociation equilibrium between the free guest and the free host and the complex. This is governed by a constant K_f, denoted formation or stability constant. The higher the K_f value is, the more stable the inclusion is, and the less dissociation that occurs. The driving forces for complex formation include release of high-energy water, solvent effects, hydrophobic interactions, hydrogen bonds, van der Waals interactions, and dipole-dipole interactions. As a general rule stated by Cramer in the 1950s, the complex is strong when there is size complementarity between the guest and the cyclodextrin cavity (Crini 2014). Depending on the respective size of the guest and host molecules, one guest molecule can interact with one or two (or more) cyclodextrin (i.e. host:guest complexes 1:1 and 2:1), or two (or more) guest molecules can interact with one cyclodextrin or two (or more) (i.e. host:guest complexes 1:2 and 2:2) as showed in Fig. 1.3. Because of their conformation, some guests can be included in one or two cyclodextrin molecules, and depending on the CD size, it is a different part of the guest molecule that can be included. Cyclodextrins themselves can self-associate to form aggregates or micelles in aqueous solutions. So, in solution, multiple inclusion equilibria can coexist. While stoichiometry is often characterized by means of Job plots, affinities are generally determined by titration, competition or solubility procedures (Connors 1997; Landy et al. 2000; Decock et al. 2006; Brewster and Loftsson 2007). Many techniques including NMR, calorimetry, fluorescence, circular dichroism and headspace gas chromatography have been used for complexation studies (Decock et al. 2008; Morin-Crini et al. 2015).

Various methods such as co-precipitation and derivatives methods such as spray drying and freeze-drying, kneading, slurry complexation, paste complexation, damp mixing, heating method (sealed-heating), extrusion method, dry mixing, microwave treatment, and supercritical carbon dioxide can be used in the formation of CD complexes. The books published by Szejtli (1982a, 1988) can be consulted on the preparation of inclusion complexes. Evidence for a guest inclusion into the cyclodextrin cavity may be proved by various analytical techniques including nuclear magnetic resonance (NMR) spectroscopy, UV-visible absorption spectroscopy, optical rotatory dispersion and circular dichroism, fluorescence, infrared (FT-IR) spectroscopy, thermo-analysis, mass spectrometry, and powder X-ray diffractometry (Morin-Crini et al. 2015).

The structure of cyclodextrins is stabilized by the formation of hydrogen bonds between C-2 and C-3 hydroxyl groups of adjacent glucose units. This phenomenon widely affects, in addition to molecular dimensions, the water solubility of cyclodextrins. The formation of a complete ring of intramolecular hydrogen bonds in β-CD counteracts its hydration and reduces its solubility as compared to other native cyclodextrins (Table 1.1) (Szejtli 1998). Disruption of intermolecular hydrogen bonds (i.e. substitution of highly reactive hydroxyl groups with either polar or apolar moieties) generally produces cyclodextrin derivatives with anomalous increased solubility. Synthesis of derivatives aims also to construct CD polymers or CDs with enhanced complexation ability and selectivity. Over than 15,000 cyclodextrin derivatives have been obtained by alkylation, hydroxyalkylation, sulfatation, phosphatation, acetylation, amination, esterification and etherification reactions (Khan et al. 1998; Szejtli 2004a; b) but only some particular one such as hydroxypropylated, methylated and sulfobutylated derivatives are marketed and widely used.

1.1.2 Fields of Application

As already stated, although they have been known for 126 years, cyclodextrins only really took off in the 1980s with the first applications in the chromatography, pharmaceutical and food industries. Cyclodextrins were used as host molecules for molecular recognition, as solubilizing agents for lipophilic drugs, excipients in formulation development, as delivery systems and complexing agents in the food and drug industries, or for masking of undesired odor and taste. Actually, applications are found in practically all sectors of industry but especially, in cosmetics, hygiene and personal care, textiles including textile finishing, functional textiles and cosmetotextiles, medicine and biomedicine, agrochemistry or pesticides, biotechnology, and catalysis (Morin-Crini et al. 2015). Cyclodextrins and their derivatives have also other practical applications in microencapsulation including aroma and fragrances, oils, analytical chemistry, organic chemistry (synthesis), macromolecular chemistry (materials), click chemistry, supramolecular chemistry, membranes, enzyme technology, environment such as remediation, decontamination and extraction, and nanotechnology (nanoparticles/nanosponges for various domains). Research on cyclodextrins is also very active in fields such as the formulation of detergents, glues and adhesives, the sector of plastics (packaging) and the industry of fibers and paper. However, the main market remains the pharmaceutical, food and cosmetic industries. It is interesting to note that the vast majority of these chemical and biological applications are based on the ability to form inclusion complexes (Szejtli 1982a, 1988; Szejtli 2004a; b; Duchêne 1987, 1991; Dodziuk 2006; Morin-Crini et al. 2015).

1.1.3 Literature Data

The literature on cyclodextrins is vast and spread across different disciplines: Chemistry, Biochemistry, Health Science, Agriculture, etc. Interesting books and book chapters on the different aspects of cyclodextrins such as the preparation, description, characterization, properties, chemistry, and derivatives for these disciplines can be consulted (Table 1.2). In the last two decades, a large number of generalist reviews has been also published in the literature on practically all these aspects of cyclodextrins (Table 1.3), so many that it would not be feasible to cite them all. Selected books and reviews are given in Tables 1.2 and 1.3, respectively, which contain the original sources of the data displayed, and also selected recent works discussed in this chapter.

Figure 1.4 demonstrates how the number of publications increased using cumulative numbers of 5-year periods. In 2015, with an average of 4.5 papers and 2.5 patent applications daily, the literature data shows that cyclodextrin research and development is still in the focus (Source: Cyclodextrin News, Cyclolab). Recently, Deorsola et al. (2014) have performed a thorough patent search on cyclodextrin-related patents using free databases. The period they evaluated was 1981-2011. The survey showed the leading role of Asia on the number patent applications filed in patent offices of different countries, being Japan and China on the first and second positions, followed by the US and the European Patent offices. The top 10 inventors among the total of 20,198 inventors have all Chinese names. The three leading companies are Procter & Gamble Co., Kao Corporation and Schering AG.

Table 1.2 Selected examples of books on cyclodextrins

Year	Title	Reference
1978	Cyclodextrin chemistry	Bender and Komiyama 1978
1982	Cyclodextrin and their inclusion complexes	Szejtli 1982a
1984	Inclusion compounds	Atwood et al. 1984
1987	Cyclodextrins and their industrial uses	Duchêne 1987
1988	Cyclodextrin technology	Szejtli 1988
1991	New trends in cyclodextrins and derivatives	Duchêne 1991
1994	Cyclodextrin in pharmacy	Frömming and Szejtli 1994
1996	Comprehensive supramolecular chemistry. Volume 3: Cyclodextrins	Szejtli and Osa 1996
2000	Principles and methods in supramolecular chemistry	Schneider and Yatsimirsky 2000
2002	Introduction to supramolecular chemistry	Dodziuk 2002
2004	Encyclopedia of supramolecular chemistry	Atwood and Steed 2004
2006	Cyclodextrins and their complexes - Chemistry, analytical methods and applications	Dodziuk 2006
2010	Use of cyclodextrin polymers in separation of organic species	Kozlowski and Sliwa 2010
2011	Cyclodextrins in pharmaceutics, cosmetics and biomedicine. current and future industrial applications	Bilensoy 2011
2012	Applications of supramolecular chemistry	Schneider 2012
2015	*Cyclodextrines - Histoire, propriétés, chimie et applications* (in French)	Morin-Crini et al. 2015
2016	Encapsulations	Grumezescu 2016

1.1.4 The Cyclodextrin Scientific Community

In 1981, Professor Szejtli organized the First International Cyclodextrin Symposium in Budapest, Hungary. This symposium was a great success, with participants coming from all over the world (more than 180 participants from 17 countries) while Professor Szejtli "*expected 25-30 participants outside Hungary*" (Szente and Szejtli 1996). The submitted manuscripts filled a 544-page volume of proceedings published by Akademiai Kiado. The second cyclodextrin symposium was organized by Professor Nagai in 1984, in Tokyo. Since 1984 and Szejtli's initiative, a broad community of researchers has met every two years to exchange and share their works on cyclodextrins. These well-established symposiums provide opportunities for scientists who work in several aspects of cyclodextrin research to meet and discuss recent advances in all cyclodextrin fields. The 19th International Cyclodextrin Symposium organized by Professor Keiko Takahashi and Professor Takashi Hayashita will be held on April 2018 in Tokyo, Japan (the next will be organized in Sicily, Italy in 2020).

Table 1.3 Selected comprehensive reviews and book chapters on cyclodextrins published in the last four decades, including selected recent works discussed in this chapter

General topic(s)	References
Aggregation - Self-assembly	Gonzalez-Gaitano et al. 2003; Loftsson et al. 2004; He et al. 2008; Fülöp et al. 2012; Macaev et al. 2013; Ryzhakov et al. 2016; Saokham and Loftsson 2017
Agrochemistry	Szente and Szejtli 1996; Morillo 2006; Luca and Grigoriu 2007; Venturini et al. 2008; Ho et al. 2014; Garrido et al. 2014; Fernandes et al. 2014; Campos et al. 2015; Yusoff et al. 2016; Letort et al. 2016
Analytical chemistry	Li and Purdy 1992; Szente 1996a; Armspach et al. 1999; Schneiderman and Stalcup 2000; Mosinger et al. 2001; Martin Del Valle 2004; Dodziuk 2006; Fakayode et al. 2007; Venturini et al. 2008; Szente et al. 2016; Prochowicz et al. 2016
Biotechnology	Szejtli 1990, 1991; Bar 1996; Singh et al. 2002; Li et al. 2011
Catalysis	Komiyama and Monflier 2006; Hapiot et al. 2006, 2011, 2014; Bhosale and Bhosale 2007; Macaev and Boldescu 2015; Hong et al. 2015
Characterization	Bender and Komiyama 1978; Szejtli 1982a; Robyt 1998; Cairo 2011; Mura 2015; Szente et al. 2016
Chemistry - Polymer science	Szejtli 1982a, 1988; Kano 1993; Luca and Grigoriu 2007; Wenz 2009; Morin-Crini et al. 2015; Rezanka 2016; Lay et al. 2016
Chromatography - LC - GC - Electrophoresis - SFC	Smolková-Keulemansová 1982; Ward and Armstrong 1988; Han and Armstrong 1989; Sybilska and Zukowski 1989; Menges and Armstrong 1991; Hilton and Armstrong 1991; Fanali 1993; Fanali et al. 1994; Schneiderman and Stalcup 2000; He and Beesley 2005; Vetter and Bester 2006; Hongdeng et al. 2011; Xiao et al. 2012; West 2014; Scriba 2016
Click chemistry	Dodziuk 2006; Bilensoy 2011; Venturini et al. 2008; Zhou and Ritter 2010; Faugeras et al. 2012; Dondoni and Marra 2012; Kushwaha et al. 2013; Schmidt et al. 2014; Hou et al. 2016
Complexing agents	Szejtli 1988; Mikus et al. 2002; Mosher and Thompson 2002; Iacovino et al. 2017
Cosmetics - Fragrances - Personal care - Toiletry - Essential oils	Szejtli 1982a, 1996; Buschmann and Schollmeyer 2002, 2004; Hashimoto 1996, 2006; Venturini et al. 2008; Cabral Marques 2010; Bhaskara-Amrit et al. 2011; Bilensoy 2011; Tarimci 2011; Auzely-Vélty 2011; Hougeir and Kircik 2012; Ammala 2013; Zarzycki et al. 2016
Derivatives	Duchêne 1987, 1991; Jicsinszky et al. 1996; Khan et al. 1998; Shen and Ji 2011; Rezanka 2016
Detergents - Surfactants	Szejtli 1988; Dodziuk 2006; Villalonga et al. 2007; Li et al. 2011; Valente and Söderman 2014
Enzymology - Enzyme technology	Komiyama 1996; Szejtli 1984; Schmid 1996
Environmental chemistry - Remediation	Crini and Morcellet 2002; Fakayode et al. 2007; Kozlowski and Sliwa 2010; Gruiz et al. 2011; Landy et al. 2012a, 2012b; Atteia et al. 2013; Morin-Crini and Crini 2014
Fluorescent cyclodextrins	Ueno 1996; Maciollek et al. 2013; Malanga et al. 2014

(continued)

Table 1.3 (continued)

General topic(s)	References
Food - Nutrition - Flavors - Aromas - Packaging	Hedges et al. 1995; Appendini and Hotchkiss 2002; Szejtli 2004b; Szente and Szejtli 2004; Han 2005; Hashimoto 2006; Astray et al. 2009; Fang and Bhandari 2010; Ko et al. 2012; Sun 2012; Martina et al. 2013; López-Nicolás et al. 2014; Calo et al. 2015; Sharma and Baldi 2016; Kfoury et al. 2016; Fenyvesi and Szente 2016; Fenyvesi et al. 2016; Zarzycki et al. 2016
Green chemistry	Venturini et al. 2008; Kushwaha et al. 2013
Hydrogels	Szejtli 1988; Li 2009; Zhao and Xu 2010; Tan et al. 2014; Pinho et al. 2014
History	Szejtli 1982a, 1988, 1998; Loeve and Normand 2011; Kurkov and Loftsson 2013; Crini 2014
Inclusion compounds	Saenger 1980; Szejtli 1982a, 1988; Bergeron 1984; Atwood et al. 1984; Clarke et al. 1988; Cram 1988; Szejtli 1988; Connors 1997; Robyt 1998; Lincoln and Easton 1998; Mikus et al. 2002; Herbstein 2005; Luca and Grigoriu 2006; Morin-Crini et al. 2015
Industrial applications	Vaution and Hutin 1987; Szejtli 1996, 1997, 1998, 2004a; Shieh and Hedges 1996; Hedges 1998; Hashimoto 2002; Sharma and Baldi 2016
Large cyclodextrins	Larsen 2002; Ueda 2002; Endo and Ueda 2004; Endo 2011; Assaf et al. 2016
Medicine - Biomedicine - Cell biology	Van de Manakker et al. 2009; Yang and Keam 2009; Caliceti et al. 2010; Dong et al. 2011; Bilensoy 2011; Kanwar et al. 2011; Moya-Ortega et al. 2012; Macaev et al. 2013; Zhang and Ma 2013; Hirakawa and Tomita 2013; Morohoshi et al. 2013; Vecsernyés et al. 2014; Miller et al. 2015; Okano et al. 2016; Brackman et al. 2016; Silva et al. 2016
Membranes	Zhou et al. 2007; Kozlowski and Sliwa 2008
Nanotechnology	Bilensoy and Hincal 2009; Goyal et al. 2011; Davis and Higson 2011; Tejashri et al. 2013, Chilajwar et al. 2014; Trotta et al. 2014, 2015; Mavridis and Yannakopoulou 2015
NMR	Schneider et al. 1998; Potrzebowski and Kazmierski 2005; Berendt et al. 2006; Pessine et al. 2012; Ferro et al. 2014, 2016
Organometallic compounds	Hapiot et al. 2006; Smaldone et al. 2010; Agostoni et al. 2015
Pharmacy - Excipients - Drug delivery	Duchêne et al. 1986; Fenyvesi 1988; Frömming and Szejtli 1994; Uekama and Irie 1996; Irie and Uekama 1997; Davis and Brewster 2004; Thompson 2006; Brewster and Loftsson 2007; Van de Manakker et al. 2009; Bilensoy and Hincal 2009; Loftsson and Brewster 2010, 2012; Bilensoy 2011; Chilajwar et al. 2014; Tejashri et al. 2013; Simoes et al. 2014; Sharma and Baldi 2016
Preparation (native CDs, inclusion compounds)	Szejtli 1982a, 1988; Kainuma 1984; Bender 1986; Szente 1996b; Robyt 1998; Li et al. 2007; Guo and Wilson 2013; Morin-Crini et al. 2015
Spectroscopy	Szejtli 1982a, 1988
Structural aspects	Szejtli 1982a, 1988; Saenger 1984; Terekhova and Kulikov 2005

(continued)

Table 1.3 (continued)

General topic(s)	References
Supramolecular chemistry	Nepogodiev and Fraser Stoddart 1998; Schneider and Yatsimirsky 2000; Chen and Jiang 2001; Atwood and Steed 2004, 2009; Wenz et al. 2006; Wenz 2009; Hapiot et al. 2006; Harada et al. 2009a, 2009b, 2014; Wang and Zhang 2010; Zhang et al. 2011; Dong et al. 2011, 2015; Schneider 2012; Schmidt et al. 2014; Karim and Loh 2016; Gontero et al. 2017
Textiles - Cosmeto-textiles	Citernesi and Sciacchitano 1995; Buschmann et al. 1998; Hebeish and El-Hilw 2001; Martel et al. 2002; Romi et al. 2005; Hashimoto 2006; Grigoriu et al. 2008; Ammayappan and Moses 2009; Abdel-Halim et al. 2010; Ripoll et al. 2010; Andreaus et al. 2010; Bhaskara-Amrit et al. 2011; Voncina 2011; Voncina and Vivod 2013; Islam et al. 2013
Toxicology	Szejtli 1982a, 1987, 1998; Gould and Scott 2005; Dodziuk 2006; Stella and He 2008; Saokham and Loftsson 2017
Wastewater	Mocanu et al. 2001; Crini 2006; Landy et al. 2012a; Morin-Crini and Crini 2013; Shen et al. 2014a; Morin-Crini et al. 2015

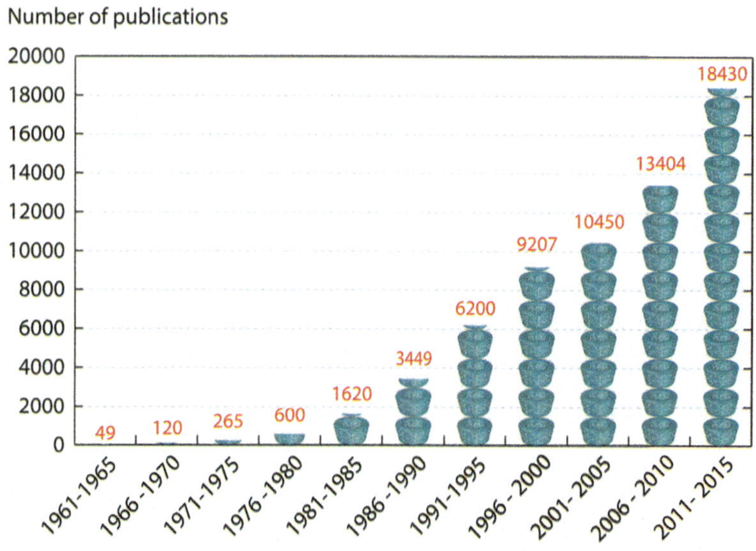

Fig. 1.4 Number of cyclodextrin-related publications including papers, patents and conferences (source: Cyclodextrin News, Cyclolab)

In Europe, the cyclodextrin scientific community is also very active. After Aalborg in 2009, Asti in 2011, Antalya in 2013 and Lille in 2015, Lisbon will welcome the 5th European conference in 2017. This is a biennial meeting where academic researchers and industrials come together to present the latest achievements in the field of cyclodextrin science and technology. The 9th and most recent Asian cyclodextrin Conference was held in 2017 in Singapore.

There are also several national groups that are very active (e.g. in Japan, France, Italy…) which enables a good quality of exchange on current research into cyclodextrins in each country. In France, the 19th annual colloquium of the *Société Française des Cyclodextrines* will be organized by Nadia Morin-Crini and Grégorio Crini in Besançon, in October 2018. This colloquium aims to provide an enabling environment for PhD students from numerous French Universities and Research Institutes, so that they can create new research collaborations. The 34th annual meeting of the Japanese community was held in 2017.

1.1.5 Szejtli Prize

The list of prestigious researchers who have contributed to the development of cyclodextrins includes Professor József Szejtli (1933–2004). Professor Szejtli made an important contribution to the chemistry of cyclodextrins, to the dissemination of results and in their industrial applications, notably by the creation of a laboratory in 1972, which became an independent company in 1992 (CYCLOLAB Ltd.) totally devoted to cyclodextrins (*"From toy to tool with industrial interest"*) and by the publication of numerous general reviews. CYCLOLAB was the first private research institute for the technological transfer between cyclodextrin research and industry (Crini 2014).

CYCLOLAB recently established Szejtli Prize (Fig. 1.5) to preserve his legacy, keep his memory alive and recognize his ground-breaking achievements in the area of cyclodextrin research, development and commercialization of related technologies. This prize aimed to award young researchers demonstrating outstanding results in the cyclodextrin science and technology. The first award was assigned to Professor

Fig 1.5 The József Szejtli medal (source: CYCLOLAB)

Keiichi Motoyama, Kumamoto University, in Saarbrücken, at the 17th International Cyclodextrin Symposium in 2014 for his works in the design of new active pharmaceutical ingredients as anticancer agents. The winner of Szejtli Prize 2016 is Professor Tamer Uyar, Bilkent University, for his studies on electrospinning of functional nanofibers with cyclodextrins.

1.1.6 Nobel Prize 2016

In 2016, the Royal Swedish Academy of Sciences has decided to award Jean-Pierre Sauvage (University of Strasbourg, France), Sir James Fraser Stoddart (Weinberg College of Arts and Sciences of Northwestern, Scotland) and Bernard Lucas Feringa (University of Groningen, The Netherlands) the Nobel Prize in Chemistry *"for the design and synthesis of molecular machines*". This was a high recognition for supramolecular chemistry utilizing building blocks of host-guest complexes including also cyclodextrin complexes.

1.2 Cyclodextrins: An update 2012-2016 – Present Situation, Trends and Outlook

Recent and continuing interest in cyclodextrins is evident from the number of papers that appears each year in the literature on various chemical and biological topics for both academic and industrial applications as showed in Fig. 1.4. In this section, we chose to highlight selected works on cyclodextrins published over the last 5 years by different research groups in order to have an update overview. The main objective is to summarize some of the developments related to the applications of cyclodextrins, based on a substantial number of relevant references published recently. Of course, this review is not exhaustive. Readers interested in cyclodextrin topics should refer to the library database "Cyclodextrin News" from CYCLOLAB, Ltd. (Hungary), which is a periodical collecting all of the cyclodextrin papers, proceedings, patents, conferences and lectures.

1.2.1 Cyclodextrins: From 1891 Until Now

Recently, Crini (2014) published a historical review of cyclodextrins, molecules discovered serendipitously by the French pharmacist Antoine Villiers in 1891 and then called "cellulosines". They were subsequently named "Schardinger dextrins" at the beginning of the 20th century, in tribute to Franz Schardinger, an Austrian chemist and microbiologist who was the first to describe their basic chemistry. Schardinger is considered the "Founding Father" of cyclodextrin chemistry. After that, they were termed "cycloamyloses" by the American chemist Dexter French at the beginning of the 1940s and finally "cyclodextrins" by the German chemist Friedrich Cramer towards the end of the 1940s. Crini's review reported their history, divided into 5 more or less clearly defined periods: The early days around their discovery from 1891 to 1911, a period of doubt between 1911 and 1935, reaching maturity from 1935 to 1950, exploration from 1950 to 1970, and the uses to which they have been put from 1970 to the present day. A series of other prestigious names has marked their history, names such as Hans Pringsheim, Paul Karrer, Karl Johann Freudenberg, Benito Casu, Wolfram Saenger, Myron Lee Bender, and József Szejtli who is considered as the "Godfather" of cyclodextrins. Each of the main periods was illustrated by precise examples from among the works published in the literature. Although this review is not exhaustive, it highlights the work of those researchers who have contributed to the development of our knowledge of cyclodextrins throughout the 120 years of their history. An historical examination of cyclodextrins can be found elsewhere (Szejtli 1998; Loftsson and Duchêne 2007; Loeve and Normand 2011).

1.2.2 Self-association of Cyclodextrins

It is now well-known that cyclodextrin molecules can adopt various types of assembly modes in aqueous solution as well as crystal structures (Gonzalez-Gaitano et al. 2003). New micro- and nanostructures including aggregates, formed by the self-assembly of cyclodextrins, have been useful, particularly in the fields of supramolecular chemistry, materials science, pharmacy including formulation and drug delivery, and medicine as reported by Kurkov and Loftsson (2013), Simoes et al. (2014), Ryzhakov et al. (2016), and Oliveri and Vecchio (2016).

The most important property of the cyclodextrins is the ability to establish specific interactions (molecular encapsulation) with various types of molecules through the formation of non-covalently bonded entities, either in the solid phase or in aqueous solution. However, cyclodextrins are not only able to form these host/guest inclusion complexes but also non-inclusion complexes (Gonzalez-Gaitano et al. 2003; Loftsson et al. 2004; Messner et al. 2010; Moya-Ortega et al. 2012; Kurkov and Loftsson 2013). The hydroxyl groups present on the outer surface can form hydrogen bonds with other molecules which makes them able, just like dextrins, non-cyclic oligosaccharides and polysaccharides, to form complexes (molecular structures) with lipophilic substances insoluble in water. In pure aqueous cyclodextrin solutions, CD molecules self-assemble to form nanoparticles with diameter from about 20 to 200 nm (Loftsson et al. 2004; He et al. 2008). At low concentrations (below 1% w/v), the fraction of CD molecules forming such aggregates is insignificant but the aggregation increases with increasing cyclodextrin concentration. Another possibility that is mentioned is the formation of aggregates able to dissolve water-insoluble lipophilic molecules (structures similar to micelles). The CD/drug complex aggregates are frequently from 100 to 4000 nm in diameter or in the nanoparticle and small microparticle size range. The complexes are kept together by weak hydrogen bonds and hydrophobic forces, and dissociate readily upon media dilution. Such systems are called cyclodextrin nanostructures, nanoassemblies or self-aggregates.

In recent years, it has been observed, in pharmaceutical applications, that other types of cyclodextrin complexes such as non-inclusion complexes, are also participating in CD solubilization of poorly soluble drugs (Bilensoy 2011; Kurkov and Loftsson 2013; Morin-Crini et al. 2015). There are some indications that formation of CD/drug complex aggregates might play an important role in CD enhancement of drug bioavailability. The cyclodextrin aggregates present the ability to form complexes, and nanosized aggregates and nanotube-type host/guest architectures can be envisaged. This is a generally unexplored domain and often causes controversies since the results obtained are closely dependent on the technique used, as pointed out by Ryzhakov et al. (2016). Another problem discussed in the literature is the stability of the cyclodextrin aggregates, in particular when these systems are used as drug delivery systems. The publications of Loftsson's group summarized the most important features and the main conclusions of these cyclodextrin-based aggregates (Fülöp et al. 2012; Kurkov and Loftsson 2013; Ryzhakov et al. 2016; Saokham and Loftsson 2017). The use of CD-based nanoaggregates in oral drug delivery could be a promising strategy to improve the bioavailability of poorly soluble drugs. Much work is necessary to study the behavior of these nanoaggregates under conditions that are representative for the gastrointestinal tract and the effects which may cause disaggregation (Ryzhakov et al. 2016).

1.2.3 Cyclodextrins as Drug Delivery Vehicles

Although cyclodextrins can be found in 56 pharmaceutical products (source: Cyclolab), they are still regarded as novel pharmaceutical excipients, drug delivery vehicles, and anti-aggregation agents (Duchêne 1987, 1991; Frömming and Szejtli 1994; Uekama and Irie 1996; Thompson 2006; Brewster and Loftsson 2007; Van de Manakker et al. 2009; Loftsson and Brewster 2010, 2012; Bilensoy 2011; Ahuja et al. 2011; Kurkov and Loftsson 2013; Chilajwar et al. 2014; Morin-Crini et al. 2015). In the last decade, as already mentioned, it has been observed that cyclodextrins and cyclodextrin complexes in particular self-assemble to form nanoparticles and that, under certain conditions, these nanoparticles can self-assemble to form microparticles, a tendency that increases upon formation of inclusion complexes with lipophilic drugs. These properties have changed the way we perform cyclodextrin pharmaceutical research and have given rise to new cyclodextrin formulation opportunities as summarized by Bilensoy (2011), and Kurkov and Loftsson (2013). The design of functional cyclodextrin nanoparticles formed by self-assembly is also a developing area in the field of nanomedicine (cancer therapy) as reported by Fülöp et al. (2012). Nanoparticle-based systems can improve bioavailability, reduce immunogenicity, modify drug metabolism, reduce toxicity, and increase the biological half-life of drugs after systemic administration (Bilensoy 2011; Fülöp et al. 2012; Kurkov and Loftsson 2013; Brandariz and Iglesias 2013; Tejashri et al. 2013; Concheiro and Alvarez-Lorenzo 2013; Simoes et al. 2014; Morin-Crini et al. 2015; Sharma and Baldi 2016).

The field of medicine using cyclodextrins and their derivatives is vast including biomedicine (wound dressing, biosensors, medical devices), nanomedicine (nanoparticles for drug delivery), antifungal treatment, antimicrobial therapy, gene therapy, tissue engineering, cancer treatment, and magnetic resonance imaging (Van de Manakker et al. 2009; Yang and Keam 2009; Caliceti et al. 2010; Dong et al. 2011; Bilensoy 2011; Kanwar et al. 2011; Moya-Ortega et al. 2012; Macaev et al. 2013; Zhang and Ma 2013; Hirakawa and Tomita 2013; Morohoshi et al. 2013; Miller et al. 2015; Okano et al. 2016; Brackman et al. 2016; Silva et al. 2016).

Interesting recent studies were devoted to the physicochemical properties of local anesthetics (LA) and their inclusion complexes with cyclodextrins in order to understand their behavior (Brandariz and Iglesias 2013). Their capacity to reach and to block sodium channels and act as anesthetics depends on their protonation state. Different studies showed that the extent of complexation with CDs varies greatly with the protonation state of the involved molecules, an interesting fact in the administration of LA, as recently reported by Brandariz and Iglesias (2013).

Another interesting example concerns the preparation of vaccines Daptacel® (Sanofi Group, Pasteur) for protection against diphtheria, tetanus and whooping cough which contain dimethyl-β-cyclodextrin.

Cyclodextrins can be used in antifungal formulations as auxiliary substances to modify and to improve physicochemical properties of the active compound such as solubility and stability. The biological effects of cyclodextrins, important for their use within antimycotic formulations, can be divided into: (i) effects based on the ability of cyclodextrins to form inclusion complexes with endogenous substances (e.g. membrane lipids, cellular cholesterol), (ii) effects based on formation of inclusion complexes with component parts of fungi cells, and (iii) effects based on the chemical nature of cyclodextrins and their derivatives. The advances in research of biological activity of CDs with focus on their properties responsible for their synergistic effect with antimycotic compounds were recently discussed by Macaev et al. (2013). The number of publications on the use of cyclodextrins in antifungal formulations is still growing.

Oliveri and Vecchio (2016) reviewed the use of cyclodextrins and their derivatives as anti-aggregation agents in a number of proteins such as insulin, prion protein, and amyloid-beta, and some multimeric enzymes. There are many diseases that are correlated to protein misfolding and amyloid formation processes affecting numerous organs and tissues. There are over 30 different amyloid proteins and a number of corresponding diseases including Alzheimer's diseases the most common neurodegenerative disease. Treatment of these diseases is still a goal to reach, and many molecules including cyclodextrins were studied in this perspective.

Onchocerciasis, also known as River Blindness, is a disease caused by infection with the parasitic worm *Onchocerca volvulus*. The parasite worm is spread by the bites of black fly. These flies live near to rivers in sub-Saharan Africa, hence the name of the disease. There is no vaccine against it. Insecticides are used to decrease the fly population. People infected are treated with ivermectin. The drug kills the larvae but not the adult worm therefore the treatment should be repeated once or twice a year. However, ivermectin is not compatible with conventional excipients (Shaw et al. 1999; Shen et al. 2014). Recently, novel formulations have been proposed: Intraperitoneal injection of cyclodextrin-conjugated ivermectin proved to be effective in inhibition of transcription factor-dependent human colon cancer xenograft in vivo (Melotti et al. 2014).

Actually, cyclodextrins are also perceived as dream molecules for the development of applications in biomedicine and nanomedicine including nanovectorization, such as, for instance, nanoparticles for drug delivery, innovative biosensors for detection of biological targets, bio-recognition events, molecular diagnosis and medical imaging, gene therapy or tissue engineering (Concheiro and Alvarez-Lorenzo 2013; Zhang and Ma 2013; Tan et al. 2014; Simoes et al. 2014; Dong et al. 2015; Wu et al. 2015b; Yuan and Zhang 2016).

Medical devices including catheters, prosthesis, vascular grafts, and bone implants can also benefit from surface grafting or thermofixation of cyclodextrins. This explains the recent increase in the number of research papers dealing with these topics (Concheiro and Alvarez-Lorenzo 2013). However, most of these studies are in the proof-of-concept stage, and only a few therapeutic nanosystems have been comprehensively investigated. The successful translation of these laboratory innovations to clinical reality remains challenging. The cost is an important factor that limits the successful translation of new materials containing cyclodextrins. Another problem concerns the lack of data on the biocompatibility evaluation and on the toxicity of these nanomaterials. In spite of these, Zhang and Ma (2013) concluded that the future of cyclodextrin-based supramolecular systems mainly in drug and gene delivery is promising in view of the notable clinical success. The cyclodextrin-based nanomaterials have found applications not only in biomedical sector, pharmacy, pharmacotherapy, biology, and biotechnology, but also in the textile industry by providing clothes for transdermal delivery.

A new generation of drug eluting stents based on the strong anchorage of a biocompatible and bioresorbable cyclodextrin-based polymer onto metallic devices has been elaborated (Perez-Anes et al. 2015). Polydopamine, a strong adhesive polymer, was applied as a first coated layer. Cyclodextrin was fixed by *in situ* polycondensation with citric acid (polycyclodextrins formation). As a third layer an amine-rich polymer, polyethyleneimine was used to stabilize the anionic cyclodextrin layer. As an alternative polycyclodextrins and chitosan layers were applied using layer-by-layer deposition technique.

Junthip et al. (2016) proposed new textiles modified with multilayer cyclodextrin polyelectrolytes (e.g. citric acid-cyclodextrin polymer and quaternary amino cyclodextrin polymer) using layer-by-layer technique to obtain textiles with antibacterial properties to be applied as wound coatings or prostheses. Sustained release of three model compounds such as *tert*-butyl-benzoic acid, methylene blue and triclosan, was presented. Biological and microbiological tests were performed to investigate the cytocompatibility and the intrinsic antibacterial activity of multilayer assemblies.

Functionalized cyclodextrins are interesting scaffolds for contrast agents used in magnetic resonance imaging as recently reported by Gouhier's group (Idriss et al. 2013; Zgani et al. 2017). For the further improvement of the sensitivity of this medical diagnostic tool, it is necessary to fully understand the role of the cyclodextrins in the efficiency of the contrast agents. The review by Mavridis and Yannakopoulou (2015) can be also consulted on the same topic.

Different methods of immobilizing cyclodextrin onto polysaccharide were discussed by Yang and Yang (2013). These materials possess the cumulative effects of inclusion, size specificity, controlled release ability and transport properties of cyclodextrin as well as the biocompatibility, non-toxicity, and biodegradability of polysaccharide. Their nanoscale association could also lead to the formation of original particles and films which pave the way to new applications in tissue engineering.

Wound dressings should be capable of mechanical wound protection and should also facilitate the healing process via maintenance of suitable environmental conditions and controlled delivery of bioactive molecules. It is known that, among macromolecular systems, hydrogels present suitable properties for wound-dressing applications such as cost-effectiveness, easy to use, good biocompatibility, together with a high-water content, the latter of which is important for the maintenance of a moist environment and ready removal from the wound with a minimal level of associated pain. As reported by Pinho et al. (2014), their properties as drug delivery systems can be improved by the use of cyclodextrins as cross-linking agents. The conjugation of cyclodextrin molecules with hydrogels may allow the achievement of an optimal wound-dressing material, because the hydrogel component will maintain the moist environment required for the healing process, and the cyclodextrin moiety has the ability to protect and modulate the release of bioactive molecules. Pinho et al. (2014) concluded that cyclodextrin-based hydrogels with target characteristics as interesting new materials for wound-dressing applications due to their powerful complexation abilities and biocompatibilities.

1.2.4 Cyclodextrins as Active Ingredients

Currently the field of medicine is closely concerned with cyclodextrin inclusion complexes. These complexes are preformed prior to be administrated. There are, however, some cases when the complexes are formed within the body. The best-known example is the one containing the active compound sugammadex (Bridion®): it is a modified γ-CD (Fig. 1.6; Table 1.4) used as an antidote to certain curare-like muscle relaxants in anesthesia since 2008 (Booij 2009; Yang and Keam 2009; Ozbilgin et al. 2016). After intravenous administration, it neutralizes steroid curare-like agents (rocuronium, vecuronium) by forming an inactive complex in the plasma which is then eliminated in the urine. Sugammadex has improved effectiveness compared with currently available methods of accelerating reversal of neuromuscular blockade. Its mechanism of action also differs from that of other commonly used

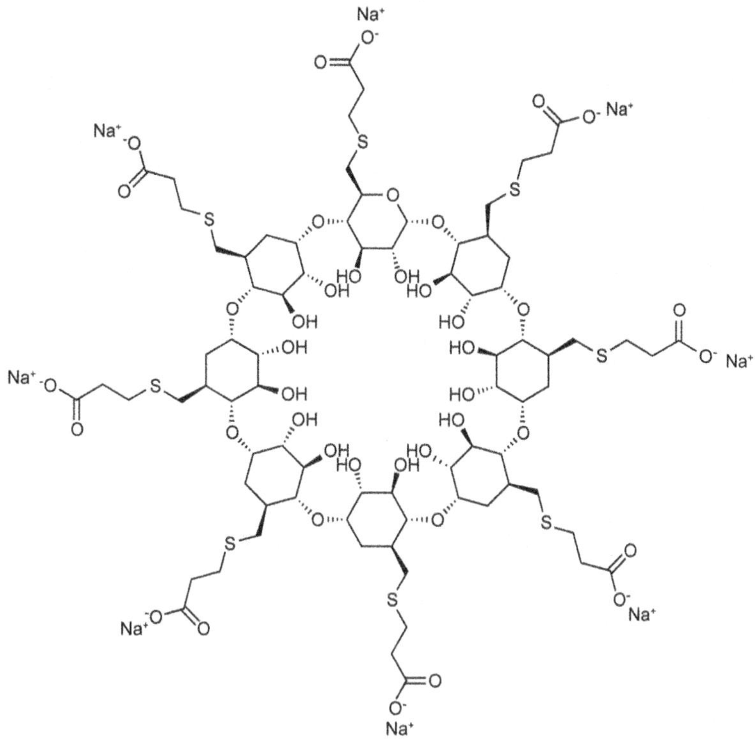

Fig. 1.6 Structure of Sugammadex

Table 1.4 Physicochemical properties of Sugammadex (Booij 2009; Yang and Keam 2009; Bilensoy 2011)

Chemical Abstract Service Registry Number	343306-71-8
Molar substitution	1
Molecular weight (g/mol)	2146
Water solubility at 25°C (g/L)	Freely soluble

reversal agents e.g. neostigmine and edrophonium. Sugammadex is biologically inactive, does not bind to plasma proteins, and appears to be safe and well tolerated although it has few side effects. Its cost is markedly higher than that of any of the other drugs used in anesthesia (Donati 2011). The literature data are actually abundant with a matter of considerable debate. Sugammadex shows an unexpected rise from 17 papers in 2000-2005 to 780 in 2010-2015 (Source: Cyclodextrin News, Cyclolab).

Application of hydroxypropyl-β-cyclodextrin (HP-β-CD) against Niemann-Pick C disease (NPC) started with a surprising observation in 2007 that this cyclodextrin aimed as excipient was more effective than allopregnanolone, the active (Walkley et al. 2016). As no alternative treatment existed at that time the NPC1 disease families pushed FDA for approval for individual access for their children and as a result of a large collaborative work in all areas of drug development, including chemistry and manufacturing, formulation, pharmacology, pharmacokinetics, toxicology, and regulatory affairs, the preclinical clinical Phase 1 and Phase 2 studies have been concluded to start Phase 3 studies in 2015. In the meantime, the research focuses on the still not completely understood mechanism how HP-β-CD can help to NPC patients (Tanaka et al. 2015; Davidson et al. 2016).

Cyclodextrins are used in autophagy, a catabolic process with an essential function in the maintenance of cellular and tissue homeostasis. It is primarily recognized for its role in the degradation of dysfunctional proteins and unwanted organelles. Actually, the range of substrates also includes lipids. Autophagy is a self-degradative process that is important for balancing sources of energy at critical times in development and in response to nutrient stress. It also plays a housekeeping role in removing misfolded or aggregated proteins, clearing damaged organelles, such as mitochondria, endoplasmic reticulum and peroxisomes, as well as eliminating intracellular pathogens. Thus, autophagy is generally thought of as a survival mechanism, although its deregulation has been linked to non-apoptotic cell death (Glick et al. 2010). As cellular membranes play important role in autophagy, their modulation by cyclodextrins modifies this important house-keeping process of the cells. However, the results of various studies seem to be controversial partly because different cell types, different cyclodextrins were used at different concentrations. The accumulation of autophagosomes, the intermediary products of autophagy, was interpreted both as a sign of activated and impaired autophagy. Hydroxypropyl-β-cyclodextrin treatment at high concentration leads to cholesterol depletion in an extent which leads to hindered fusion of the cellular membranes and this way to the diminished fusion of autophagosomes with lysosomes or to reduced expulsion of the autophagolysosomes. At low HP-β-CD concentrations, however, the autophagy is not perturbed or might be even improved. A combination of cholesterol removal by HP-β-CD and autophagy stimulation by rapamycin or carbamazepine seems to be a promising strategy in the treatment of impaired autophagy in lysosomal storage disorders, such as Niemann-Pick type C disease (Maetzel et al. 2014; Ward et al. 2016). Other interesting works have been published on cyclodextrin role (Yokoo et al. 2015; Tamura and Yui 2015; Motoyama et al. 2015, 2016; Manchon et al. 2016). Hydroxypropyl-β-cyclodextrin was found effective in inhibition of leukemic cell proliferation at various leukemic cell lines as recently reported by Arima's group (Yokoo et al. 2015). Their results demonstrated that hydroxypropyl-β-cyclodextrin was a potential anticancer agent in leukemia. Further studies are needed to understand the effect of cyclodextrins on autophagy and specially to learn how these effects can be utilized in the therapy of various illnesses including neurodegenerative diseases and cancers.

Cyclodextrins are also used for prevention and treatment of atherosclerosis as recently reported by Zimmer et al. (2016). These authors published a comprehensively study on the anti-atherosclerotic and anti-inflammatory effects of hydroxypropyl-β-cyclodextrin used as solubilizing agent to increase cholesterol solubility. Atherosclerosis is a chronic inflammatory disease driven primarily by a continuous retention of cholesterol within the subendothelial space where it precipitates to form cholesterol crystals (CC). Despite ongoing advances in the prevention and treatment of atherosclerosis, cardiovascular disease remains the leading cause of death worldwide. Continuous retention of apolipoprotein B-containing lipoproteins in the subendothelial space causes a local overabundance of free cholesterol. Cyclodextrin treatment of murine atherosclerosis reduced atherosclerotic plaque size and CC load and promoted plaque regression even with a continued cholesterol-rich diet. Mechanistically, cyclodextrins increased oxysterol production in both macrophages and human atherosclerotic plaques and promoted liver X receptor-mediated transcriptional reprogramming to improve cholesterol efflux and exert anti-inflammatory effects. In vivo, this cyclodextrin-mediated liver X receptor agonist was required for the anti-atherosclerotic and anti-inflammatory effects of cyclodextrins as well as for augmented reverse cholesterol transport. Because cyclodextrin treatment in humans is safe and cyclodextrin beneficially affects key mechanisms of atherogenesis, it may therefore be used clinically to prevent or treat human atherosclerosis. The authors are optimistic as hydroxypropyl-β-cyclodextrin is a drug (it has received the orphan drug status against NPC).

A recent review on the effect of cyclodextrins on blood brain barrier (BBB) summarizes the findings of several research groups on various animal models of the central nervous system (CNS) diseases (Vecsernyés et al. 2014). The CD-mediated cholesterol modulations change the action of various proteins in the membrane of epithelial cells in BBB. These proteins play significant role in pathogenesis of stroke, cerebral hypoxia and ishemia, Alzheimer, Parkinson and Huttington disease, epilepsia, CNS infections, and brain tumor. Extensive research is going on to translate these effects into therapy.

The alarming spread of bacterial resistance to traditional antibiotics has warranted the study of alternative antimicrobial agents. Quorum sensing (QS) is a chemical cell-to-cell communication mechanism utilized by bacteria to coordinate group behaviors and establish infections (Hirakawa and Tomita 2013; Morohoshi et al. 2013; Miller et al. 2015; Okano et al. 2016). It is known that cyclodextrins can interact with *N*-acyl-L-homoserine lactones (AHLs), the main signaling molecules for the bacterial cell-to-cell communication QS system. Many bacteria regulate their cooperative activities through releasing, sensing and responding to small signaling molecules. This mechanism called quorum sensing makes possible for a population of bacteria to behave as a multi-cellular organism in host colonization, formation of biofilms, defense against competitors and adaptation to changing environment. Quorum sensing is a new target for the development of antibiotic agents. The QS system is regulated by three different-type of signals: AHLs and small peptides (called auto-inducing peptides, AIP) are produced in gram-negative and gram-positive bacteria, respectively, while production of the autoinducer-2 (AI-2) was known as a universal signal for responding to bacterial populations in both gram-negative and gram-positive bacteria. Inclusion complexes between cyclodextrins and bacterial signal molecules are responsible for inhibitory effects on quorum sensing. Since many bacteria have QS system for controlling gene expression in response to cell population density by means of signal molecules, an intercept of the quorum sensing signal onto the cyclodextrins can be a general method to control transcription of the QS-regulated genes. The concept of complexation of signal molecules (AHLs or peptides) by cyclodextrins does not aim to kill the bacteria but to control their growth and to decrease their virulence (Hirakawa and Tomita 2013). The development of novel strategies in the prevention and treatment of biofilm infections are expected in the next years. This will be useful not only in medicine but also in cosmetic, textile and packaging fields (Brackman et al. 2016; Silva et al. 2016).

Fluorophore-functionalized cyclodextrins have been extensively studied and successfully utilized in photodynamic therapy (PDT) (Mazzaglia et al. 2012). This minimally invasive therapeutic approach has proven to be efficient for the treatment of cancer and bacterial infections. In PDT, the photosensitizer gets excited and transfers the energy to nearby molecular oxygen photodonor. This leads to an *in situ* generation of singlet oxygen ($^1O_2$) offering advantages over conventional drugs (e.g. no drug resistance is developed). In the combined chemotherapy and PDT, the cavity of cyclodextrin is utilized as carrier for a drug with anticancer or antibiotic effect while the cavity entrance serves as molecular scaffold for attaching the photosensitizer (porphyrin or xanthene dye), nitric oxide releasing moiety, etc. (Králová et al. 2010; Fraix et al. 2015, 2016). The fluorescence of such nanoassemblies is utilized in *in vitro* experiments for imaging (Kirejev et al. 2014).

1.2.5 Cyclodextrins and Nanotechnology

Cyclodextrins and their derivatives have been successfully employed to create novel nanomaterials, often called nanosponges (Bilensoy and Hincal 2009; Goyal et al. 2011; Davis and Higson 2011; Tejashri et al. 2013; Chilajwar et al. 2014; Trotta et al. 2014, 2015; Mavridis and Yannakopoulou 2015). A broad spectrum of cyclodextrin-containing materials with versatile supramolecular architectures such as nanoparticles, nanosponges, nanomicelles, and nanovesicles has been synthetized to assemble functional platforms. For pharmaceutical and biomedical applications (Bilensoy and Hincal 2009; Goyal et al. 2011), nanomaterials can be formulated as oral, parenteral, topical or inhalation dosage forms. These materials have also found applications in nanomedicine.

At present, nanotechnology is receiving considerable acknowledgment due to its potential to combine features that are difficult to achieve by making use of a drug alone. Cyclodextrin-based nanomaterials are a contemporary approach for highlighting the advancements which could be brought about in a drug delivery system. Chilajwar et al. (2014) recently reported that statistical analyses have shown that around 40% of currently marketed drugs and about 90% of drugs in their developmental phase encounter solubility-related problems. Cyclodextrin-based nanosponges have the capacity to emerge as an innovative approach over conventional cycodextrins by overcoming the disadvantages associated with the latter. These novel class structures have been also developed since their use can improve a drug's bioavailability by modifying the pharmacokinetic parameters of actives. Nanosponges offer high drug loading compared to other nanocarriers and are thus suitable for solving issues related to solubility, stability, and delayed release of actives (Tejashri et al. 2013). Although the methods of preparation are well-known, more information on their characterization is required to optimize their performance for therapeutic purposes and to demonstrate the role of the nanocavity in the complexation. Neutral, cationic and/or anionic amphiphilic cyclodextrins have been also proposed to increase interactions of cyclodextrins with biological membranes.

Cyclodextrin-based nanosponges have been also developed as tool for the delivery of anticancer drugs, e.g. paclitaxel, doxorubicin, 5-fluorourcil and tamoxifen, as reported by Trotta et al. (2014, 2015). These innovative materials can be considered as a challenging technology for the development of innovative formulations, suitable for various administration routes for anti-cancer drugs.

Metal organic frameworks (MOF) based on cyclodextrin have been developed by the group of 2016 Nobel laureate, Sir Stoddart (Smaldone et al. 2010). Mixing γ-cyclodextrin with alkali metal salts and alcohol (all edible constituents) robust, nanoporous MOF are obtained. On the other hand, nanoMOF of iron polycarboxylates were coated with cyclodextrin phosphates to obtain engineered nanoMOF for targeted drug delivery, catalysis, and sensing (Agostoni et al. 2015).

1.2.6 Cyclodextrins and Foods

In recent years, the growth of the functional foods industry has increased research into new compounds with high added value for use in the fortification of traditional products (Martina et al. 2013; López-Nicolás et al. 2014; Calo et al. 2015; Sharma and Baldi 2016; Kfoury et al. 2016; Fenyvesi and Szente 2016; Zarzycki et al. 2016). One of the most promising functional food groups is those enriched in antioxidant compounds of a lipophilic nature. In spite of the numerous advantages reported for such antioxidant molecules, they may also have disadvantages that impede their use in functional foods, although these problems may be avoided by the use of encapsulant agents such as cyclodextrins (Hedges et al. 1995; Appendini and Hotchkiss 2002; Szejtli 2004b; Szente and Szejtli 2004; Hashimoto 2006; Astray et al. 2009; Fang and Bhandari 2010; Fenyvesi et al. 2016).

It has been recently recognized that α-cyclodextrin being non-digestible is a dietary fiber with beneficial effects on digestion of fat and carbohydrates (Artiss et al. 2006). It has been marketed for body weight control in several countries. The antidiabetic effect has been recognized by European Food Safety Administration (EFSA) approving the health claim: "Consumption of alpha-cyclodextrin contributes to the reduction of the blood glucose rise after starch-containing meals" (EFSA 2012). The role of cyclodextrin molecules played in food technology and human nutrition have been recently reviewed (Fenyvesi et al. 2016).

An excellent review of the most recent studies on the complexes formed between several important types of antioxidant compounds and cyclodextrins was published by López-Nicolás et al. (2014). This comprehensive review focus on the contradictory data reported in the literature concerning the antioxidant activity of the host/guest molecule complexes, the different complexation constants reported for identical complexes, the bioavailability of the antioxidant compound in the presence of cyclodextrins and the recommendations concerning the use of natural or modified cyclodextrins. The authors also concluded that cyclodextrins will act as secondary antioxidants, enhancing the ability of traditional antioxidants to prevent enzymatic browning in different foods. Another interesting review on the encapsulation of antioxidants such as flavonoids and phenolic acids and their applications in food products, food supplements and also packaging has been published by Zarzycki et al. (2016).

Effects of encapsulation, such as solubility enhancement, protection, controlled release, improved organoleptic behavior and masked off-flavors, active packaging as well as improved handling and dosage were recently discussed in details in the review by Kfoury et al. (2016). The authors also gave a detailed evaluation of the experimental techniques used for studying the cyclodextrin/flavor complexes including static headspace gas chromatography, UV-visible and fluorescence spectroscopy, isothermal titration calorimetry, phase solubility studies, NMR spectroscopy, thermoanalytical methods and microscopy. There was an interesting large table listing the methods of complexation and analysis for over 100 aroma compounds and essential oils with significant references.

A general overview on the nano-encapsulation of flavors and aromas was also given in a chapter published by Fenyvesi and Szente (2016). Starting with the history dating back to the sixties of the last century, the advantages were summarized, the approval status of cyclodextrins in food was evaluated, and the methods of preparation and analysis were shortly outlined. Flavor complexes in food processing to reduce the loss in color, odor and taste were revealed. Several examples were given to illustrate the advantages of nanoencapsulation of flavors in aroma preserving food packaging.

An interesting effect of cyclodextrin encapsulation is in the container and wrapping materials for foods. Wasabi (Japanese horseradish) is a plant extract used for commercial antimicrobial packaging (Ko et al. 2012). The main active antimicrobial ingredient in wasabi extract is allyl-isothiocyanate (AIT). This substance is volatile and decomposes rapidly by oxidation. The AIT gas inhibits many fungi and bacteria. This extract has been encapsulated in cyclodextrin to control the volatility of AIT and to stabilize it in wrapping materials. The AIT in the encapsulated powder becomes volatile when the AIT-CD complex is exposed to high humidity conditions after the packaging of the food product. The evaporated AIT then migrates to the food surface, and inhibits the growth of aerobic bacteria. The AIT-CD complex powder has been incorporated in packaging materials of drip sheets, polymer films (polyethylene, nylon), and tablets (Han 2005; Sun 2012). These products are used for rice lunch boxes, meats, fresh produce, and raw fish (Sashimi) to keep them fresh, and are commercially available in Japan (under trade name Wasapower®).

1.2.7 Cyclodextrins and Cosmetics

Encapsulation techniques using cyclodextrins as (nano)-encapsulating agents are increasingly used not only by food, aroma, and pharmaceutical industries but also by cosmetic, fragrance and flavor industries, toiletry and personal care sectors for improving the efficiency of odorant and aroma substances, odor control in perfumes, masking unpleasant smells and tastes of some compounds, improving the physical and/or stability of essential oils and volatile compounds, stabilizing volatiles by reducing or eliminating any losses through evaporation, modifying the physicochemical and/or biological properties of the guest to afford a protective effect, or transforming of liquid compounds into crystalline form (Bilensoy 2011; Ammala 2013; Hougeir and Kircik 2012; Kfoury et al. 2016; Zarzycki et al. 2016; Fenyvesi and Szente 2016). The fragrance and flavor industry is a large and innovate sector of the chemical industry. Fragrance chemicals are added to consumer products such as personal care products, perfumes, deodorants, laundry detergents, etc. Encapsulation techniques using cyclodextrins are increasingly used by this industry for protecting fragile molecules (eye-drop solutions) and improving the efficiency of odorant and aroma substances, but also to avoid the degradation of flavors by processing or, on storage, allowing the use of minor amounts of flavors. Encapsulation of flavors by cyclodextrins is also an essential process that ensure controlled release, reduce volatility, increase solubility, dissolution and bioavailability, and decrease the allergic reactions (Szejtli 1982a, 1982b, 1996; Buschmann and Schollmeyer 2002, 2004; Hashimoto 1996, 2006; Venturini et al. 2008; Cabral Marques 2010; Bhaskara-Amrit et al. 2011; Bilensoy 2011; Tarimci 2011; Auzely-Vélty 2011). The use of these host compounds is also promising for various emerging fields such as aromatherapy and cosmeto-textiles – these new products also called 'cosmeceuticals' (Bilensoy 2011).

1.2.8 Cyclodextrins and Textiles

Textile finishing is crucial for giving textiles new functionalities and making them appropriate for special applications such as antimicrobial resistance and flame retardancy (Citernesi and Sciacchitano 1995; Buschmann et al. 1998; Hashimoto 2006; Grigoriu et al. 2008; Ammayappan and Moses 2009; Ripoll et al. 2010; Andreaus et al. 2010). In recent years, the use of low-environmental impact technologies based on sustainable substances such as polysaccharides (chitosan, alginate), cyclodextrins and others compounds (sericin protein) has been proposed as a novel possible route for large scale development of bioactive textiles (Bhaskara-Amrit et al. 2011; Voncina 2011; Voncina and Vivod 2013; Islam et al. 2013). Traditionally, chemicals such as inorganic salts (zinc pyrithione), phenolic derivatives, antibiotics, formaldehyde derivatives, dyes, and other compounds (triclosan) have been employed to impart antimicrobial activity to textile fibers. However, these compounds are not environmental friendly and some have toxic effects. With the public's enhanced awareness of ecosafety, it is necessary to develop new strategies including both the use of natural products and the concept of green chemistry. Cyclodextrin molecules can be an ecofriendly alternative and used as finishing agents. In this sector, their main characteristics are: ecofriendly nature, cost-effectiveness and ease of production at large-scale, physicochemical (e.g. inclusion complex forming ability, solubilizing activity, chelating activity, slow release of fragrances) and biological (e.g. biocompatibility, biodegradability, drug carrier ability, insecticidal delivery) properties. These properties can be applied to different areas of applications such as deodorant (odor absorption, stabilization of active ingredients), fragrance/aroma, UV protection, water resistance, antimicrobial resistance, flame retardancy, and also insect repellent. In general, cyclodextrins are grafted in materials using binding and cross-linking agents. The comprehensive reviews of Islam et al. (2013) and Voncina and Vivod (2013), and the previous works by Hebeish and El-Hilw (2001), Martel et al. (2002), Romi et al. (2005), and Abdel-Halim et al. (2010) can be consulted. Cyclodextrin is considered as a promising reagent in textile finishing.

More and more research and practical use results also indicate that cyclodextrins might also act as active compounds in functionalized textiles called cosmeto-textiles. A cosmeto-textile allows the administration of active natural substances, like vitamins, oils and therapeutic extracts, simply and controllably. Neither in cosmetics nor textiles, the microencapsulated ingredients on cosmeto-textiles ensure their slimming, hydrating or perfuming progressive effect on the skin. It can also be used to change the surface properties of a fabric in order to make it self-cleaning, hydrophobic or lipophobic. Cyclodextrins may function as encapsulating, dispersing and levelling agents in the dyeing and washing of materials. Furthermore, as already mentioned, they may be anchored to polymers and fibers in order to impart target properties such as odor reduction, UV protection, or for the controlled release of perfumes, aromas, substances with therapeutical effects like antimicrobial properties, and also mosquito repellents. The techniques used to functionalize materials include microencapsulation, plasma, and sol-gel. The adhesion of the substances that functionalize textiles can be achieved physically or chemically. Voncina and Vivod (2013) reviewed the current state of the art concerning functionalization techniques and the methods used to characterize a functionalized fabric.

Cosmeto-textiles meet an increasing demand on the market and the cosmetic and textile industries are on the forefront of the research on this topic. These innovative materials are increasingly used not only in these sectors but also in pharmaceutical and medical industries, and in food packaging. Development of textiles with an antimicrobial activity can be also useful for water and air treatment.

1.2.9 Cyclodextrins and Separation Techniques

Chirality is an important modulator of the effects and properties of chiral substances not only in pharmacology, agrochemistry, food and environmental chemistry but also in all biological systems. As additives and/or chiral selectors, cyclodextrins and their derivatives have been used extensively in separation science because they have been shown to discriminate between positional isomers, functional groups, homologues and enantiomers. This property makes them one of the most useful agents for a wide variety of separations (Smolková-Keulemansová 1982; Ward and Armstrong 1988; Han and Armstrong 1989; Sybilska and Zukowski 1989; Menges and Armstrong 1991; Hilton and Armstrong 1991; Fanali 1993; Fanali et al. 1994; Schneiderman and Stalcup 2000; He and Beesley 2005; Vetter and Bester 2006). Cyclodextrin derivatives are more interesting than natives due to their higher solubility and increased selectivity introduced by the presence of non-ionic or ionic substituents on the cyclodextrin molecules. Cyclodextrins are used in liquid chromatography (LC), gas chromatography (GC), thin-layer chromatography, gel electrophoresis, capillary electrophoresis, electrokinetic chromatography, dialysis, separation on hollow fibers, solid- and liquid-phases extractions, liquid, gas and supercritical fluid chromatography, separation through liquid and composite membranes, and also molecularly imprinted polymers.

Although an important number of works have been published, the sector of cyclodextrin-based chromatography and electrophoresis continues to interest the scientific community (Hongdeng et al. 2011; Xiao et al. 2012; West 2014; Scriba 2016). The number of publications continues to grow not only in high performance LC, ultra-high performance LC, capillary chromatography and GC, but also in supercritical fluid chromatography (SFC). Actually, SFC is considered as a green separation technique, as it avoids the use of organic mobile phase, and is an ideal alternative technique with fast and efficient separation for the preparation and separation of pure substances e.g. enantioselective separation, chiral extraction, etc.).

Xiao et al. (2012) reported various synthetic and functional groups immobilization strategies of novel cyclodextrin chiral stationary phases for chromatography, in particular for supercritical fluid chromatography. The authors concluded that the use of high performance CD-based stationary phases for preparative supercritical fluid chromatography will likely play an important role for future pharmaceutical industry. The interest of supercritical fluid chromatography for screening methods and for the preparative scale was also showed by West (2014). While the pharmaceutical sector is the main application area of enantioselective separations, other industries such as agrochemicals and fragrances are also concerned.

Scriba (2016) recently published an interesting update review on the contributions to the understanding of the binding mechanism between chiral selectors and cyclodextrins in analytical enantioseparations dating between 2012 and early 2016. The author showed that many tools are available nowadays to study the mechanism of enantiorecognition including spectroscopic techniques (NMR) as well as molecular modeling for the visualization and analysis of the dynamics of the process.

Selectors such as cyclodextrins appeared advantageous due to the much broader range of applications for structurally diverse analytes. Either random substituted cyclodextrins, which are mixtures of isomers with similar structure, or single isomer cyclodextrins are used. This high versatility makes cyclodextrins the first-choice selectors. The single isomers have the advantage of uniform structure, while the random substituted cyclodextrins often suffer from the batch-to-batch reproducibility (Li and Vigh 2004; Benkovics et al. 2016).

Lay et al. (2016) illustrated the exotic applications of imprinting techniques (IT) employing cyclodextrins. The exploitation of IT could produce products of molecularly imprinted polymers (MIP), which are very robust with long-term stability, reliability, cost-efficiency, and selectivity. MIP containing either cyclodextrin or its derivatives demonstrate superior binding effects for a target molecule e.g. steroidals, amino acids, polysaccharides, drugs, plant hormones, proteins, pesticides, and plastic additives. The authors concluded that cyclodextrins and their derivatives as emerging single or binary functional monomers are a versatile tool in separation science.

1.2.10 Cyclodextrins and Catalysis

Cyclodextrins have long been known to be good contributors to the development of catalytic processes. They have been used for mass-transfer catalytic reactions and for the production of new catalysts mimicking enzymatic activity (Macaev and Boldescu 2015; Hong et al. 2015). The following classification for catalytic applications of cyclodextrins was proposed by Monflier's group (Hapiot et al. 2006, 2011, 2014): (i) to significantly increase the rate and selectivity of reactions catalyzed by water-soluble organometallic complexes; (ii) to design new water-soluble ligands for aqueous organometallic catalysis; (iii) to stabilize catalytically-active noble metal nanoparticles in water; and (iv) to facilitate reactions catalyzed by supported metals or metallic powder in water. Cyclodextrins were used in organic synthesis as promoters or catalysts of different reactions, as components of artificial enzymatic systems, as stabilizers for the nanodimensional metallic catalysts. A comprehensive collection of recent breakthroughs in aqueous cyclodextrin-assisted supramolecular catalysis can be found in the review by Hapiot et al. (2014).

As already mentioned, cyclodextrins have been used as separating agents for racemic mixtures or in chiral resolution of enantiomers. The formation of an inclusion complex between the β-CD unit of a catalytic system and a substrate leads to an increase of the local concentration of the latter and immobilizes it near the catalytically-active center. As a result, acceleration in the reaction rate, higher substrate selectivity, enantioselectivity and regioselectivity can often be achieved. Recently, Macaev and Boldescu (2015) published a state-of-the-art of cyclodextrins in asymmetric and stereospecific synthesis. Three topics were summarized: (1) cyclodextrins' complexes with transition metals as asymmetric and stereospecific catalysts; (2) cyclodextrins' non-metallic derivatives as asymmetric and stereospecific catalysts; and (3) cyclodextrins promoting asymmetric and stereospecific catalysis in water. The authors concluded that cyclodextrins and their derivatives can be a feasible alternative to traditional catalysts in a variety of reactions. Particularly, cyclodextrins can play the role of nanoreactors in which the asymmetric or stereospecific synthesis is facilitated. A wide range of stereospecific and asymmetric reactions can be performed with the application of non-metallic derivatives of cyclodextrins as catalysts. These include reactions of halogenation and hydrohalogenation, oxidation, reduction, hydrogenation, aldol reactions, photolysis, addition and substitution reactions. Finally, an interesting discussion was made on the effects of different factors such as correlation between the substrate molecular dimensions and the cyclodextrin cavity size, geometry in the approach of the substrate to the "active site", influence of pH and temperature, on the enantioselectivity of these reactions.

1.2.11 Cyclodextrin-based Supramolecular Architectures

Supramolecular chemistry was defined as "*the chemistry of non-covalent interactions*" by the 1987 Nobel Prize Jean-Marie Lehn. These interactions can be used to form new nanoassemblies for multiple applications. The state of the art reviews of the design of complex macromolecular architectures based on cyclodextrin were presented by Schmidt et al. (2014) and Dong et al. (2015), and comprehensively discussed. In particular, by an elegant combination of dynamic/reversible structures with exceptional functions, functional supramolecular polymers are attracting increasing attention in various fields such as biomedical, e.g. gene transfection, protein delivery, bio-imaging and diagnosis, tissue engineering, and biomimetic chemistry, and material science (polymer science, nanotechnology). Although extensive work has been done on cyclodextrin-based supramolecular architectures, future research needs to take into account their precise physicochemical characterization Wenz et al. 2006; Wenz 2009; Li 2009; Hapiot et al. 2006; Harada et al. 2009a, 2009b; Dong et al. 2011, 2015; Schmidt et al. 2014; Agostoni et al. 2015; Karim and Loh 2016; Gontero et al. 2017; Valetti et al. 2017). Indeed, the understanding and the design of supramolecular systems require a detailed characterization with respect to stoichiometry, affinity, structure, heterogeneity, and supramolecular dynamics.

1.2.12 Cyclodextrins and Sugar-based Surfactants

Carbohydrate-based surfactants are today an important class of amphiphilic compounds which play an important role with both fundamental and practical applications. The growing interest in such compounds is due to, inter alia, their preparation from renewable raw materials, their ready biodegradability and biocompatibility, as well as other more basic reasons of practical, economic and environmental order. When complexed with cyclodextrins, carbohydrate-based surfactants considerably increase their performance and potential application range (Dodziuk 2006; Villalonga et al. 2007; Li et al. 2011). The formation of CD/surfactant host-guest compounds leads to an increase in the critical micelle concentration and in the solubility of surfactants. The use of these new systems is promising as reported by Valente and Söderman (2014).

1.2.13 Cyclodextrins and Click Chemistry

Click chemistry describes a family of modular, efficient, versatile and reliable reactions which have acquired a pivotal role as one of the most useful synthetic tools for functionalization of cyclodextrins with a potentially broad range of applications (Faugeras et al. 2012; Dondoni and Marra 2012; Schmidt et al. 2014; Hou et al. 2016). Cyclodextrins modified by the click reaction are building blocks of superstructures used for drug delivery systems (Venturini et al. 2008; Zhou and Ritter 2010; Nielsen et al. 2010), for the modification of macromolecular surfaces (Celebioglu et al. 2014a), for the preparation of cyclodextrin dimers and trimers (Mourer et al. 2008; Tungala et al. 2013), and for the generation of various glycoconjugates (glycopeptides, glycodendrimers, etc.) (Kushwaha et al. 2013; Dodziuk 2006; Bilensoy 2011). A tutorial review with the recent development in thiol-ene coupling for peptide glycosylation was published by Dondoni and Marra (2012). Cyclodextrins will play a very important role in all these new developments.

1.2.14 Cyclodextrins and Agrochemistry

An interesting sector for cyclodextrins is agrochemistry (Ho et al. 2014; Campos et al. 2015; Yusoff et al. 2016). Development of environmentally-friendly pesticides for sustainable agriculture is a key focus in the agrochemical industry. Identifying novel active ingredients and improving the delivery system of an active ingredient are the main challenges in developing new agrochemical formulations. Conventional formulations of agrochemicals are likely to contaminate the environment and there is a need for controlled-release formulations of agrochemicals to reduce pollution and health hazards. Strategies for the control of delivery systems for slow and sustained release of agrochemicals are of great interest to environmental scientists. In this context, the selection of materials to be used as carrier agents in agrochemical formulations is crucial. In recent years, the efficacy of cyclodextrins for pesticide formulations has been evaluated. Cyclodextrins can effectively encapsulate or bind the pesticide's active ingredients in the material's matrix with a sustained release profile and slow mobility in soil (Yusoff et al. 2016). It is well-known that cyclodextrin molecules form complexes with a wide variety of agricultural substances such as insecticides, fungicides, herbicides, repellents, pheromones, and growth regulators (Szente and Szejtli 1996; Morillo 2006; Luca and Grigoriu 2007; Venturini et al. 2008). This is at the origin of many benefits: modification of the physicochemical properties of the included guest, i.e., physical state, stability, solubility, and bioavailability, stabilization against the effects of light or biochemical degradation, and reduction of volatility. All these benefits are interesting during the preparation of the commercial formulations. It is important to note that most of the pesticides-CD inclusion complexes studied in the literature have used β-CD because of its lower price, and most of the published papers related to pesticides are not directly practice-oriented (Garrido et al. 2014; Fernandes et al. 2014).

1.2.15 Electrospinning of Functional Nanofibers with Cyclodextrins

Electrospinning (ES) is one of the most useful techniques for nanofibers (NF) production due to its versatility and cost-effectiveness. Discovered in the early 1900's, ES is a long-known polymer processing technique that has recently been rediscovered to produce nanofibers/nanowebs in order to create many products for medical (medical devices, tissue engineering scaffolds), textile (clothing) and environmental (filtration media, membranes) applications. Nanofibers also represent high potential in other fields such as sensors, electronics, energy and biotechnology. The principle is simple: ES is an electro-hydrodynamic process in which a charged polymer jet is collected on a grounded collector; a rapidly rotating collector results in aligned nanofibers while stationary collectors result in randomly oriented fiber mats. The polymer jet is formed when an applied electrostatic charge overcomes the surface tension of the solution. There is a minimum concentration for a given polymer, termed the critical entanglement concentration, below which a stable jet cannot be achieved and no nanofiber will form - although nanoparticles may be achieved (electrospray). The characteristics of nanofibers such as high surface-to-volume, porous structure, controllable fiber diameter and variety of morphologies (core-shell, hollow, porous) can be manipulated by changing the parameters of the process (i.e. viscosity, conductivity, flow rate, concentration, solvent and needle type). The ultra-fine fibers are 1000 times smaller than a single human hair and range from about ten nanometers to few microns in diameter. ES nanofibers possess several remarkable characteristics such as very large surface-to-volume ratio, high porosity within the nanoscale, and unique physical and mechanical properties along with the flexibility for chemical/physical (multi-)functionalization. These properties and multi-functionality make them favorable candidates in many areas including textiles, energy, sensors, electronics, healthcare, environmental (filtration), food, packaging, and also agriculture.

A literature data survey shows that studies mostly focused on polymeric nanofibers and their functionalization. Indeed, electrospinning of nanofibers from polymer solution is very common since entanglements and overlapping between macromolecular chains play crucial role for uniform fiber formation. Polymers have to possess high molecular weight and high polymer concentrations are desirable. Numerous polymers including polystyrene, polyester, cellulose acetate, polylactic acid, polycaprolactone, and polyethylene oxide were studied. However, polymer-based nanofibers production requires organic solvents and this causes some problems in specific applications.

Recently, electrospinning of functional nanofibers with cyclodextrin molecules was proposed, and an important contribution was made by Uyar's group on this topic (Celebioglu and Uyar 2012, 2013a, 2013b; Kayaci and Uyar 2012, 2014; Kayaci et al. 2013a, 2013b, 2013c, 2014, 2015; Celebioglu et al. 2014a, 2014b, 2016; Keskin et al. 2015a, 2015b; Aytac et al. 2015, 2016a, 2016b; Aytac and Uyar 2016, 2017; Senthamizhan et al. 2016). Functional nanofibers incorporating cyclodextrin were developed via ES using different polymers and reactions (cross-linking, grafting, click chemistry). Cyclodextrin molecules were also proposed to produce nanofibers without using any polymer carrier because cyclodextrin can form aggregates via intermolecular hydrogen bonding in their concentrated solutions or polymeric structures through cross-linked reactions. The functionalization of nanofibers with cyclodextrins and cyclodextrin inclusion complexes (denoted CD-ICs) is extremely attractive since electrospun nanofibers/nanowebs containing cyclodextrins and/or CD-ICs have unique characteristics that can potentially improve and broaden the application areas of cyclodextrins and electrospun nanofibers. Uyar and his research team proposed various materials: CD functionalized electrospinning polymeric nanofibers, molecular filters based on cyclodextrin functionalized electrospinning polymeric nanofibers, electrospinning of nanofibers from cyclodextrins, and electrospinning of CD-ICs nanofibers. Various applications were target of including functional materials for textile and medical textile, drug delivery, control release of antibacterials, packaging and food applications, complexation and microencapsulation (essential oils, antioxidants), and environmental purposes (e.g. filtration, purification and separation processes).

By means of the electrospinning technique, Uyar's group have successfully produced cyclodextrin functionalized polymeric nanofibers using polymers such as polystyrene, polymethylmethacrylate and polyethylene oxide. These polymers were blended with α-, β- and γ-cyclodextrins and electrospinning into uniform nanofibers. The authors noted that the addition of cyclodextrin in the polymer solutions facilitated the ability to produce electrospun nanofibers from polymer solutions at low polymer concentration yielding in bead-free and uniform nanofibers. Cyclodextrin molecules were homogeneously distributed within the polymer matrix without forming phase separated crystalline aggregates. However, the choice of polymeric matrix is important in which polymer chains should not make inclusion complexation with cyclodextrin cavity and therefore cyclodextrin will be available for further complexation depending on the application target. Uyar's group also demonstrated that the combination of cyclodextrins and the electrospinning nanofibers can potentially increase the efficiency of filters by facilitating complex formation with organic compounds and the very high surface area of the nanofibers.

1.2.16 Cyclodextrins and Remediation

In the last two decades, cyclodextrins as complexing agents have attracted considerable attention in environmental science in terms of removal of pollutants present in all environmental compartments, soils, air, waters and wastewaters, and sediments. Cyclodextrins can be used in soluble or insoluble form. In general, the removal of pollutants is mainly accounted for by the formation of inclusion processes due to cyclodextrin molecules and the specific structure of the material containing CDs used. Different materials have been proposed such as insoluble polymers, e.g. cross-linked structures, hydrogels and nanosponges, functionalized materials, e.g. polymers, silica and organic resins, organic-inorganic systems, membranes, and nanofibers (Crini and Morcellet 2002; Fakayode et al. 2007; Kozlowski and Sliwa 2010; Landy et al. 2012a, 2012b; Atteia et al. 2013; Morin-Crini and Crini 2014; Morin-Crini et al. 2015).

Soil flushing using cyclodextrin-based aqueous solutions was employed to solubilize pollutants. Cyclodextrin molecules are used as additives to enhance efficiencies and reduce the treatment time compared to the use of water alone or conventional surfactants. The review by Atteia et al. (2013) can be consulted on this topic.

Cyclodextrins found their role in the environmental risk management. Both risk assessment and risk reduction technologies may benefit from the CD's ability of forming inclusion complexes with the typical organic contaminants in soils as reported by Gruiz et al. (2011) and Fenyvesi et al. (2016). The solubility enhancement of these compounds of usually poor water solubility can be utilized. Extraction of soils with aqueous solutions of well soluble cyclodextrin derivatives (HP-β-CD or RAMEB, randomly methylated-β-CD) gives information on the contaminant fraction easy to mobilize. As this fraction is the most available for the soil microbes, extracting the soil with HP-β-CD solution is considered as a measure of the bioavailable, biodegradable fraction. Nowadays, this method has become a part of the everyday protocol to predict the microbial bioavailability of organic pollutants of soils.

The aim of the risk reduction technologies is not only the removal of the contaminant from the soil but also destroying the polluting compounds by chemical or biological means. These technologies have improved economic parameters when performed *in situ* (no need of transporting the soil or its extract). Both *in situ* chemical oxidation and microbial degradation show improved efficiency when combined with CD flushing. The lab-scale experiments with various pollutants including monoaromatics such as benzene, toluene, ethylbenzene and xylenes, and BTEX, polycyclic aromatic hydrocarbons (PAH), polychlorinated biphenyls (PCB), and motor oil proved that the bioavailability of the contaminants in bioremediation technologies is enhanced. RAMEB has a catalytic effect in *in situ* chemical oxidation of chlorinated solvents such as trichloroethylene. Several strictly controlled field demonstrations of the technologies showed that CD flushing can be adjusted to the current running technologies without special technical difficulties. The benefits are clear: saving time compared to the traditional technologies which may pay off the price of cyclodextrin. The vivid soil life proved that cyclodextrins were nontoxic adjuvants enhancing the bioavailability of the nutrients as well. The biodegradability of the cyclodextrin derivatives used in soil remediation is pre-requisite of their application. The cyclodextrin derivatives were degraded in a lower rate compared to the underivatized ones. However, even the less degradable RAMEB was slowly degraded by the microflora of contaminated soils.

Cyclodextrin-based materials for water and wastewater treatment include cross-linked polymers and nanosponges, membranes, nanofibers, and functionalized systems such as polymers, silica, and organic resins. Morin-Crini and Crini (2013) reviewed the developments in the use of cross-linked cyclodextrin-based polymers as complexing polymeric matrices for pollutant removal by oriented-adsorption processes. The authors summarized the features of these polymers and how they were used in decontamination applications. Numerous interesting studies on the treatment of real effluents can be found in the literature (Nagy et al. 2014; Khaoulani et al. 2015; Morin-Crini et al. 2015; Charles et al. 2016; Euvrard et al. 2016; Han et al. 2016; Alsbaiee et al. 2016). Imprinting technology (IT) makes possible to develop sorbents with extremely high specificity (Lay et al. 2016). Cyclodextrin-containing molecular imprinting polymers and cyclodextrins grafted to the surface of silica or other support *via* IT offer multiple binding sites with tunable surface properties for sorption of selected chemicals, e.g. perfluorinated compounds (Karoyo and Wilson 2015).

Nanofibers containing cyclodextrin molecules are another interesting material for pollutant removal. All the results published by Uyar's group demonstrated that these new materials may have high potential to be used as air filters for the removal of organic vapor wastes, adsorbents/filters for pollutant removal from aqueous solutions, and also drug delivery system. The pollutants studied include dyes (Keskin et al. 2015a, 2015b), volatile organic compounds (VOC) (Celebioglu et al. 2016; Kayaci et al. 2015; Kayaci and Uyar 2014), aromatic derivatives such as PAH (Celebioglu et al. 2014a; Kayaci et al. 2013a; b; c), triclosan (Celebioglu et al. 2014b), metals (Senthamizhan et al. 2016) and drugs (Aytac et al. 2015). Electrospun nanofibers without having cyclodextrin were ineffective for entrapment of pollutants whereas cyclodextrin-based nanofibrous membranes can effectively entrap pollutants by taking advantage of the high surface-volume ratio of nanofibers with the added advantage of inclusion complexation capability of cyclodextrins presenting on the nanofiber surface. Uyar's group proposed cyclodextrin functionalized cellulose acetate nanofibers prepared by combining electrospinning and click reaction as efficient adsorbent for the purpose of water purification and wastewater treatment containing PAH molecules (Celebioglu et al. 2014a). First, cyclodextrin molecules and cellulose acetate nanofibers were modified to obtain azide-CD and propargyl-terminated nanofibers, respectively. Then, click chemistry reaction was performed between these two derivatives to obtain permanent grafting of cyclodextrin molecules onto nanofibers surface. Click chemistry can be an interesting alternative to modify the surface of nanofibers because these reactions showed high yields. The adsorption capacity of cyclodextrin-grafted electrospun cellulose acetate nanofibers were determined by removing phenanthrene as guest model present in aqueous solutions. The material exhibited high adsorption properties and all the results were interpreted using adsorption surface due to the high surface area of nanofibers and chemisorption via inclusion complexation. Another alternative proposed was to pre-polymerize cyclodextrin onto electrospun polyester nanofibers using citric acid as cross-linking agent. However, although, nanofibers have shown enhanced mechanical and thermal properties, their surface areas were less important (Kayaci and Uyar 2012). Uyar's group also published a series of papers on the electrospinning of nanofibers from non-polymeric systems such as native cyclodextrin molecules (Celebioglu and Uyar 2013a, 2013b, Celebioglu and Uyar 2012; Celebioglu et al. 2016). They have successfully performed electrospinning of nanofibers from α-CD, β-CD and γ-CD molecules and their derivatives (such as hydroxypropyl or methyl derivatives) without using a carrier polymer matrix. Similar to polymeric systems, the electrospinning of cyclodextrin solutions resulted in different morphologies and average fiber diameters depending on the cyclodextrin type and cyclodextrin concentration. Also, the success of the electrospinning of the cyclodextrin-nanofibers strongly depends on type of solvent (water, DMF) and intermolecular interactions between the cyclodextrin molecules. The existence of cyclodextrin aggregates via hydrogen bonding and very high solution viscosity and viscoelastic solid-like behavior of cyclodextrin solutions were found to be the key factors for obtaining bead-free nanofibers from cyclodextrin. The addition of urea disrupted cyclodextrin aggregates and lowered the viscosity significantly, and therefore, the urea-added cyclodextrin solutions yielded beaded fibers and/or beads. Using XRD data, the authors showed that electrospun cyclodextrin-nanofibers have amorphous characteristic without showing any particular orientation or crystalline aggregation of cyclodextrin molecules. Celebioglu and Uyar (2013b) demonstrated that γ-CD nanofibers were efficient for entrapping of aniline and toluene as guest VOC models by inclusion complexation whereas γ-CD in powder form did not show any entrapment ability. The driving forces of inclusion complex formation proposed were electrostatic interactions, Van der Waals contributions and hydrogen bonding. The inclusion complexation ability of cyclodextrin molecules was combined with very high surface area and versatile features of nanofibers. The authors concluded that electrospun cyclodextrin-nanofibers could serve as useful filtering material for air filtration purposes due to their molecular entrapment capability of volatile organic compounds.

There are several other interesting works for synthesizing membranes and nanofibers containing cyclodextrins (Meng et al. 2014; Xiao et al. 2014; Ghemati and Aliouche 2014; Zhao et al. 2015; Zhang et al. 2015; Wu et al. 2015a; Norena-Caro and Alvarez-Lainez 2016; Wei et al. 2016; Costoya et al. 2017). For example, Norena-Caro and Alvarez-Lainez (2016) proposed two different methods for the production of electrospun polyacrylonitrile (PAN) nanofibers containing cyclodextrin capable of capturing formaldehyde, a common indoor pollutant. The former comprised the addition of cyclodextrins to PAN/dimethyl sulfoxide solutions and the subsequent electrospinning of the mixture. The latter involved the crosslinking of cyclodextrins on electrospun PAN fibers by alkaline hydrolysis and esterification with citric acid. The results showed that both functionalized nanofibers might be used for indoor air purification. However, functionalized fibers obtained by addition of cyclodextrins were more effective for capturing formaldehyde than fibers obtained by crosslinking of cyclodextrins.

Cyclodextrin-functionalized silica networks used as adsorbents/filters for environmental applications has recently received a lot of attention as reported by Gibson (2014), Samiey et al. (2014), Lee and Park (2015), Dinker and Kulkarni (2015), Vunain et al. (2016), Mahmud and Wilson (2016), and Yamamoto and Kuroda (2016). Two main class of materials have been proposed: cyclodextrin-functionalized silicas prepared through grafting reactions and cyclodextrin-silica hybrid systems prepared through sol-gel or self-assembly process. These materials could be applied in the elimination, enrichment and detection of environmental pollutants in air and water samples.

Actually, fundamental research is also focusing on cyclodextrin-based nanoparticles for environmental applications (Taka et al. 2017). Indeed, nanosponges have not only been explored for their pharmaceutical applications but for water purification and wastewater treatment (Landy et al. 2012a; Tong and Chen 2013; Taka et al. 2017). This emerging technology of cyclodextrin-based nanosponges is expected to provide technical solutions to water treatment.

To conclude, the environmental application of cyclodextrins was one of the focuses of the 18th International Cyclodextrin Symposium held in Gainesville on May 19-21, 2016. Two of the 4 invited lectures, 7 of 24 oral presentations and 6 of 31 posters were related to this topic. However, in spite of numerous results, publications, and patents, cyclodextrins are still not used for soil remediation and water treatment. Why? It is difficult to respond.

1.2.17 Cyclodextrins: Other Interesting Selected Reviews and Works

Szente et al. (2016) published a comprehensive overview on the methods used for analysis of cyclodextrins and their derivatives. The paper intends to act as a guide for the readers in looking around the classical and modern instrumental analytical methods suitable for identification, characterization and determination of cyclodextrins themselves, cyclodextrins in finished products or even in biological samples. A thorough analytical characterization of cyclodextrin complexes is of fundamental importance to provide an adequate support in selection of the most suitable cyclodextrin for each guest molecule, and also in view of possible future marketing of drug-cyclodextrin formulations for example. However, the analytical characterization of drug-cyclodextrin solid systems and the assessment of the actual inclusion complex formation is not a simple task and involves the combined use of several analytical techniques as reported by Mura (2015). In this review, the author also presented a general prospect of the principal analytical techniques which can be employed for a suitable characterization of drug-cyclodextrin systems in the solid state, evidencing their respective potential advantages and limits. The applications of each examined technique were described and discussed by pertinent examples from literature.

Curcumionoids such as curcumin and related compounds are well-known, in particular in South Asia for their antioxidant, antibacterial, antifungal, anticancer and anti-inflammatory effects. Curcumin/CD formulations for application as food supplement are on the market. Recently, the benefits in cystic fibrosis and Alzheimer disease have also been proved in cellular and animal models (Higdon and Drake 2014). Curcumin is an oil-soluble polyphenol pigment, practically insoluble in water at acidic and neutral pH, but soluble in alkali. It has low bioavailability owing to the poor aqueous solubility. Recent studies showed that natural cyclodextrin molecules can enhance notably the solubility of curcumin. This result in enhanced bioavailability and improved clinical effects (e.g. aggregation of curcumin/CD complex in spherical nanoparticles form helps the cell penetration). Complexation with cyclodextrins also protects from decomposition upon UV-exposure. Further results on prostate cancers are expected in the near future (Boztas et al. 2013; Zhang et al. 2016).

Rezanka (2016) recently reviewed monosubstituted cyclodextrin derivatives with particular focus on the synthesis of allyl, cinnamyl, propargyl, formylmethyl, carboxymethyl, azido and amino derivatives. Synthesis and modification methods of magnetic chitosan materials containing cyclodextrins were reviewed along with some interesting applications in analytical separations by Tong and Chen (2013). Magnetic nanoparticles grafted with cyclodextrin were developed for various applications, such as solid phase extraction that is for sample concentration prior analysis, for magnetoresponsive drug delivery (Banerjee and Chen 2008) and for magnetic manipulation of cholesterol removal (Horák et al. 2017).

The advances in the developments of supramolecular hydrogels based on the polypseudorotaxanes and polyrotaxanes formed by inclusion complexes of cyclodextrins threading onto polymer chains were detailed by Li (2009). Both physical and chemical hydrogels of many different types were discussed with respect to their preparation, structure, property, and gelation mechanism. A large number of physical supramolecular hydrogels were formed induced by self-assembly of densely packed cyclodextrin rings threaded on polymer or copolymer chains acting as physical crosslinking points. The thermo-reversible and thixotropic properties of these hydrogels had inspired their applications as injectable drug delivery systems. Slide-ring gels of extreme swelling capacity are formed by threading cyclodextrin polymers onto polymer backbones (Kali et al. 2015; Murakami et al. 2017). Chemical supramolecular hydrogels synthesized from polypseudorotaxanes and polyrotaxanes were based on the chemical crosslinking of either the cyclodextrin molecules or the included polymer chains. The chemical supramolecular hydrogels were often made biodegradable through incorporation of hydrolyzable threading polymers, end caps, or cross-linkers, for their potential applications as biomaterials.

While interactions of cyclodextrins with metal ions have been studied for decades, structurally well-defined systems are relatively rare and the application of cyclodextrin-based functional materials is in its initial phase. The review published by Prochowicz et al. (2016) focuses on the synthesis, reactivity and structural diversity of well-defined metal complexes derived essentially from native cyclodextrins. Various structural motifs for metal complexes based on cyclodextrins were delineated ranging from monomeric species, dinuclear systems, homo- and heterometallic sandwich-type complexes to cylindrical, extended structures. The reported examples are discussed with an emphasis placed on how the character of used metal or auxiliary ions, and the formation of intra- and/or intermolecular hydrogen bonds can influence the mode of aggregation and supramolecular arrangement of the resulting metal complexes.

Letort et al. (2016) provided an update on the current use of cyclodextrins against organophosphorus compound intoxications. Organophosphorus pesticides and nerve agents play a determinant role in the inhibition of cholinesterases. The cyclic structure of cyclodextrin molecules and their toroidal shape are perfectly suitable to design new chemical scavengers able to trap and hydrolyze the organophosphorus compounds before they reach their biological target.

Cyclodextrin-based nanosponges found a broad range of applications. However, their structural and dynamic characterization is still a challenge, mainly due to their amorphous state. Mele's group demonstrated that NMR spectroscopy, especially high resolution magic angle spinning techniques, is an interesting tool to obtain information not only on their structure but also on the dynamics of organic molecules encapsulated in these polymeric hydrogels (Mallard et al. 2015; Ferro et al. 2014, 2016).

1.3 Conclusion

This chapter gives an overview of recent selected works on cyclodextrins used in various fields. Large-scale applications of cyclodextrins have been made possible by their ease of production and low cost, in particular for the native molecules, non-toxicity, biocompatibility and their ability to form inclusion complexes. This is the basis for their applications in the pharmaceutical, food, and cosmetic industries. Other practical and industrial applications have been reported in other sectors such as chromatography and biomedicine. The increasing number of publications on cyclodextrins shows that there is always an increasing interest in these molecules and their applications from academic and practical point of views. The emergence of cyclodextrins as active and smart molecules rather than complexing molecules in numerous cosmetic, textile, therapeutic and biomedical products seems to be the next step in the development of cyclodextrin technology. Further studies, contributions, and industrial developments are expected in the near future in the following domains: catalysis, bacterial resistance, anticancer drugs (chemotherapy), magnetic resonance imaging, biotechnology, e.g. biotransformation, fermentation processing, enzyme models, peptide and protein delivery, material science, e.g. wrapping materials and packaging, cosmeto-textiles, agrochemistry, and sensor applications.

References

Abdel-Halim ES, Fouda MMG, Hamdy I, Abdel-Mohdy FA, El-Sawy SM (2010) Incorporation of chlorohexidin diacetate into cotton fabrics grafted with glycidyl methacrylate and cyclodextrin. Carbohydr Polym 79:47–53. https://doi.org/10.1016/j.carbpol.2009.07.050

Agostoni V, Horcajada P, Noiray M, Malanga M, Aykaç A, Jicsinszky L, Vargas-Berenguel A, Semiramoth N, Daoud-Mahammed S, Nicolas V, Martineau C, Taulelle F, Vigneron J, Etcheberry F, Serre C, Gref R (2015) A "green" strategy to construct non-covalent, stable and bioactive coatings on porous MOF nanoparticles. Sci Rep 5:7925. https://doi.org/10.1038/srep07925

Ahuja A, Baboota S, Ali J, Mustafa G (2011) Cyclodextrins as potential excipients in pharmaceutical formulations: solubilizing and stabilizing effects. In: Bilensoy E (ed) Cyclodextrins in pharmaceutics, cosmetics and biomedicine: current and future industrial. John Wiley. Chapter 2, Hoboken, pp 19–43. https://doi.org/10.1002/9780470926819

Alsbaiee A, Smith BJ, Xiao L, Ling YH, Helbling DE, Dichtel WR (2016) Rapid removal of organic micropollutants from water by a porous beta-cyclodextrin polymer. Nature 529:190–U146. https://doi.org/10.1038/nature16185

Ammala A (2013) Biodegradable polymers as encapsulation materials for cosmetics and personal care markets. Int J Cosmet Sci 35:113–124. https://doi.org/10.1111/ics.12017

Ammayappan L, Moses JJ (2009) An overview on application of cyclodextrins in textile product enhancement. J Text Assoc 70:9–18

Andreaus J, Dalmolin MC, De Oliveira IB, Barcellos IO (2010) Application of cyclodextrins in textile processes. Quim Nova 33:929–937. https://doi.org/10.1590/S0100-40422010000400031

Appendini P, Hotchkiss JH (2002) Review of antimicrobial food packaging. Innovative Food Sci Emerg Technol 3:113–126. https://doi.org/10.1016/S1466-8564(02)00012-7

Armspach D, Gattuso G, Koeniger R, Stoddart JF (1999) Cyclodextrins. In: Hecht SM (ed) Bioorganic chemistry: carbohydrates. Oxford University Press, New York, pp 458–488

Artiss JD, Brogan K, Brucal M, Moghaddam M, Jen KL (2006) The effects of a new soluble dietary fiber on weight gain and selected blood parameters in rats. Metabolism 55:195–202

Assaf KI, Gabel D, Zimmermann W, Nau WM (2016) High-affinity host-guest chemistry of large-ring cyclodextrins. Org Biomol Chem 14:7702–7706. https://doi.org/10.1039/c6ob01161f

Astray G, Gonzalez-Barreiro C, Mejuto JC, Rial-Otero R, Simal-Gandara J (2009) A review on the use of cyclodextrins in foods. Food Hydrocoll 23:1631–1640. https://doi.org/10.1016/j.foodhyd.2009.01.001

Atteia O, Estrada ED, Bertin H (2013) Soil flushing: a review of the origin of efficiency variability. Rev Environ Sci Biotechnol 12:379–389. https://doi.org/10.1007/s11157-013-9316-0

Atwood JL, Steed JW (2004) Encyclopedia of supramolecular chemistry. Marcel Dekker, Inc., New-York

Atwood JL, Steed JW (2009) Supramolecular chemistry, 2nd edn. John Wiley, Chichester

Atwood JL, Davies EE, MacNicol DD (1984) Inclusion compounds. Academic Press, Michigan

Auzely-Vélty R (2011) Self-assembling polysaccharide systems based on cyclodextrin complexation: synthesis, properties and potential applications in the biomaterials field. CR Chim 14:167–177. https://doi.org/10.1016/j.crci.2010.04.019

Aytac Z, Uyar T (2016) Antioxidant activity and photostability of α-tocopherol/β-cyclodextrin inclusion complex encapsulated electrospun polycaprolactone nanofibers. Eur Polym J 79:140–149. https://doi.org/10.1016/j.eurpolymj.2016.04.029

Aytac Z, Uyar T (2017) Core-shell nanofibers of curcumin/cyclodextrin inclusion complex and polylactic acid: enhanced water solubility and slow release of curcumin. Int J Pharm 518:177–184. https://doi.org/10.1016/j.ijpharm.2016.12.061

Aytac Z, Sen HS, Durgun E, Uyar T (2015) Sulfisoxazole/cyclodextrin inclusion complex incorporated in electrospun hydroxypropyl cellulose nanofibers as drug delivery system. Colloids Surf B Biointerfaces 128:331–338. https://doi.org/10.1016/j.colsurfb.2015.02.019

Aytac Z, Yildiz ZI, Kayaci F, San NO, Kusku SI, Durgun E, Tekinay T, Uyar T (2016a) Fast-dissolving, prolonged release and antibacterial cyclodextrin/limonene-inclusion complex nanofibrous webs via polymer-free electrospinning. J Agric Food Chem 64:7325–7334. https://doi.org/10.1021/acs.jafc.6b02632

Aytac Z, Yildiz ZI, Kayaci F, San NO, Tekinay T, Uyar T (2016b) Electrospinning of polymer-free cyclodextrin/geraniol-inclusion complex nanofibers: enhanced shelf-life of geraniol with antibacterial and antioxidant properties. RSC Adv 6:46089–46099. https://doi.org/10.1039/C6RA07088D

Banerjee SS, Chen DH (2008) Cyclodextrin conjugated magnetic colloidal nanoparticles as a nanocarrier for targeted anticancer drug delivery. Nanotechnology 19:265602. https://doi.org/10.1088/0957-4484/19/26/265602

Bar R (1996) Applications of cyclodextrins in biotechnology. In: Szejtli J, Osa T (eds) Comprehensive supramolecular chemistry, Volume 3. Pergamon Oxford, London, pp 423–440

Bender H (1986) Production, characterization and applications of cyclodextrins. In: Liss AR (ed) Advances in biotechnological processes. Wiley. Chapter 6, New York, pp 31–71

Bender ML, Komiyama M (1978) Cyclodextrins chemistry. Springer Verlag, Berlin

Benkovics G, Fejős I, Darcsi A, Varga E, Malanga M, Fenyvesi É, Sohajda T, Szente L, Sz B, Szemán J (2016) Single-isomer carboxymethyl-γ-cyclodextrin as chiral resolving agent for capillary electrophoresis. J Chromatogr A 1467:445–453. https://doi.org/10.1016/j.chroma.2016.06.083

Berendt RT, Sperger DM, Isbester PK, Munson EJ (2006) Solid-state NMR spectroscopy in pharmaceutical research and analysis. TrAC Trends Anal Chem 25:977–984. https://doi.org/10.1016/j.trac.2006.07.006

Bergeron RJ (1984) Cycloamylose-substrate binding. In: Inclusion compounds - physical properties and applications. Atwood JL, Davies JED, MacNicol DD, eds. London: Academic Press. Volume 3, chapter 12, pp 391-443. ISBN-13: 978-0120671038

Bhaskara-Amrit UR, Agrawal PB, Warmoeskerken MMCG (2011) Applications of β-cyclodextrins in textiles. Autex Res J 11:94–101. http://autexrj.org/No4-2011-/0020_11.pdf

Bhosale SV, Bhosale SV (2007) Beta-cyclodextrin as a catalyst in organic synthesis. Mini-Rev Org Chem 4:231–242. https://doi.org/10.2174/157019307781369922

Bilensoy E (2011) Cyclodextrins in pharmaceutics, cosmetics and biomedicine: current and future industrial applications. John Wiley, Hoboken, 395 p. https://doi.org/10.1002/9780470926819

Bilensoy E, Hincal AA (2009) Recent advances and future directions in amphiphilic cyclodextrin nanoparticles. Expert Opin Drug Deliv 6:1161–1173. https://doi.org/10.1517/17425240903222218

Booij LHDJ (2009) Cyclodextrins and the emergence of Sugammadex. Anaesthesia 64:31–37. https://doi.org/10.1111/j.1365-2044.2008.05868.x

Boztas AO, Karakuzu O, Galante G, Ugur Z, Kocabas F, Altuntas CZ, Yazaydin AO (2013) Synergistic interaction of paclitaxel and curcumin with cyclodextrin polymer complexation in human cancer cells. Mol Pharm 10:2676–2683. https://doi.org/10.1021/mp400101k

Brackman G, Garcia-Fernandez MJ, Lenoir J, De Meyer L, Remon JP, De Beer T, Concheiro A, Alvarez-Lorenzo C, Coenye T (2016) Dressings loaded with cyclodextrin-hamamelitannin complexes increase *Staphylococcus aureus* susceptibility toward antibiotics both in single as well as in mixed biofilm communities. Macromol Biosci 16:859–869. https://doi.org/10.1002/mabi.201500437

Brandariz I, Iglesias E (2013) Local anesthetics: acid-base behavior and inclusion with cyclodextrins. Curr Org Chem 10:3050–3063. https://doi.org/10.2174/13852728113179990023

Brewster ME, Loftsson T (2007) Cyclodextrins as pharmaceutical solubilizers. Adv Drug Deliv Rev 59:645–666. https://doi.org/10.1016/j.addr.2007.05.012

Buschmann HJ, Schollmeyer EJ (2002) Applications of cyclodextrins in cosmetic products: a review. J Cosmet Sci 53:185–191

Buschmann HJ, Schollmeyer E (2004) Cosmetic textiles: A new functionality of clothes. Cosmet Toiletries 11:105–112

Buschmann HJ, Denter U, Knittel D, Schollmeyer E (1998) The use of cyclodextrins in textile processes - an overview. J Text Inst 89:554–561. https://doi.org/10.1080/00405009808658641
Cabral Marques HM (2010) A review on cyclodextrin encapsulation of essential oils and volatiles. Flavour Fragr J 25:313–326. https://doi.org/10.1002/ffj.2019
Cairo MR (2011) Structural aspects of crystalline derivatized cyclodextrins and their inclusion complexes. Curr Org Chem 15:815–830. https://doi.org/10.2174/138527211794518862
Caliceti P, Salmaso S, Bersani S (2010) Polysaccharide-based anticancer prodrugs. In: Reddy LH, Couvreur P (eds) Macromolecular anticancer therapeutics. Springer, New York, pp 163–166
Calo JR, Crandall PG, O'Bryan CA, Ricke SC (2015) Essential oils as antimicrobials in food systems - a review. Food Control 54:111–119. https://doi.org/10.1016/j.foodcont.2014.12.040
Campos EVR, de Oliveira JL, Fraceto LF, Singh B (2015) Polysaccharides as safer release systems for agrochemicals. Agron Sustain Dev 35:47–66. https://doi.org/10.1007/s13593-014-0263-0
Celebioglu A, Uyar T (2012) Electrospinning of nanofibers from non-polymeric systems: polymer-free nanofibers from cyclodextrin derivatives. Nanoscale 4:621–631. https://doi.org/10.1039/c1nr11364j
Celebioglu A, Uyar T (2013a) Electrospinning of nanofibers from non-polymeric systems: electrospun nanofibers from native cyclodextrins. J Colloid Interface Sci 404:1–7. https://doi.org/10.1016/j.jcis.2013.04.034
Celebioglu A, Uyar T (2013b) Electrospun gamma-cyclodextrin (gamma-CD) nanofibers for the entrapment of volatile organic compounds. RSC Adv 3(45):22891–22895. https://doi.org/10.1039/C3RA44870C
Celebioglu A, Demirci S, Uyar T (2014a) Cyclodextrin-grafted electrospun cellulose acetate nanofibers via click reaction for removal of phenanthrene. Appl Surf Sci 305:581–588. https://doi.org/10.1016/j.apsusc.2014.03.138
Celebioglu A, Umu OCO, Tekinay T, Uyar T (2014b) Antibacterial electrospun nanofibers from triclosan/cyclodextrin inclusion complexes. Colloids Surf B Biointerfaces 116:612–619. https://doi.org/10.1016/j.colsurfb.2013.10.029
Celebioglu A, Sen HS, Durgun E, Uyar T (2016) Molecular entrapment of volatile organic compounds (VOCs) by electrospun cyclodextrin nanofibers. Chemosphere 144:736–744. https://doi.org/10.1016/j.chemosphere.2015.09.029
Charles J, Bradu C, Morin-Crini N, Sancey B, Winterton P, Torri G, Badot PM, Crini G (2016) Pollutant removal from industrial discharge water using individual and combined effects of adsorption and ion-exchange processes: Chemical abatement. J Saudi Chem Soc 20:185–194. https://doi.org/10.1016/j.jscs.2013.03.007
Chen G, Jiang M (2001) Cyclodextrin-based inclusion complexation bridging supramolecular chemistry and macromolecular self-assembly. Chem Soc Rev 40:2254–2266. https://doi.org/10.1039/c0cs00153h
Chilajwar SV, Pednekar PP, Jadhav KR, Gupta GJC, Kadam VJ (2014) Cyclodextrin-based nanosponges: A propitious platform for enhancing drug delivery. Expert Opin Drug Deliv 11:111–120. https://doi.org/10.1517/17425247.2014.865013
Citernesi U, Sciacchitano M (1995) Cyclodextrins in functional dermocosmetics. Cosmetics Toiletries 110:53–61
Clarke RJ, Coates JH, Lincoln SF (1988) Inclusion complexes of the cyclomalto-oligosaccharides (cyclodextrins). Adv Carbohydr Chem Biochem 46:205–249. https://doi.org/10.3998/ark.5550190.0006.e14
Concheiro A, Alvarez-Lorenzo C (2013) Chemically cross-linked and grafted cyclodextrin hydrogels: From nanostructures to drug-eluting medical devices. Adv Drug Deliv Rev 65:1188–1203. https://doi.org/10.1016/j.addr.2013.04.015
Connors KA (1997) The stability of cyclodextrin complexes in solution. Chem Rev 97:1325–1357. https://doi.org/10.1021/cr960371r
Costoya A, Concheiro A, Alvarez-Lorenzo C (2017) Electrospun fibers of cyclodextrins and poly(cyclodextrins). Molecules 22:230. https://doi.org/10.3390/molecules22020230

Cram DJ (1988) The design of molecular hosts, guests and their complexes. Angew Chem Int Ed 27(8):1009–1020. https://doi.org/10.1002/anie.198810093

Crini G (2006) Non-conventional adsorbents for dye removal: A review. Bioresour Technol 97:1061–1085. https://doi.org/10.1016/j.biortech.2005.05.001

Crini G (2014) Review: a history of cyclodextrins. Chem Rev 114:10940–10975. https://doi.org/10.1021/cr500081p

Crini G, Morcellet M (2002) Synthesis and applications of adsorbents containing cyclodextrins. J Sep Sci 25:789–813. https://doi.org/10.1002/1615-9314(20020901)25:13<789::AID-JSSC789>3.0.CO;2-J

Davidson CD, Fishman YI, Puskás I, Szemán J, Sohajda T, McCauliff LA, Sikora J, Storch J, Vanier MT, Szente L, Walkley SU, Dobrenis K (2016) Efficacy and ototoxicity of different cyclodextrins in Niemann-Pick C disease. Ann Clin Transl Neuro 3:366–380. https://doi.org/10.1002/acn3.306

Davis ME, Brewster ME (2004) Cyclodextrin-based pharmaceutics: past, present and future. Nat Rev Drug Discov 3:1023–1035. https://doi.org/10.1038/nrd1576

Davis F, Higson S (2011) Cyclodextrins. In: Davis F, Higson S (eds) Macrocycles: construction, chemistry and nanotechnology applications. John Wiley, United Kingdom. Chapter 6

Decock G, Fourmentin S, Surpateanu GG, Landy D, Decock P, Surpateanu G (2006) Experimental and theoretical study on the inclusion compounds of aroma components with β-cyclodextrins. Supramol Chem 18:477. https://doi.org/10.1080/10610270600665749

Decock G, Landy D, Surpateanu G, Fourmentin S (2008) Study of the retention of aroma components by cyclodextrins by static headspace gas chromatography. J Incl Phenom Macrocycl Chem 62:297. https://doi.org/10.1007/s10847-008-9471-z

Deorsola AC, Mothé CG, de Oliviera LG, Deorsola AB (2014) Technological monitoring of cyclodextrin – World panorama. World Patent Inf 39:41–49. https://doi.org/10.1016/j.wpi.2014.06.004

Dinker MK, Kulkarni PS (2015) Recent advances in silica-based materials for the removal of hexavalent chromium: a review. J Chem Eng Data 60:2521–2540. https://doi.org/10.1021/acs.jced.5b00292

Dodziuk H (2002) Introduction to supramolecular chemistry. Kluwer Academic Publishers, Dordrecht. 10.1007/0-306-47587-1 (e-book)

Dodziuk H (ed) (2006) Cyclodextrins and their complexes: chemistry, analytical methods, applications. Wiley-VCH Verlag GmbH & Co. KGaA, Weinheim, 489p. https://doi.org/10.1002/3527608982

Donati F (2011) Sugammadex: a cyclodextrin-based novel formulation and marketing story. In: Bilensoy E (ed) Cyclodextrins in pharmaceutics, cosmetics and biomedicine: current and future industrial applications. John Wiley. Chapter 19, Hoboken, pp 636–370. https://doi.org/10.1002/9780470926819

Dondoni A, Marra A (2012) Recent applications of thiol-ene coupling as a click process for glycoconjugation. Chem Soc Rev 41:573–586. https://doi.org/10.1039/c1cs15157f

Dong HQ, Li YY, Li L, Shi DL (2011) Cyclodextrins/polymer based (pseudo)polyrotaxanes for biomedical applications. Prog Chem 23:914–922

Dong RJ, Zhou YF, Huang XH, Zhu XY, Lu YF, Shen J (2015) Functional supramolecular polymers for biomedical applications. Adv Mater 27:498–526. https://doi.org/10.1002/adma.201402975

Duchêne D (ed) (1987) Cyclodextrins and their industrial uses. Éditions de Santé, Paris

Duchêne D (ed) (1991) New trends in cyclodextrins and derivatives. Éditions de Santé, Paris

Duchêne D, Vaution C, Glomot F (1986) Cyclodextrins, their value in pharmaceutical technology. Drug Dev Ind Pharm 12:2193–2215

EFSA (2012) Scientific Opinion on the substantiation of health claims related to alpha-cyclodextrin and reduction of post prandial glycaemic responses (ID 2926, further assessment) pursuant to Article 13(1) of Regulation (EC) No 1924/2006. EFSA J 10:2713–2730. https://doi.org/10.2903/j.efsa.2012.2713

Endo T (2011) Large-ring cyclodextrins. Trends Glycosci Glycotechnol 23:79–92. https://doi.org/10.4052/tigg.23.79
Endo T, Ueda H (2004) Large ring cyclodextrins - recent progress. Fabad. J Pharm Sci 29:27–38. https://doi.org/10.1002/chin.200638261
Euvrard E, Morin-Crini N, Druart C, Bugnet J, Martel B, Cosentino C, Moutarlier V, Crini G (2016) Cross-linked cyclodextrin-based material for treatment of metals and organic substances present in industrial discharge waters. Beilstein J Org Chem 12:1826–1838. https://doi.org/10.3762/bjoc.12.172
Fakayode SO, Lowry M, Fletcher KA, Huang XD, Powe AM, Warner IM (2007) Cyclodextrins host-guest chemistry in analytical and environmental chemistry. Curr Anal Chem 3:171–181. https://doi.org/10.2174/157341107781023811
Fanali S (1993) Use of cyclodextrins in capillary electrophoresis. In: Capillary electrophoresis technology. Guzman NA, ed. New York: Marcel Dekker Inc. Chromatographic Science Series, Volume 64, part V, pp. 731–752. ISBN: 0-8247-9042-1
Fanali S, Cristalli M, Vespalec R, Bocek P (1994) Chiral separations in capillary electrophoresis. In: Chrambach A, Dunn MJ, Radola BJ (eds) Advances in electrophoresis. VCH Verlagsgesellschaft mbH. Chapter 7, Weinheim, pp 1–88
Fang Z, Bhandari B (2010) Encapsulation of polyphenols- A review. Trends Food Sci Technol 21:510–523. https://doi.org/10.1016/j.tifs.2010.08.003
Faugeras PA, Boens B, Elchinger PH, Brouillette F, Montplaisir D, Zerrouki R, Lucas R (2012) When cyclodextrins meet click chemistry. Eur J Org Chem (22):4087–4105. https://doi.org/10.1002/ejoc.201200013
Fenyvesi É (1988) Cyclodextrin polymers in the pharmaceutical industry. J Incl Phenom 6:537–545. https://doi.org/10.1007/BF00660751
Fenyvesi É, Szente (2016) Nanoencapsulation of flavors and aromas by cyclodextrins. In: Encapsulations: nanotechnology in the agri-food industry. Grumezescu A, ed. 1st edition. Volume 2, chapter 18, pp. 769–792. ISBN: 978-0-12-804378-3
Fenyvesi É, Vikmon MA, Szente L (2016) Cyclodextrins in food technology and human nutrition: benefits and limitations. Crit Rev Food Sci Nutr 56:1981–2004. https://doi.org/10.1080/10408398.2013.809513
Fernandes C, Encarnação I, Gaspar A, Garrido J, Borges F, Garrido EM (2014) Influence of hydroxypropyl-cyclodextrin on the photostability of fungicide pyrimethanil. Int J Photoenergy 2014:1–8. https://doi.org/10.1155/2014/489873
Ferro M, Castiglione F, Punta C, Meloni L, Panzeri W, Rossi B, Trotta F, Mele A (2014) Anomalous diffusion of ibuprofen in cyclodextrin nanosponge hydrogels: an HRMAS NMR study. Beilstein J Org Chem 10:2715–2723. https://doi.org/10.3762/bjoc.10.286
Ferro M, Castiglione F, Punta C, Meloni L, Panzeri W, Rossi B, Trotta F, Mele A (2016) Transport properties of ibuprofen encapsulated in cyclodextrin nanosponge hydrogels: a proton HR-MAS NMR spectroscopy study. J Vis Exp 114:53769. https://doi.org/10.3791/53769
Fraix A, Marino N, Sortino S (2015) Phototherapeutic release of nitric oxide with engineered nanoconstructs. In: Sortino S (ed) Light-responsive nanostructured systems for applications in nanomedicine. Springer, New York. https://doi.org/10.1007/978-3-319-22942-3_8
Fraix A, Marino N, Sortino S (2016) Phototherapeutic release of nitric oxide with engineered nanoconstructs. Top Curr Chem 370:225–257. https://doi.org/10.1007/978-3-319-22942-3_8
Frömming KH, Szejtli J (1994) Cyclodextrins in pharmacy. Kluwer Academic Publishers, Dordrecht
Fülöp Z, Kurkov SV, Nielsen TT, Larsen KL, Loftsson T (2012) Self-assembly of cyclodextrins: formation of cyclodextrin polymer based nanoparticles. J Drug Delivery Sci Technol 22:215–221. https://doi.org/10.1016/S1773-2247(12)50032-8
Garrido J, Cagide F, Melle-Franco M, Borges F, Garrido EM (2014) Microencapsulation of herbicide MCPA with native β-cyclodextrin and its methyl and hydroxypropyl derivatives: an experimental and theoretical investigation. J Mol Struct 1061:76–81. https://doi.org/10.1016/j.molstruc.2013.12.067

Ghemati D, Aliouche D (2014) Dye adsorption behavior of polyvinyl alcohol/glutaraldehyde/beta-cyclodextrin polymer membranes. J Appl Spectrosc 81(2):257–263. https://doi.org/10.1007/s10812-014-9919-4

Gibson LT (2014) Mesosilica materials and organic pollutant adsorption: Part B - Removal from aqueous solution. Chem Soc Rev 43(15):5173–5182. https://doi.org/10.1039/C3CS60095E

Glick D, Barth S, MacLeod KF (2010) Autophagy: cellular and molecular mechanisms. J Pathol 221:3–12. https://doi.org/10.1002/path.2697

Gontero D, Lessard-Viger M, Brouard D, Bracamonte AG, Boudreau D, Veglia AV (2017) Smart multifunctional nanoparticles design as sensors and drug delivery systems based on supramolecular chemistry. Microchem J 130:316–328. https://doi.org/10.1016/j.microc.2016.10.007

Gonzalez-Gaitano G, Rodriguez P, Isasi JR, Fuentes M, Tardajos G, Sanchez M (2003) The aggregation of cyclodextrins as studied by photon correlation spectroscopy. J Incl Phenom Macrocycl Chem 44:101–105. https://doi.org/10.1023/A:1023065823358

Gould S, Scott RC (2005) 2-Hydroxypropyl-beta-cyclodextrin (HP-beta-CD): a toxicology review. Food Chem Toxicol 43:1451–1459. https://doi.org/10.1016/j.fct.2005.03.007

Goyal AK, Johal ES, Rath G (2011) Nanotechnology for water treatment. Curr Nanosci 7:640–654. https://doi.org/10.2174/157341311796196772

Grigoriu AM, Luca C, Grigoriu A (2008) Cyclodextrins applications in the textile industry. Cellulose. Chem Technol 42:103–112

Gruiz K, Molnár M, Fenyvesi É, Cs H, Atkári Á, Barkács K (2011) Cyclodextrins in innovative engineering tools for risk-based environmental management. J Incl Phenom Macrocycl Chem 70:299–306. https://doi.org/10.1007/s10847-010-9909-y

Grumezescu AM (2016) Encapsulations. In: Nanotechnology in the agri-food industry. Academic Press Elsevier. 1st edition. vol 2, 924 p. ISBN: 978-0-12-804378-3

Guo R, Wilson LD (2013) Cyclodextrin-based microcapsule materials - their preparation and physiochemical properties. Curr Org Chem 17:14–21. https://doi.org/10.2174/138527213805289204

Han JH (2005) Innovations in food packaging. Food science and technology, international series. ISBN: 0-12-311632-5. 503 p

Han SM, Armstrong DW (1989) HPLC separation of enantiomers and other isomers with cyclodextrin-bonded phases: rules for chiral recognition. In: Krstulovic AM (ed) Chiral separations by HPLC. Ellis Horwood Limited, John Wiley & Sons. Chapter 10, New York, p 208

Hapiot F, Tilloy S, Monflier E (2006) Cyclodextrins as supramolecular hosts for organometallic complexes. Chem Rev 106:767–781. https://doi.org/10.1021/cr050576c

Hapiot F, Ponchel A, Tilloy S, Monflier E (2011) Cyclodextrins and their applications in aqueous-phase metal-catalyzed reactions. CR Chimie 14:149–166. https://doi.org/10.1016/j.crci.2010.04.003

Hapiot F, Bricout H, Menuel S, Tilloy S, Monflier E (2014) Recent breakthroughs in aqueous cyclodextrin-assisted supramolecular catalysis. Cat Sci Technol 4:1899–1908. https://doi.org/10.1039/C4CY00005F

Harada A, Takashima Y, Yamaguchi H (2009a) Cyclodextrin-based supramolecular polymers. Chem Soc Rev 38:875–882. https://doi.org/10.1039/B705458K

Harada A, Takashima Y, Yamaguchi H (2009b) Polymeric rotaxanes. Chem Rev 109:5974–6023. https://doi.org/10.1021/cr9000622

Harada A, Takashima Y, Nakahata M (2014) Supramolecular polymeric materials via cyclodextrin-guest interactions. Acc Chem Res 47:2128–2140. https://doi.org/10.1021/ar500109h

Hashimoto HJ (1996) Cyclodextrins in foods, cosmetics, and toiletries. In: Szejtli J, Osa T (eds) Comprehensive supramolecular chemistry, vol 3. Pergamon Oxford, London, pp 483–502

Hashimoto H (2002) Present status of industrial application of cyclodextrins in Japan. J Incl Phenom 44:57–62. https://doi.org/10.1023/A:1023036406829

Hashimoto H (2006) Cyclodextrin applications in food, cosmetic, toiletry, textile and wrapping materiel fields. In: Dodziuk H (ed) Cyclodextrins and their Complexes. Chemistry, Analytical Methods, Applications. Wiley-VCH, Verlag GmbH & Co. KGaA, Weinheim., Chapter 16, pp 452–459. https://doi.org/10.1002/3527608982.ch16

He LF, Beesley TE (2005) Applications of enantiomeric gas chromatography: A review. J Liq Chromatogr Relat Technol 28:1075–1114. https://doi.org/10.1081/JLC-200052997
He Y, Fu P, Shen X, Gao H (2008) Cyclodextrin-based aggregates and characterization by microscopy. Micron 39:495–516. https://doi.org/10.1016/j.micron.2007.06.017
Hebeish A, El-Hilw ZH (2001) Chemical finishing of cotton using reactive cyclodextrin. Color Technol 117:104–110. https://doi.org/10.1111/j.1478-4408.2001.tb00343.x
Hedges AR (1998) Industrial applications of cyclodextrins. Chem Rev 98:2035–2044. https://doi.org/10.1021/cr970014w
Hedges AR, Shieh WJ, Sikorski CT (1995) Use of cyclodextrins for encapsulation in the use and treatment of food products. In: Risch SJ, Reineccius GA. (eds) ACS Sym SerEncapsulation and controlled release of food ingredients, vol 590. American Chemical Soc, Washington, DC, pp 60–71
Herbstein FH (2005) Cyclodextrins, and some analogs, as hosts. In: Crystalline molecular complexes and compounds. Oxford Science Publications, vol 1. Oxford University Press, New York, p 73
Higdon J, Drake JV (2014) Curcumin. Available from: Linus Pauling Institute, Micronutrient Research for Optimum Health, Oregon State University. http://lpi.oregonstate.edu/infocenter/phytochemicals/curcumin
Hilton ML, Armstrong DW (1991) Contribution of cyclodextrins and derivatives to liquid chromatography. In: Duchêne D (ed) New trends in cyclodextrins and derivatives. Éditions de Santé, Paris, pp 517–549
Hirakawa H, Tomita H (2013) Interference of bacterial cell-to-cell communication: a new concept of antimicrobial chemotherapy breaks antibiotic resistance. Front Microbiol 4:114. https://doi.org/10.3389/fmicb.2013.00114
Ho TM, Howes T, Bhandari BR (2014) Encapsulation of gases in powder solid matrices and their applications: a review. Powder Technol 259:87–108. https://doi.org/10.1016/j.powtec.2014.03.054
Hong SB, Liu MY, Zhang W, Deng W (2015) Organic reactions catalyzed by cyclodextrin and its derivatives. Chin J Org Chem 35:325–336. https://doi.org/10.6023/cjoc201409001
Hongdeng Q, Xiaojing L, Min S, Shengxiang J (2011) Development of silica-based stationary phases for high-performance liquid chromatography. Anal Bioanal Chem 399:3307–3322. https://doi.org/10.1007/s00216-010-4611-x
Horák D, Beneš M, Procházková Z, Trchová M, Borysov A, Pastukhov A, Paliienko K, Borisova T (2017) Effect of O-methyl-β-cyclodextrin-modified magnetic nanoparticles on the uptake and extracellular level of l-glutamate in brain nerve terminals. Colloids Surf B Biointerfaces 149:64–71. https://doi.org/10.1016/j.colsurfb.2016.10.007
Hou XS, Ke CF, Stoddart JF (2016) Cooperative capture synthesis: yet another playground for copper-free click chemistry. Chem Soc Rev 45:3766–3780. https://doi.org/10.1039/c6cs00055j
Hougeir EG, Kircik L (2012) A review of delivery systems in cosmetics. Dermatol Therm 25:234–237. https://doi.org/10.1111/j.1529-8019.2012.01501.x
Iacovino R, Caso JV, Di Donato C, Malgieri G, Palmieri M, Russo L, Isernia C (2017) Cyclodextrins as complexing agents: preparation and applications. Curr Org Chem 21:1–15. https://doi.org/10.2174/1385272820666160909111842
Idriss H, Estour F, Zgani I, Barbot C, Biscotti A, Petit S, Galaup C, Hubert-Roux M, Nicol L, Mulder P, Gouhier G (2013) Effect of the second coordination sphere on new contrast agents based on cyclodextrin scaffolds for MRI signals. RSC Adv 3:4531. https://doi.org/10.1039/C3RA40314A
Irie T, Uekama K (1997) Pharmaceutical applications of cyclodextrins. III. Toxicological issues and safety evaluation. J Pharm Sci 86:147–162. https://doi.org/10.1021/js960213f
Islam SU, Shahid M, Mohammad F (2013) Green chemistry approaches to develop antimicrobial textiles based on sustainable biopolymers - a review. Ind Eng Chem Res 52:5245–5260. https://doi.org/10.1021/ie303627x

Jicsinszky L, Fenyvesi É, Hashimoto H (1996) Cyclodextrin derivatives. In: Szejtli J, Osa T (eds) Comprehensive supramolecular chemistry, vol 3, London, Pergamon Oxford, pp 57–188
Junthip J, Tabary N, Chai F, Leclercq L, Maton M, Cazaux F, Neut C, Paccou L, Guinet Y, Staelens JN, Bria M, Landy D, Hedoux A, Blanchemain N, Martel B (2016) Layer-by-layer coating of textile with two oppositely charged cyclodextrin polyelectrolytes for extended drug delivery. J Biomed Mater Res A 104:408–424. https://doi.org/10.1002/jbm.a.35674
Kainuma K (1984) Starch oligosaccharides: Linear, branched, and cyclic. In: Whistler RL, JN BM, Paschall EF (eds) Starch: chemistry and technology, 2nd edn. Academic Press, London, pp 125–152. 978-0-12-746270-7
Kali G, Eisenbarth H, Wenz G (2015) One pot synthesis of a polyisoprene polyrotaxane and conversion to a slide-ring gel. Macromol Rapid Commun. https://doi.org/10.1002/marc.201500548
Kano K (1993) Selectivities in cyclodextrin chemistry. In: Dugas H, Schmidtchen FP (eds) Bioorganic Chemistry Frontiers, vol 3. Springer-Verlag, Berlin, pp 1–23. 978-3-642-78110-0
Kanwar JR, Long BM, Kanwar RK (2011) The use of cyclodextrins nanoparticles for oral delivery. Curr Med Chem 18:2079–2085
Karim AA, Loh XJ (2016) Towards cyclodextrin-based supramolecular materials. In: Polymers for personnal care products and cosmetics. Chapter 9, pp 154–177. https://doi.org/10.1039/9781782623984-00154
Karoyo AH, Wilson LD (2015) Nano-sized cyclodextrin-based molecularly imprinted polymer adsorbents for perfluorinated compounds - a mini-review. Nanomaterials 5:981–1003. https://doi.org/10.3390/nano5020981
Kayaci F, Uyar T (2012) Encapsulation of vanillin/cyclodextrin inclusion complex in electrospun polyvinyl alcohol (PVA) nanowebs: Prolonged shelf-life and high temperature stability of vanillin. Food Chem 133:641–649. https://doi.org/10.1016/j.foodchem.2012.01.040
Kayaci F, Uyar T (2014) Electrospun polyester/cyclodextrin nanofibers for entrapment of volatile organic compounds. Polym Eng Sci 54(12):2970–2978. https://doi.org/10.1002/pen.23858
Kayaci F, Aytac Z, Uyar T (2013a) Surface modification of electrospun polyester nanofibers with cyclodextrin polymer for the removal of phenanthrene from aqueous solution. J Hazard Mater 261:286–294. https://doi.org/10.1016/j.jhazmat.2013.07.041
Kayaci F, Umu OCO, Tekinay T, Uyar T (2013b) Antibacterial electrospun polylactic acid (PLA) nanofibrous webs incorporating triclosan/cyclodextrin inclusion complexes. J Agric Food Chem 61:3901–3908. https://doi.org/10.1021/jf400440b
Kayaci F, Ertas Y, Uyar T (2013c) Enhanced thermal stability of eugenol by cyclodextrin inclusion complex encapsulated in electrospun polymeric nanofibers. J Agric Food Chem 61:8156–8165. https://doi.org/10.1021/jf402923c
Kayaci F, Sen HS, Durgun E, Uyar T (2014) Functional electrospun polymeric nanofibers ıncorporating geraniol-cyclodextrin ınclusion complexes: high thermal stability and enhanced durability of geraniol. Food Res Int 62:424–431. https://doi.org/10.1016/j.foodres.2014.03.033
Kayaci F, Sen HS, Durgun E, Uyar T (2015) Electrospun nylon 6,6 nanofibers functionalized with cyclodextrins for removal of toluene vapor. J Appl Polym Sci 132:41941. https://doi.org/10.1002/app.41941
Keskin NOS, Celebioglu A, Uyar T, Tekinay T (2015a) Microalgae immobilized by nanofibrous web for removal of reactive dyes from wastewater. Ind Eng Chem Res 54:5802–5809. https://doi.org/10.1021/acs.iecr.5b01033
Keskin NOS, Celebioglu A, Sarioglu OF, Ozkan AD, Uyar T, Tekinay T (2015b) Removal of a reactive dye and hexavalent chromium by a reusable bacteria attached electrospun nanofibrous web. RSC Adv 5:86867–86874. https://doi.org/10.1039/C5RA15601G
Kfoury M, Hădărugă NG, Hădărugă DI, Fourmentin S (2016) Cyclodextrins as encapsulation material for flavors and aroma. In: Encapsulations: nanotechnology in the agri-food industry, vol 2, 1st edn, pp 127–192. 978-0-12-804378-3
Khan AR, Forgo P, Stine KJ, D'Souza VT (1998) Methods for selective modifications of cyclodextrins. Chem Rev 98:1977–1996. https://doi.org/10.1021/cr970012b

Khaoulani S, Chaker H, Cadet C, Bychkov E, Cherif L, Bengueddach A, Fourmentin S (2015) Wastewater treatment by cyclodextrin polymers and noble metal/mesoporous TiO_2 photocatalysts. CR Chimie 18:23–31. https://doi.org/10.1016/j.crci.2014.07.004

Kirejev V, Gonçalves AR, Aggelidou C, Manet I, Mårtensson J, Yannakopoulou K, Ericson MB (2014) Photophysics and ex vivo biodistribution of β-cyclodextrin-meso-tetra(m-hydroxyphenyl)porphyrin conjugate for biomedical applications. Photochem Photobiol Sci 13:1185–1191. https://doi.org/10.1039/c4pp00088a

Ko JA, Jeon JY, Park HJ (2012) Preparation and characterization of allyl isothiocyanate microcapsules by spray drying. J Food Biochem 36:255–261. https://doi.org/10.1111/j.1745-4514.2010.00533.x

Komiyama M (1996) Cyclodextrins as enzyme models. In: Szejtli J, Osa T (eds) Comprehensive supramolecular chemistry, vol 3. Pergamon Oxford, London, pp 401–422

Komiyama M, Monflier E (2006) Cyclodextrin catalysis. In: Dodziuk H (ed) Cyclodextrins and their complexes: chemistry, analytical methods, applications. Wiley-VCH, Verlag GmbH & Co. KGaA, Weinheim. Chapter 4, pp 93–105

Kozlowski CA, Sliwa W (2008) The use of membranes with cyclodextrin units in separation processes: recent advances. Carbohydr Polym 74:1–9. https://doi.org/10.1016/j.carbpol.2008.01.010

Kozlowski CA, Sliwa W (2010) Use of cyclodextrin polymers in separation of organic species. In: Polymer science and technology series. Nova Science Publishers, New York

Králová J, Kejík Z, Bríza T, Poucková P, Král A, Martásek P, Král V (2010) Porphyrin-cyclodextrin conjugates as a nanosystem for versatile drug delivery and multimodal cancer therapy. J Med Chem 53:128–138. https://doi.org/10.1021/jm9007278

Kurkov SV, Loftsson T (2013) Cyclodextrins. Int J Pharm 453:167–180. https://doi.org/10.1016/j.ijpharm.2012.06.055

Kushwaha D, Dwivedi P, Kuanar SK, Tiwari VK (2013) Click reaction in carbohydrate chemistry: recent developments and future perspective. Curr Org Synth 10:90–135. https://doi.org/10.2174/1570179411310010005

Landy D, Mallard I, Ponchel A, Monflier E, Fourmentin S (2012a) Cyclodextrins for remediation technologies. In: Lichtfouse E, Schwarzbauer J, Robert D (eds) Environmental chemistry for a sustainable world: nanotechnology and health risk, vol 1. Springer Verlag, Berlin, pp 47–81

Landy D, Mallard I, Ponchel A, Monflier E, Fourmentin S (2012b) Remediation technologies using cyclodextrins: an overview. Environ Chem Lett 10:225–237. https://doi.org/10.1007/s10311-011-0351-1

Larsen KL (2002) Large cyclodextrins. J Incl Phenom 43:1–13. https://doi.org/10.1023/A:1020494503684

Lay S, Ni XF, Yu HN, Shen SR (2016) State-of-the-art applications of cyclodextrins as functional monomers in molecular imprinting techniques: a review. J Sep Sci 39:2321–2331. https://doi.org/10.1002/jssc.201600003

Lee SY, Park SJ (2015) A review on solid adsorbents for carbon dioxide capture. J Ind Eng Chem 23:1–11. https://doi.org/10.1016/j.jiec.2014.09.001

Letort S, Balieu S, Erb W, Gouhier G, Estour F (2016) Interactions of cyclodextrins and their derivatives with toxic organophosphorus compounds. Beilstein J Org Chem 12:204–228. https://doi.org/10.3762/bjoc.12.23

Li J (2009) Cyclodextrin inclusion polymers forming hydrogels. In: Wenz G (ed) Inclusion polymers, Advances in polymer science, vol 222, pp 79–112. https://doi.org/10.1007/12_2008_9

Li S, Purdy WC (1992) Cyclodextrins and their applications in analytical chemistry. Chem Rev 92:1457–1470. https://doi.org/10.1021/cr00014a009

Li S, Vigh G (2004) Single-isomer sulfated alpha-cyclodextrins for capillary electrophoresis: Hexakis(2,3-di-O-methyl-6-O-sulfo)-alpha-cyclodextrin, synthesis, analytical characterization, and initial screening tests. Electrophoresis 25:2657–2670. https://doi.org/10.1002/elps.200405839

Li ZF, Wang M, Wang F, Gu ZB, Du GC, Wu J, Chen J (2007) gamma-Cyclodextrin: a review on enzymatic production and applications. Appl Microbiol Biotechnol 77:245–255. https://doi.org/10.1007/s00253-007-1166-7

Li JJ, Zhao F, Li J (2011) Polyrotaxanes for applications in life science and biotechnology. Appl Microbiol Biotechnol 90:427–443. https://doi.org/10.1007/s00253-010-3037-x

Lincoln SF, Easton CJ (1998) Inclusion complexes of the cyclodextrins. In: Polysaccharides: structural diversity and functional versatility. Dumitriu S, ed. New York: Marcel Dekker, Inc.. Chapter 14, pp 473–521. ISBN: 0-8247-0127-5

Loeve S, Normand M (2011) How to trust a molecule? The case of cyclodextrins entering the naorealm. In: Zülsdorf TB, Coenen C, Fiedeler U, Ferrari A, Milburn C, Wienroth M (eds) Quantum engagements: social reflections of nanoscience and emerging techologies. IOS Press AKA Verlag, Heidelberg, pp 195–216

Loftsson T, Brewster ME (2010) Pharmaceutical applications of cyclodextrins: basic science and product development. J Pharm Pharmacol 62:1607–1621. https://doi.org/10.1111/j.2042-7158.2010.01030.x

Loftsson T, Brewster ME (2012) Cyclodextrins as functional excipients: methods to enhance complexation efficiency. J Pharm Sci 101:3019–3032. https://doi.org/10.1002/jps.23077

Loftsson T, Duchêne D (2007) Cyclodextrins and their pharmaceutical applications: historical perspectives. Int J Pharm 329:1–11. https://doi.org/10.1016/j.ijpharm.2006.10.044

Loftsson T, Masson M, Brewster ME (2004) Self-association of cyclodextrins and cyclodextrin complexes. J Pharm Sci 93:1091–1099. https://doi.org/10.1002/jps.20047

López-Nicolás JM, Rodríguez-Bonilla P, García-Carmona F (2014) Cyclodextrins and antioxidants. Crit Rev Food Sci Nutr 54:251–276. https://doi.org/10.1080/10408398.2011.582544

Luca C, Grigoriu AM (2006) Cyclodextrins inclusion compounds. Rev Chim 57:248–252

Luca C, Grigoriu AM (2007) Cyclodextrins inclusion compounds in macromolecular chemistry. Cellul Chem Technol 41:1–12

Macaev F, Boldescu V (2015) Cyclodextrins in asymmetric and stereospecific synthesis. Symmetry 7:1699–1720. https://doi.org/10.3390/sym7041699

Macaev F, Boldescu V, Geronikaki A, Sucman N (2013) Recent advances in the use of cyclodextrins in antifungal formulations. Curr Top Med Chem 21:2677–2683. https://doi.org/10.2174/15680266113136660194

Maciollek A, Ritter H, Beckert R (2013) Superstructures of fluorescent cyclodextrin via click-reaction. Beilstein J Org Chem 9:827–831. https://doi.org/10.3762/bjoc.9.94

Maetzel D, Sarkar S, Wang H, Abi-Mosleh L, Xu P, Cheng AW, Gao Q, Mitalipova M, Jaenisch R (2014) Genetic and chemical correction of cholesterol accumulation and impaired autophagy in hepatic and neural cells derived from Niemann-Pick Type C patient-specific iPS cells. Stem Cell Rep 2:866–880. https://doi.org/10.1016/j.stemcr.2014.03.014

Mahmud ST, Wilson LD (2016) Synthesis and characterization of surface-modified mesoporous silica materials with beta-cyclodextrin. Cogent Chem 2:1132984. https://doi.org/10.1080/23312009.2015.1132984

Malanga M, Bálint M, Puskás I, Tuza K, Sohajda T, Jicsinszky L, Szente L, Fenyvesi É (2014) Synthetic strategies for the fluorescent labeling of epichlorohydrin-branched cyclodextrin polymers. Beilstein J Org Chem 10:3007–3018. https://doi.org/10.3762/bjoc.10.319

Mallard I, Baudelet D, Castiglione F, Ferro M, Panzeri W, Ragg E, Mele A (2015) Polydisperse methyl beta-cyclodextrin-epichlorohydrin polymers: variable contact time C-13 CP-MAS solid state NMR characterization. Beilstein J Org Chem 11:2785–2794. https://doi.org/10.3762/bjoc.11.299

Manchon JFM, Uzor NE, Kesler SR, Wefel JS, Townley DM, Nagaraja AS, Pradeep S, Mangala LS, Sood AK, Tsvetkov AS (2016) TFEB ameliorates the impairment of the autophagy-lysosome pathway in neurons induced by doxorubicin. Aging 8:3507–3519. https://doi.org/10.18632/aging.101144

Martel B, Morcellet M, Ruffin D, Vinet F, Weltrowski L (2002) Capture and controlled release of fragrances by CD finished textiles. J Incl Phenom Macrocycl Chem 44:439–442. https://doi.org/10.1023/A:1023028105012

Martin Del Valle EM (2004) Cyclodextrins and their uses. Process Biochem 39:1033–1046. https://doi.org/10.1016/S0032-9592(03)00258-9

Martina K, Binello A, Lawson D, Jicsinszky L, Cravotto G (2013) Recent applications of cyclodextrins as food additives and in food processing. Curr Nutr Food Sci 9:167–179. https://doi.org/10.2174/1573401311309030001

Mavridis IM, Yannakopoulou K (2015) Anionic cyclodextrins as versatile hosts for pharmaceutical nanotechnology: synthesis, drug delivery, enantioselectivity, contrast agents for MRI. Int J Pharma 492:275–290. https://doi.org/10.1016/j.ijpharm.2015.06.004

Mazzaglia A, Sciortino T, Kandoth N, Sortino S (2012) Cyclodextrin-based nanoconstructs for photoactivated therapies. J Drug Deliv Sci Technol 22:235–242. https://doi.org/10.1016/S1773-2247(12)50034-1

Melotti A, Mas C, Kuciak M, Lorente-Trigos A, Borges I, Ruiz i, Altaba A (2014) The river blindness drug ivermectin and related macrocyclic lactones inhibit WNT-TCF pathway responses in human cancer. EMBO Mol Med 6:1263–1278. https://doi.org/10.15252/emmm.201404084

Meng QR, Bai J, Li CP, Huang YR, Liu H, Li HQ (2014) Electrospun functional cyclodextrins/polystyrene (PS) composite nanofibers and their applications for sorption of Cu (II) ions under aqueous solution. Nanosci Nanotechnol Lett 6:289–294. https://doi.org/10.1166/nnl.2014.1768

Menges RA, Armstrong DW (1991) Chiral separations using native and functionalized cyclodextrin-bonded stationary phases in high-pressure liquid chromatography. In: Ahuja S (ed) Chiral Separations by Liquid Chromatography, ACS symposium series 471. Chapter 4, Washington, p 67. isbn:0841221162

Messner M, Kurkov SV, Jansook T, Loftsson T (2010) Self-assembled cyclodextrin aggregates and nanoparticles. Int J Pharm 387:199–208. https://doi.org/10.1016/j.ijpharm.2009.11.035

Mikus P, Sebesta R, Kaniansky D, Salisova M (2002) Cyclodextrins and their complexes - structure and interactions. Chem List 96:693–697

Miller KP, Wang L, Chen YP, Pellechia PJ, Benicewicz BC, Decho AW (2015) Engineering nanoparticles to silence bacterial communication. Front Microbiol 6:189. https://doi.org/10.3389/fmicb.2015.00189

Mocanu G, Vizitiu D, Carpov A (2001) Cyclodextrins polymers. J Bioact Compat Polym 16:315–342

Morillo E (2006) Application of cyclodextrins in agrochemistry. In: Dodziuk H (ed) Cyclodextrins and their Complexes: chemistry, analytical methods, applications. Wiley-VCH, Verlag GmbH & Co. KGaA. Chapter 16, Weinheim, pp 459–467

Morin-Crini N, Crini G (2013) Environmental applications of water-insoluble β-cyclodextrin-epichlorohydrin polymers. Prog Polym Sci 38:344–368. https://doi.org/10.1016/j.progpolymsci.2012.06.005

Morin-Crini N, Fourmentin S, Crini G (2015) Cyclodextrines (in French). Besançon: PUFC. 370 p. ISBN: 978-2-84867-520-6

Morohoshi T, Tokita K, Ito S, Saito Y, Maeda S, Kato K, Ikeda T (2013) Inhibition of quorum sensing in gram-negative bacteria by alkylamine-modified cyclodextrins. J Biosci Bioeng 116:175–179. https://doi.org/10.1016/j.jbiosc.2013.01.022

Mosher G, Thompson DO (2002) Complexation and cyclodextrins. In: Swarbrick J, JCE B (eds) Encyclopedia of pharmaceutical technology, 2nd edn. Marcel Dekker, New York, pp 531–558

Mosinger J, Tomankova V, Nemcova I, Zyka J (2001) Cyclodextrins in analytical chemistry. Anal Lett 34:1979–2004. https://doi.org/10.1081/AL-100106834

Motoyama K, Hirai Y, Nishiyama R, Maeda Y, Higashi T, Ishitsuka Y, Kondo Y, Irie T, Era T, Arima H (2015) Cholesterol lowering effects of mono-lactose-appended beta-cyclodextrin in Niemann-Pick type C disease-like HepG2 cells. Beilstein J Org Chem 11:2079–2086. https://doi.org/10.3762/bjoc.11.224

Motoyama K, Nishiyama R, Maeda Y, Higashi T, Kawaguchi Y, Futaki S, Ishitsuka Y, Kondo Y, Irie T, Era T, Arima H (2016) Cholesterol-lowering effect of octaarginine-appended beta-cyclodextrin in Npc1-Trap-CHO cells. Biol Pharm Bull 39:1823–1829. https://doi.org/10.1248/bpb.b16-00369

Mourer M, Hapiot F, Monflier E, Menuel S (2008) Click chemistry as an efficient tool to access β-cyclodextrin dimers. Tetrahedron 64:7159–7163. https://doi.org/10.1016/j.tet.2008.05.095

Moya-Ortega M, Alvarez-Lorenzo C, Concheiro A, Loftsson T (2012) Cyclodextrin-based nanogels for pharmaceuticals and biomedical applications. Int J Pharm 428:152–163. https://doi.org/10.1016/j.ijpharm.2012.02.038

Mura P (2015) Analytical techniques for characterization of cyclodextrin complexes in the solid state: a review. J Pharm Biomed Anal 113:226–238. https://doi.org/10.1016/j.jpba.2015.01.058

Murakami T, Schmidt BVKJ, Brown HR, Hawker CJ (2017) Structural versatility in slide-ring gels: influence of co-threaded cyclodextrin spacers. J Polym Sci A Polym Chem 55:1156–1165. https://doi.org/10.1002/pola.28490

Nagy ZM, Molnár M, Fekete-Kertész I, Molnár-Perl I, Fenyvesi É, Gruiz K (2014) Removal of emerging micropollutants from water using cyclodextrins. Sci Total Environ 485-486:711–719. https://doi.org/10.1016/j.scitotenv.2014.04.003

Nepogodiev S, Fraser Stoddart J (1998) Cyclodextrin-based catenanes and rotaxanes. Chem Rev 98:1959–1576. https://doi.org/10.1021/cr970049w

Nielsen TT, Wintgens V, Amiel C, Wimmer R, Larsen KL (2010) Facile synthesis of β-cyclodextrin-dextran polymers by “click” chemistry. Biomacromol 11:1710–1715. https://doi.org/10.1021/bm9013233

Norena-Caro D, Alvarez-Lainez M (2016) Functionalization of polyacrylonitrile nanofibers with beta-cyclodextrin for the capture of formaldehyde. Mater Des 95:632–640. https://doi.org/10.1016/j.matdes.2016.01.106

Okano C, Nasuno E, Iimura K, Kato N (2016) Cyclodextrin-immobilized microspheres for uptake of the quorum-sensing signaling molecule N-acylhomoserine lactone. J Appl Polym Sci 133:43198. https://doi.org/10.1002/app.43198

Oliveri V, Vecchio G (2016) Cyclodextrins as protective agents of protein aggregation: an overview. Chem Asian J 11:1648–1657. https://doi.org/10.1002/asia.201600259

Ozbilgin S, Yilmaz O, Ergur BU, Hanci V, Ozbal S, Yurtlu S, Gunenc SF, Kuvaki B, Kucuk BA, Sisman AR (2016) Effectiveness of sugammadex for cerebral ischemia/reperfusion injury. Kaohsiung J Med Sci 32:292–301. https://doi.org/10.1016/j.kjms.2016.05.002

Perez-Anes A, Gargouri M, laure W, Van Den Berghe H, Courcot E, Sobocinski J, Tabary N, Chai F, Blach JF, Addad A, Woisel P, Douroumis D, Martel B, Blanchemain N, Lyskawa J (2015) Bioinspired titanium drug eluting platforms based on a poly-beta-cyclodextrin-chitosan layer-by-layer self-assembly targeting infections. ACS Appl Mater Interfaces 7:12882–12893. https://doi.org/10.1021/acsami.5b02402

Pessine FBT, Calderini A, Alexandrino GL (2012) Review: cyclodextrin inclusion complexes probed by NMR techniques. In: Dong-Hyun K. (ed) Magnetic resonance spectroscopy. In TECH, Rijeak, chapter 2, pp 237–264. https://doi.org/10.5772/32029

Pinho E, Grootveld M, Soares G, Henriques M (2014) Cyclodextrin-based hydrogels toward improved wound dressings. Crit Rev Biotechnol 34:328–337. https://doi.org/10.3109/07388551.2013.794413

Potrzebowski MJ, Kazmierski S (2005) High-resolution solid-state NMR studies of inclusion complexes. In: Klinowski J (ed) New Techniques in solid-state NMR, Topics in current chemistry, vol 246, pp 91–140. https://doi.org/10.1007/b98649

Prochowicz D, Kornowicz A, Justyniak I, Lewinski J (2016) Metal complexes based on native cyclodextrins: synthesis and structural diversity. Coord Chem Rev 306:331–345. https://doi.org/10.1016/j.ccr.2015.07.016

Rezanka M (2016) Monosubstituted cyclodextrins as precursors for further use. Eur J Org Chem (32):5322–5334. https://doi.org/10.1002/ejoc.201600693

Ripoll L, Bordes C, Etheve S, Elaissari A, Fessi H (2010) Cosmeto-textile from formulation to characterization: an overview. e-Polymers:040. https://doi.org/10.1515/epoly.2010.10.1.409
Robyt JF (1998) Cyclodextrins. In: Essentials of carbohydrate of chemistry. New York: Springer-Verlag, p. 245–261. ISBN: 978-1-4612-7220-5 (print). DOI https://doi.org/10.1007/978-1-4612-1622-3
Romi R, Nostro PL, Bocci E, Ridi F, Baglioni P (2005) Bioengineering of a cellulosic fabric for insecticide delivery via grafted cyclodextrin. Biotechnol Prog 21:1724–1730. https://doi.org/10.1021/bp050276g
Ryzhakov A, Thi TD, Stappaerts J, Bertoletti L, Kimpe K, Couto ARS, Saokham P, Van den Mooter G, Augustijns P, Somsen GW, Kurkov S, Inghelbrecht S, Arien A, Jimidar MI, Schrijnemakers K, Loftsson T (2016) Self-assembly of cyclodextrins and their complexes in aqueous solutions. J Pharm Sci 105:2556–2569. https://doi.org/10.1016/j.xphs.2016.01.019
Saenger W (1980) Cyclodextrin inclusion compounds in research and industry. Angew Chem Int Ed 19:344–362. https://doi.org/10.1002/anie.198003441
Saenger W (1984) Structural aspects of cyclodextrins and their inclusion complexes. In: Inclusion compounds - structural aspects of inclusion compounds formed by organic host lattices. Atwood JL, Davies JED, MacNicol DD, eds. London: Academic Press. Volume 2, pp. 231–259. ISBN-13: 978-0120671021
Samiey B, Cheng CH, Wu JG (2014) Organic-inorganic hybrid polymers as adsorbent for removal of heavy metal ions from solutions: a review. Materials 7:673–726. https://doi.org/10.3390/ma7020673
Saokham P, Loftsson T (2017) gamma-Cyclodextrin. Int J Pharm 516:278–292. https://doi.org/10.1016/j.ijpharm.2016.10.062
Schmid G (1996) Enzymology of cyclodextrins. In: Szejtli J, Osa T (eds) Comprehensive supramolecular chemistry, vol 3. Pergamon Oxford, London, pp 615–626
Schmidt BVKJ, Hetzer M, Ritter H, Barner-Kowollik C (2014) Complex macromolecular architecture design via cyclodextrin host/guest complexes. Prog Polym Sci 39:235–249. https://doi.org/10.1016/j.progpolymsci.2013.09.006
Schneider HJ (2012) Applications of supramolecular chemistry. CRC Press, Boca Raton
Schneider HJ, Yatsimirsky AK (2000) Principles and methods in supramolecular chemistry. John Wiley, Chichester
Schneider HJ, Hacket F, Rüdiger V, Ikeda H (1998) NMR Studies of cyclodextrins and cyclodextrin complexes. Chem Rev 98:1755–1785. https://doi.org/10.1021/cr970019t
Schneiderman E, Stalcup AM (2000) Cyclodextrins: a versatile tool in separation science. J Chromatogr B 745:83–102. https://doi.org/10.1016/S0378-4347(00)00057-8
Scriba GKE (2016) Chiral recognition in separation science - an update. J Chromatogr A 1467:56–78. https://doi.org/10.1016/j.chroma.2016.05.061
Senthamizhan A, Balusamy B, Celebioglu A, Uyar T (2016) Nanotraps in porous electrospun fibers for effective removal of lead(II) in water. J Mater Chem A 4:2484–2493. https://doi.org/10.1039/C5TA09166G
Sharma N, Baldi A (2016) Exploring versatile applications of cyclodextrins: an overview. Drug Deliv 23:739–757. https://doi.org/10.3109/10717544.2014.938839
Shaw NO, De Villiers MM, Lotter AP (1999) Preformulation stability screening of ivermectin with non-ionic emulsion excipients. Pharmazie 54:372–376
Shen H, Ji H (2011) Application of cyclodextrin derivatives in liquid-phase organic synthesis. Chin J Org Chem 31:791–803
Shen HM, Ji HB, Wu HK, Shi HX (2014a) Recent advances in the immobilization of beta-cyclodextrin and their application. Chin J Org Chem 34:1549–1572. https://doi.org/10.6023/cjoc201402024
Shen W, Zhang GH, Guo N, Li YT (2014b) Study on the inclusion compound of avermectin by infrared spectroscopy. Guang Pu Xue Yu Guang Pu Fen Xi 34:1201–1205. https://doi.org/10.3964/j.issn.1000-0593(2014)05-1201-05

Shieh WJ, Hedges AR (1996) Properties and applications of cyclodextrins. J Macromol Sci Pure Appl Chem A33:673–683. https://doi.org/10.1080/10601329608010886
Silva A, Duarte A, Sousa S, Ramos A, Domingues FC (2016) Characterization and antimicrobial activity of cellulose derivatives films incorporated with a resveratrol inclusion complex. LWT-Food Sci Technol 73:481–489. https://doi.org/10.1016/j.lwt.2016.06.043
Simoes SMN, Veiga F, Torres-Labandeira JJ, Ribeiro ACF, Concheiro A, Alvarez-Lorenzo C (2014) Syringeable self-assembled cyclodextrin gels for drug delivery. Curr Top Med Chem 14:494–509. https://doi.org/10.2174/1568026613666131219124308
Singh M, Sharma R, Banerjee UC (2002) Biotechnological applications of cyclodextrins. Biotechnol Adv 20:341–359. https://doi.org/10.1016/S0734-9750(02)00020-4
Smaldone RA, Forgan RS, Furukawa H, Gassensmith JJ, Slawin AMZ, Yaghi OM Stoddart JF (2010) Metal-organic frameworks from edible natural products. Angew Chem Int Ed 49:8630–8634. https://doi.org/10.1002/anie.201002343
Smolková-Keulemansová E (1982) Cyclodextrins as stationary phases in chromatography. J Chromatogr 251:17–34. https://doi.org/10.1002/jhrc.1240120113
Stella VJ, He Q (2008) Cyclodextrins. Toxicol Pathol 36:30–42. https://doi.org/10.1177/0192623307310945
Sun DW (2012) Handbook of frozen food processing and packaging. Boca Raton: CRC Press. ISBN: 978-1-4398-3604-0. 936 p
Sybilska D, Zukowski J (1989) Cyclodextrin additives. In: Krstulovic AM (ed) Chiral separations by HPLC. Ellis Horwood Limited, John Wiley. Chapter 7, New York, p 147
Szejtli J (1982a) Cyclodextrins and their inclusion complexes. Akademiai Kiado, Budapest
Szejtli J (1982b) Cyclodextrins in food, cosmetics and toiletries. Starch/Stärke 34:379–385. https://doi.org/10.1002/star.19820341106
Szejtli J (1984) Industrial applications of cyclodextrins. In: Atwood JL, JED D, DD MN (eds) Inclusion compounds, vol 3. Academic Press, London, pp 331–390
Szejtli J (1987) The metabolism, toxicity and biological effects of cyclodextrins. In: Duchêne D (ed) Cyclodextrins and their industrial uses. Editions de Santé, Paris. Chapter 5, pp 173–210
Szejtli J (1988) Cyclodextrin technology. Kluwer Academic Publishers, Dordrecht
Szejtli J (1990) The cyclodextrins and their applications in biotechnology. Carbohydr Polym 12:375–392. https://doi.org/10.1016/0144-8617(90)90088-A
Szejtli J (1991) The use of cyclodextrins in biotechnological operations. In: Duchêne D (ed) New trends in cyclodextrins and derivatives. Éditions de Santé, Paris, pp 595–626
Szejtli J (1996) Use of cyclodextrins in chemical products and processes. In: Szejtli J, Osa T (eds) Comprehensive supramolecular chemistry, vol 3. Pergamon Oxford, London, pp 603–614
Szejtli J (1997) Utilization of cyclodextrins in industrial products and processes. J Mater Chem 7:575–587. https://doi.org/10.1039/A605235E
Szejtli J (1998) Introduction and general overview of cyclodextrin chemistry. Chem Rev 98:1743–1753. https://doi.org/10.1021/cr970022c
Szejtli J (2004a) Past, present and future of cyclodextrin research. Pure Appl Chem 76:1825–1845
Szejtli J (2004b) Cyclodextrins. In: Chemical and functional properties of food saccharides. Tomasik P, ed. New York: CRC Press. Chapter 17, p. 272–290. ISBN 0-8493-1486-0
Szente L (1996a) Analytical methods for cyclodextrins, cyclodextrin derivatives, and cyclodextrin complex. In: Szejtli J, Osa T (eds) Comprehensive supramolecular chemistry, vol 3. Pergamon Oxford, London, pp 253–278
Szente L (1996b) Preparation of cyclodextrin complexes. In: Szejtli J, Osa T (eds) Comprehensive supramolecular chemistry, vol 3. Pergamon Oxford, London, pp 243–252
Szente L, Szejtli J (1996) Cyclodextrins in pesticides. In: Szejtli J, Osa T (eds) Comprehensive supramolecular chemistry, vol 3. Pergamon Oxford, London, pp 503–514
Szente L, Szejtli J (2004) Cyclodextrins as food ingredients. Trends Food Sci Technol 15:137–142. https://doi.org/10.1016/j.tifs.2003.09.019

Szente L, Szeman J, Sohajda T (2016) Analytical characterization of cyclodextrins: history, official methods and recommended new techniques. J Pharm Biomed Anal 130:347–365. https://doi.org/10.1016/j.jpba.2016.05.009

Taka AL, Pillay K, Mbianda XY (2017) Nanosponge cyclodextrin polyurethanes and their modification with nanomaterials for the removal of pollutants from waste water: a review. Carbohydr Polym 159:94–107. https://doi.org/10.1016/j.carbpol.2016.12.027

Tamura A, Yui N (2015) beta-Cyclodextrin-threaded biocleavable polyrotaxanes ameliorate impaired autophagic flux in Niemann-Pick type C disease. J Biol Chem 290:9442–9454. https://doi.org/10.1074/jbc.M114.636803

Tan S, Ladewig K, Fu Q, Blencowe A, Qiao GG (2014) Cyclodextrin-based supramolecular assemblies and hydrogels: recent advances and future perspectives. Macromol Rapid Commun 35:1166–1184. https://doi.org/10.1002/marc.201400080

Tanaka Y, Yamada Y, Ishitsuka Y, Matsuo M, Shiraishi K, Wada K, Uchio Y, Kondo Y, Takeo T, Nakagata N, Higashi T, Motoyama K, Arima H, Mochinaga S, Higaki K, Ohno K, Irie T (2015) Efficacy of 2-hydroxypropyl-β-cyclodextrin in Niemann-pick disease type C model mice and its pharmacokinetic analysis in a patient with the disease. Biol Pharm Bull 38:844–851. https://doi.org/10.1248/bpb.b14-00726

Tarimci N (2011) Cyclodextrins in the cosmetic field. In: Bilensoy E (ed) Cyclodextrins in pharmaceutics, cosmetics and biomedicine. John Wiley, London., Chapter 7, pp. 131–144. https://doi.org/10.1002/9780470926819.ch7

Tejashri G, Amrita B, Darshana J (2013) Cyclodextrin based nanosponges for pharmaceutical use: a review. Acta Pharm 63:335–358. https://doi.org/10.2478/acph-2013-0021

Terekhova IV, Kulikov OV (2005) Cyclodextrins: physical-chemical aspects of formation complexes host-guest and molecular selectivity in relation to biologically active compounds. In: Chemistry of polysaccharides. Zaikov GE, ed. London: CRC Press. Chapter 2, pp. 38–70. ISBN: 9789067644198

Thompson DO (2006) Cyclodextrins – Enabling excipients: a case study of the development of a new excipient – Sulfobutylether β-cyclodextrin (CAPTISOL®). In: Excipient development for pharmaceutical, biotechnology, and drug delivery systems. Katdare A, Chaubal MV, eds. New York: CRC Press. pp. 51–68. ISBN:978-1-4200-0413-7. (e-book - pdf)

Tong J, Chen LG (2013) Review: preparation and application of magnetic chitosan derivatives in separation processes. Anal Lett 46:2635–2656. https://doi.org/10.1080/00032719.2013.807815

Trotta F, Dianzani C, Caldera F, Mognetti B, Cavalli R (2014) The application of nanosponges to cancer drug delivery. Expert Opin Drug Deliv 11(6):931–941. https://doi.org/10.1517/17425247.2014.911729

Trotta F, Caldera F, Dianzani C, Argenziano M, Barrera G, Cavalli R (2015) New glutathione bio-responsive cyclodextrin nanosponges. ChemPlusChem 81:439–443. https://doi.org/10.1002/cplu.201500531

Tungala K, Adhikary P, Krishnamoorthi S (2013) Trimerization of β-cyclodextrin through the click reaction. Carbohydr Polym 95:295–298. https://doi.org/10.1016/j.carbpol.2013.02.074

Ueda H (2002) Physicochemical properties and complex formation abilities of large-ring cyclodextrins. J Incl Phenom Macrocycl Chem 44:53–56. https://doi.org/10.1023/A:1023055516398

Uekama K, Irie T (1996) Pharmaceutical use of cyclodextrins in various drug formulations. In: Szejtli J, Osa T (eds) Comprehensive supramolecular chemistry, vol 3. Pergamon Oxford, London, pp 451–482

Ueno A (1996) Review: fluorescent cyclodextrins for molecule sensing. Supramol Sci 3:31–36. https://doi.org/10.1016/0968-5677(96)00016-8

Valente AJM, Söderman O (2014) The formation of host-guest complexes between surfactants and cyclodextrins. Adv Colloid Int Sci 205:156–176. https://doi.org/10.1016/j.cis.2013.08.001

Valetti S, Xia X, Costa-Gouveia J, Brodin P, Bernet-Camard MF, Andersson M, Feiler A (2017) Clofazimine encapsulation in nanoporous silica particles for the oral treatment of antibiotic-resistant mycobacterium tuberculosis infections. Nanomedicine (Lond) 12(8):831–844. https://doi.org/10.2217/nnm-2016-0364

Van de Manakker F, Vermonden T, Van Nostrum CF, Hennink WE (2009) Cyclodextrin-based polymeric materials: synthesis, properties, and pharmaceutical/biomedical applications. Biomacromol 10:3157–3175. https://doi.org/10.1021/bm901065f

Vaution C, Hutin M (1987) The use of cyclodextrins in various industries. In: Duchêne D (ed) Cyclodextrins and their industrial uses. Éditions de Santé, Paris, pp 297–350

Vecsernyés M, Fenyvesi F, Bácskay I, Deli MA, Szente L, Fenyvesi É (2014) Cyclodextrins, blood-brain barrier, and treatment of neurological diseases. Arch Med Res 45:711–729. https://doi.org/10.1016/j.arcmed.2014.11.020

Venturini CDG, Nicolini J, Machado C, Machado VG (2008) Properties and recent applications of cyclodextrins. Quim Nova 31:360–368. https://doi.org/10.1590/S0100-40422008000200032

Vetter W, Bester K (2006) Gas chromatographic enantioseparation of chiral pollutants - techniques and results. In: Busch KW, Busch MA (eds) Chiral separation. Elsevier, The Netherlands. Chapter 6, pp 131–228

Villalonga R, Cao R, Fragoso A (2007) Supramolecular chemistry of cyclodextrin in enzyme technology. Chem Rev 107:3088–3116. https://doi.org/10.1021/cr050253g

Voncina, B (2011). Application of cyclodextrins in textile dyeing. In: Textile dyeing, Hauser P, ed. Rijeka: InTech. Chapter 17, pp 373–392. ISBN: 978-953-307-565-5

Voncina B, Vivod V (2013) Cyclodextrins in textile finishing. In: Günay M (ed) Textile dyeing. InTech. Chapter 3, Rijeka, pp 53–75. https://doi.org/10.5772/53777

Vunain E, Mishra AK, Mamba BB (2016) Dendrimers, mesoporous silicas and chitosan-based nanosorbents for the removal of heavy-metal ions: a review. Int J Biol Macromol 86:570–586. https://doi.org/10.1016/j.ijbiomac.2016.02.005

Walkley SU, Davidson CD, Jacoby J, Marella PD, Ottinger EA, Austin CP, Porter FD, Vite CH, Ory DS (2016) Fostering collaborative research for rare genetic disease: the example of niemann-pick type C disease. Orphanet J Rare Dis 11:161. https://doi.org/10.1186/s13023-016-0540-x

Wang HF, Zhang LM (2010) Molecularly imprinted functional materials based on polysaccharides. Prog Chem 22:2165–2172

Ward TJ, Armstrong DW (1988) Cyclodextrin-stationary phases. In: Zief M, Crane LJ (eds) Chromatographic chiral separations. Marcel Dekker Inc., Chapter 5, New York, p 131

Ward C, Martinez-Lopez N, Otten EG, Carroll B, Maetzel D, Singh R, Sarkar S, Korolchuk VI (2016) Autophagy, lipophagy and lysosomal lipid storage disorders. Biochim Biophys Acta 1861:269–284. https://doi.org/10.1016/j.bbalip.2016.01.006

Wei ZZ, Liu YL, Hu HM, Yu JY, Li FX (2016) Biodegradable poly(butylene succinate-co-terephthalate) nanofibrous membranes functionalized with cyclodextrin polymer for effective methylene blue adsorption. RSC Adv 6:108240–108246. https://doi.org/10.1039/C6RA22941G

Wenz G (2009) Recognition of monomers and polymers by cyclodextrins. Incl Polym 222:1–54. https://doi.org/10.1007/12_2008_13

Wenz G, Han BH, Müller A (2006) Cyclodextrin rotaxanes and polyrotaxanes. Chem Rev 106:782–817. https://doi.org/10.1021/cr970027+

West C (2014) Enantioselective separations with supercritical fluids – review. Curr Anal Chem 10:99–120. https://doi.org/10.2174/1573411011410010009

Wu HQ, Kong JH, Yao XY, Zhao CY, Dong YL, Lu XH (2015a) Polydopamine-assisted attachment of beta-cyclodextrin on porous electrospun fibers for water purification under highly basic condition. Chem Eng J 270:101–109. https://doi.org/10.1016/j.cej.2015.02.019

Wu ZL, Song N, Menz R, Pingali B, Yang YW, Zheng YB (2015b) Nanoparticles functionalized with supramolecular host-guest systems for nanomedicine and healthcare. Nanomedicine 10:1493–1514. https://doi.org/10.2217/NNM.15.1

Xiao Y, Ng SC, Tan TTY, Wang Y (2012) Recent development of cyclodextrin chiral stationary phases and their applications in chromatography. J Chromatogr A 1269:52–68. https://doi.org/10.1016/j.chroma.2012.08.049

Xiao N, Wen Q, Liu QW, Yang QB, Li YX (2014) Electrospinning preparation of beta-cyclodextrin/glutaraldehyde crosslinked PVP nanofibrous membranes to adsorb dye in aqueous solution. Chem Res Chin Univ 30:1057–1062. https://doi.org/10.1007/s40242-014-4203-y

Yamamoto E, Kuroda K (2016) Colloidal mesoporous silica nanoparticles. Bull Chem Soc Jpn 89:501–539. https://doi.org/10.1246/bcsj.20150420

Yang LPH, Keam SJ (2009) Sugammadex: a review of its use in anaesthetic practice. Drugs 69:919–942. https://doi.org/10.2165/00003495-200969070-00008

Yang JS, Yang L (2013) Preparation and application of cyclodextrin immobilized polysaccharides. J Mater Chem B 1:909–918. https://doi.org/10.1039/c2tb00107a

Yokoo M, Kubota Y, Motoyama K, Higashi T, Taniyoshi M, Tokommaru H, Nishiyama R, Tabe Y, Mochinaga S, Sato A, Sueoka-Aragane N, Sueoka E, Arima H, Irie T, Kimura S (2015) 2-Hydroxypropyl-β-cyclodextrin acts as a novel anticancer agent. PLoS One 10:e0141946. https://doi.org/10.1371/journal.pone.0141946

Yuan Z, Zhang L (2016) Photoinduced controlled-release drug delivery systems for applications in nanomedicine. Curr Org Chem 20:1768–1785. https://doi.org/10.2174/1385272820666160112001944

Yusoff SNM, Kamari A, Aljafree NFA (2016) A review of materials used as carrier agents in pesticide formulations. Int J Environ Sci Technol 13:2977–2994. https://doi.org/10.1007/s13762-016-1096-y

Zarzycki PK, Fenert BE, Głód BK (2016) Cyclodextrins-based nanocomplexes for encapsulation of bioactive compounds in food, cosmetics, and pharmaceutical products: Principles of supramolecular complexes formation, their influence on the antioxidative properties of target chemicals, and recent advances in selected industrial applications. In: Encapsulations. nanotechnology in the agri-food industry. Grumezescu A (ed). Chapter 17, pp 717–767. ISBN: 978-0-12-804378-3

Zgani I, Idriss H, Barbot C, Djedaïni-Pilard F, Petit S, Hubert-Roux M, Estour F, Gouhier G (2017) Positive variation of the MRI signal via intramolecular inclusion complexation of a C-2 functionalized β-cyclodextrin. Org Biomol Chem 15:564–569. https://doi.org/10.1039/c6ob02583h

Zhang J, Ma PX (2013) Cyclodextrin-based supramolecular systems for drug delivery: Recent progress and future perspective. Adv Drug Deliv Rev 65:1215–1233. https://doi.org/10.1016/j.addr.2013.05.001

Zhang L, Li LL, Cheng LH, Chen HL (2011) Preparation and application of cyclodextrin-based molecular tube. Prog Chem 23:1936–1944

Zhang XM, Li HZ, Cao ML, Shi L, Chen CY (2015) Adsorption of basic dyes on beta-cyclodextrin functionalized poly (styrene-alt-maleic anhydride). Sep Sci Technol 50:947–957. https://doi.org/10.1080/01496395.2014.978461

Zhang LL, Man SL, Qiu HN, Liu Z, Zhang M, Ma L, Gao WY (2016) Curcumin-cyclodextrin complexes enhanced the anti-cancer effects of curcumin. Environ Toxicol Pharmacol 48:31–38. https://doi.org/10.1016/j.etap.2016.09.021

Zhao SP, Xu W (2010) Cyclodextrin-containing supramolecular hydrogels. Prog Chem 22:916–926

Zhao R, Wang Y, Li X, Sun BL, Wang C (2015) Synthesis of beta-cyclodextrin-based electrospun nanofiber membranes for highly efficient adsorption and separation of methylene Blue. ACS Appl Mater Interfaces 7:26649–26657. https://doi.org/10.1021/acsami.5b08403

Zhou JW, Ritter H (2010) Cyclodextrin functionalized polymers as drug delivery systems. Polym Chem 1:1552–1559. https://doi.org/10.1039/c0py00219d

Zhou ZJ, Cai RX, Liu NG, Zhang L, Chen HL (2007) Preparation and application of membrane with cyclodextrins. Prog Chem 19:1436–1442

Zimmer S, Grebe A, Bakke SS, Bode N, Halvorsen B, Ulas T, Skjelland M, De Nardo D, Labzin LI, Kerksiek A, Hempel C, Heneka MT, Hawxhurst V, Fitzgerald ML, Trebicka J, Bjorkhem I, Gustafsson JA, Westerterp M, Tall AR, Wright SD, Espevik T, Schultze JL, Nickenig G, Lutjohann D, Latz E (2016) Cyclodextrin promotes atherosclerosis regression via macrophage reprogramming. Sci Transl Med 8:333ra50. https://doi.org/10.1126/scitranslmed.aad6100

Chapter 2
Synthesis of Cyclodextrin Derivatives

Michal Řezanka

Contents

Abstract Cyclodextrins are naturally occurring cyclic oligosaccharides, which are non-toxic and biodegradable. The main feature of cyclodextrins is the ability to encapsulate lipophilic compounds, and thus many applications have been developped in various disciplines. Although many cyclodextrin derivatives have become available in the market, their price is in the range of fine chemicals, and thus they are still often synthesised in laboratories. The actual number of cyclodextrin derivatives exceeds 11,000, but new cyclodextrin derivatives are still needed for more advanced applications. Therefore, many newcomers or beginners in cyclodextrin chemistry struggle with a reliable choice of a synthetic route.

This chapter reviews the synthesis cyclodextrin derivatives that are able to be subsequently modified. Indeed, the modification of a cyclodextrin already substituted with a suitable functional group is much easier than the optimisation of the substitution for every new cyclodextrin derivative desired. This chapter describes the synthesis of different types of cyclodextrine derivatives: persubstituted, randomly substituted, persubstituted at selected positions, selectively substituted and monosubstituted.

M. Řezanka (✉)
Institute for Nanomaterials, Advanced Technologies and Innovation,
Technical University of Liberec, Liberec, Czech Republic
e-mail: michal.rezanka@tul.cz

S. Fourmentin et al. (eds.), *Cyclodextrin Fundamentals, Reactivity and Analysis*,
Environmental Chemistry for a Sustainable World,
https://doi.org/10.1007/978-3-319-76159-6_2

2.1 Introduction

The history of cyclodextrin derivatives goes back to the beginning of the twentieth century when cyclodextrin properties were examined and the first derivatives were prepared (Crini 2014). Since then, cyclodextrin derivatives have come a long way, and they can currently find their use in many kinds of human activities: in pharmaceutical and biomedical applications (Jambhekar and Breen 2016; Sharma and Baldi 2016; Oliveri and Vecchio 2016; Coisne et al. 2016; Leclercq 2016; di Cagno 2017; Saokham and Loftsson 2017); nanotherapeutics (Bonnet et al. 2015; Swaminathan et al. 2016; Antoniuk and Amiel 2016; Mejia-Ariza et al. 2017; Venuti et al. 2017); cosmetics, toiletries and personal care (Sharma and Baldi 2016); nutrition industry (Astray et al. 2009; Fenyvesi et al. 2016; Sharma and Baldi 2016); textile and packing industry (Radu et al. 2016; Sharma and Baldi 2016); separation techniques (Řezanka et al. 2014; Zhou et al. 2015; Adly et al. 2016; Saz and Marina 2016; Zhu and Scriba 2016); and as artificial enzymes or catalysts (Kryjewski et al. 2015; Macaev and Boldescu 2015; Aghahosseini and Ramazani 2016; Letort et al. 2016).

Probably the most known application of cyclodextrin derivatives in daily life is the use of octakis[6-(2-carboxyethylthio)-6-deoxy]-γ-cyclodextrin sodium salt (sugammadex) in Bridion® (Donati 2008). When an individual undergoes surgery, anaesthesia must be provided. Drugs like rocuronium bromide are administered as they block transmission at the cholinergic nicotinic receptor at the neuromuscular junction. After the surgery the patient is transited from the anaesthetised to the fully recovered state. Therefore, sugammadex is administered. It is designed to bind the neuromuscular blocking agent instead of relying on rocuronium pharmacokinetic properties or on the inhibition of acetylcholine breakdown with a reversal agent.

2.1.1 Inclusion Complexes of Cyclodextrin Derivatives

As it was illustrated in the previous paragraph, the key property of cyclodextrin derivatives lies in their ability to complex compounds in their cavity. These compounds are usually denoted as “guests”. There are several driving forces that could lead to the formation of cyclodextrin-guest complexes: van der Waals interactions, electrostatic interactions, hydrogen bonding, hydrophobic interactions, release of conformational strain, exclusion of “high-energy” water from the cavity, and charge-transfer interactions. Liu and Guo (Liu and Guo 2002) have shown that with the exception of the release of conformation strain and exclusion of water from the cyclodextrin cavity, the other interactions contribute to the complex formation. They concluded that hydrophobic and van der Waals interactions are the main driving forces, whereas hydrogen bonding and electrostatic interaction can significantly affect the stability of the inclusion complex.

The shape and size of guests forming complexes with cyclodextrin or its derivatives are variable and therefore the strength of a complex depends on the cavity size (Szente and Fenyvesi 2017). In the case of cyclodextrin derivatives, the substituents present on a cyclodextrin skeleton should not be overlooked – as they could significantly affect the formation of the inclusion complex. For example, an association constant of the abovementioned sugammadex with rocuronium bromide is 10^7 M^{-1} (Bom et al. 2002), while native γ-cyclodextrin has the constant only in the order of 10^4 M^{-1} (Cameron et al. 2002). The association constants for complexes of cyclodextrin derivatives with a wide variety of guests can be found in the review by Rekharsky and Inoue (Rekharsky and Inoue 1998).

Cyclodextrin derivative-guest complexes are prepared by several methods: co-precipitation, freeze-drying, kneading, melting, neutralization, grinding, sealed-heating and others (Iacovino et al. 2017). Cyclodextrin complexes usually have a cyclodextrin derivative:guest ratio of 1:1. However many other cyclodextrin derivative:guest ratios are known (Song et al. 2009b; Lima et al. 2016), and cyclodextrins can even form a wide variety of other supramolecular structures including catenanes, rotaxanes (Harada et al. 2009; Garcia-Rio et al. 2014) and supramolecular polymers (Wang et al. 2016).

2.1.2 Cyclodextrin Derivatives Properties

Nomenclature usually used by researchers for the description of cyclodextrin derivatives is depicted in Fig. 2.1. A glucose unit possesses substitution sites at positions 2, 3 and 6. Glucose units are named A, B, C etc. respectively in superscript. Sometimes the glucose units are numbered with roman numerals in superscript – i. e. $^{I, II, III}$ etc. Simplicity also often wins over precision. For example, per-*O*-methyl-2^A-*O*-allyl-β-cyclodextrin refers to β-cyclodextrin, where allyl substituent is at position 2 at one glucose unit and all 20 the other hydroxyl groups are protected with methyl groups.

As it was already shown in the previous section, substituents present on a cyclodextrin skeleton play a key role in its properties. For example cyclodextrins with hydrophobic chains could form micelles or vesicles in water environment (Sallas and Darcy 2008), other cyclodextrins are capable of catalysing decomposition of organophosphorus compounds (Letort et al. 2016), and positively charged cyclodextrin derivatives are used in chiral separations in capillary electrophoresis as they are able to interact with carboxylic acids (Tang and Ng 2008a).

The dependence of solubility on substituents could be demonstrated in the following example. Only 1.85 g of β-cyclodextrin is soluble in 100 ml water at room temperature (Szejtli 1998), while almost 24 g of per-*O*-methyl-β-cyclodextrin could be dissolved in the same amount of water (Szente and Szejtli 1999). Moreover, when heated, the solubility of the β-cyclodextrin increases, while in the case of the latter its solubility decreases. Permethylated cyclodextrins are also well soluble in most organic solvents.

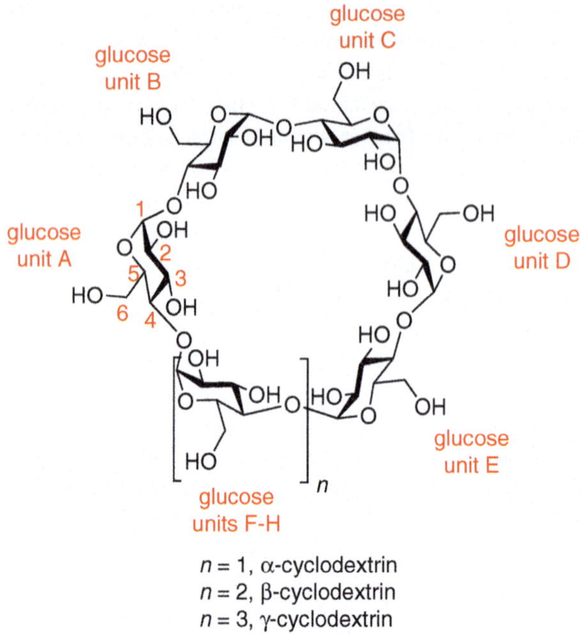

Fig. 2.1 Structure and numbering of α-, β-, and γ-cyclodextrin. Glucose units are named A, B, C etc. respectively and carbon atoms in each unit are numbered as usual

Cyclodextrin derivatives toxicity depends on the substituents present as well. The vast majority of toxicity tests are performed only on cyclodextrins randomly substituted, persubstituted or persubstituted at 2-, 3-, or 6-positions – as they are of interest of the pharmaceutical industry. Cytotoxic and hemolytic properties of these derivatives were compared by Kiss et al. on Caco-2 cells. Cell toxicity of several methylated β-cyclodextrin was the highest, while ionic derivatives were observed to be less toxic. The authors concluded cholesterol-solubilising properties could be a predictive factor for cyclodextrin derivatives cell toxicity (Kiss et al. 2010).

Randomly substituted sulfobutyl-β-cyclodextrin sodium salt with a degree of substitution 6.0–7.1 is well tolerated in male volunteers when administered intravenously at doses up to 200 mg/kg. Only a mild toxicity in the kidneys and liver was observed in rats at the dose of 3 g/kg, which is approximately 50-fold greater than the dose usually administered to men (Luke et al. 2010).

Randomly substituted 2-hydroxypropyl-β-cyclodextrin is well tolerated in rats, mice and dogs, particularly when administered orally. This cyclodextrin derivative is also well tolerated in humans, with the main adverse effect being diarrhoea. Moreover, intraperitoneal single dose of 10 g/kg in mice was neither lethal nor produced any toxicity (Gould and Scott 2005).

When speaking about randomly substituted derivatives, it should be taken into account there could be differences between derivatives with different degree of substitution (Li et al. 2016). Three randomly substituted 2-hydroxypropyl-β-cyclodextrin with degree of substitution 4.55, 6.16 and 7.76 were compared when administered intravenously to rats once daily at a dose of 500 mg/kg for 7 days. It was concluded cyclodextrin derivative with lower degree of substitution resulted in more changes in hematological and biochemical parameters. However, the effects were reversible at the end of recovery.

2.2 Cyclodextrin Derivatives

When native cyclodextrins are not suitable for a given application, their derivatives come into play. What type of a reaction should be used depends on the purpose of the product. For example, if a more soluble cyclodextrin derivative is required for the pharmaceutical application, then a random conversion of hydroxyl groups e.g. to sulphate (or another hydrophilic) group is carried out. However, such a product is a mixture of single compounds – usually characterised by degree of substitution. Moreover, the mixture differs in the detailed representation of individual isomers from batch to batch, even when degree of substitution remains constant (Estrada III and Vigh 2012).

When a single isomer cyclodextrin derivative is desired, the most straightforward way is to synthesise a persubstituted derivative. The direct synthesis of other single isomer derivatives is more challenging due to the number of theoretically possible isomers (Wenz 1994). The number of possible isomers is three for monosubstituted, dozens for disubstituted, and more than one hundred for trisubstituted derivatives.

The number of known cyclodextrin derivatives is huge. A search in SciFinder® for any substituted cyclodextrin skeleton revealed there are more than 2000 derivatives for α-cyclodextrin, almost 8000 for β-cyclodextrin, and more than 1000 for γ-cyclodextrin. Large ring cyclodextrins, i.e. having more than 8 glucose units (Endo 2011) or pre-α-cyclodextrin (Nakagawa et al. 1994) derivatives are not covered herein as their use is scarce. However, the reactivity of these cyclodextrins is expected to be similar to the three basic ones, with the exception of reactions where the reagent interferes with the cavity – see below. The large ring cyclodextrins' cavity is more flexible as well as more spacious, while pre-α-cyclodextrin has the smallest one.

Listing of more than 11,000 α-, β-, and γ-cyclodextrin derivatives would exceed the possibilities of this chapter. Thus the aim of this chapter is to provide a comprehensive view on the synthesis of favourite or interesting cyclodextrin derivatives, especially on those which are suitable for further modifications. The reason for this approach is obvious – the modification of a cyclodextrin already substituted with one or more suitable functional groups is much easier than the optimisation of substitution for every new cyclodextrin derivative desired. The chapter is divided into several sections, each focused on one type of cyclodextrin derivatives: persubstituted cyclodextrins, randomly substituted cyclodextrins, cyclodextrins persubstituted at selected positions, selectively substituted

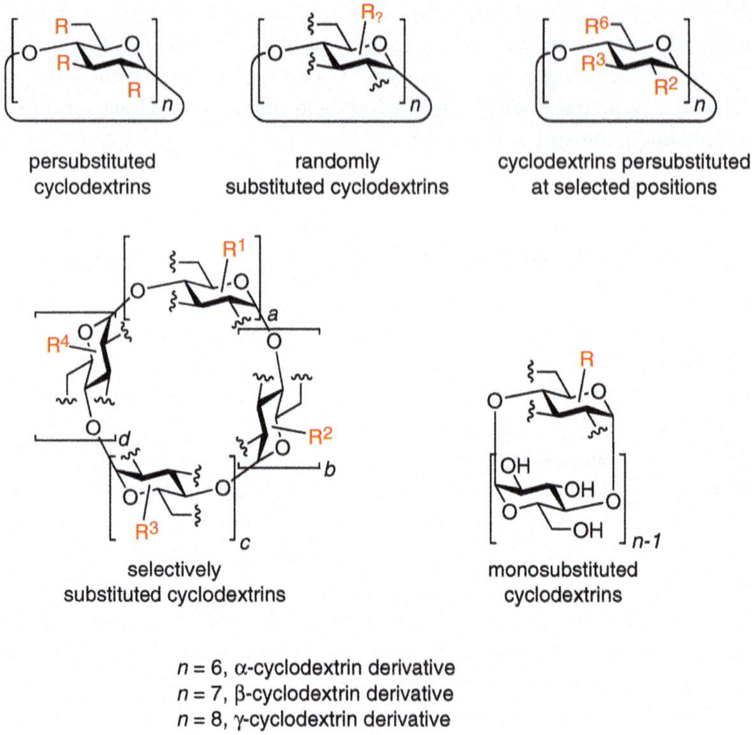

Fig. 2.2 Schematic representation of the cyclodextrin derivatives types

cyclodextrins, and monosubstituted cyclodextrins (Fig. 2.2). A mini-summary could be found at the end of these sections as a help for busy readers.

For the purpose of this review, the term monosubstituted cyclodextrins or selectively substituted cyclodextrins also refers to corresponding cyclodextrin derivatives with all the remaining hydroxyl groups protected – methylated, acetylated, benzylated etc.

Preparation of other cyclodextrin derivatives not described herein could be found in the previous reviews (Table 2.1).

2.2.1 *Reactivity of Cyclodextrins*

The modification reactions on cyclodextrins take place at the hydroxyl groups. As the hydroxyl groups are nucleophiles, the reaction proceeds *via* an electrophilic attack. However, a selective substitution of cyclodextrins is a great challenge for chemists as there are three types of hydroxyl groups present in one glucose unit (at position 2, 3 or 6). Moreover, several glucose units of which cyclodextrins are composed make the process rather difficult. Hydroxyl groups at positions 2, 3 and 6

Table 2.1 Reviews focused on preparation of cyclodextrin derivatives

Description	Reference
Cyclodextrin derivatives used against organophosphorus compounds	(Letort et al. 2016)
Cyclodextrins monosubstituted by several functional groups suitable for further synthesis	(Řezanka 2016)
Phosphorus-containing cyclodextrins	(Grachev 2013)
Cyclodextrin derivatives prepared by click chemistry	(Faugeras et al. 2012)
A few selectively functionalised cyclodextrins	(Bellia et al. 2009)
Monosubstituted positively charged cyclodextrins for chiral separations	(Tang and Ng 2008a)
Amphiphilic cyclodextrin derivatives	(Sallas and Darcy 2008)
Cyclodextrin derivatives bearing bridging substituents	(Engeldinger et al. 2003)
Systematic analysis of available methods for modification of cyclodextrins at position 2, 3 or 6	(Khan et al. 1998)
Cyclodextrin derivatives listed by functional groups present	(Croft and Bartsch 1983)

compete against each other during the reaction. Fortunately, there are at least some differences among them. Hydroxyl groups at positions 6 are primary and at positions 2 and 3 are secondary. Hydroxyl groups at positions 6 are the least acidic and the most accessible, at positions 2 are the most acidic, and at positions 3 are the least accessible (Khan et al. 1998).

When all the hydroxyl groups are deprotonated with an excess of a base, an electrophilic reagent reacts at position 6, because it is the most accessible. The more reactive agents attack the hydroxyl groups at position 6 less selectively, and therefore they react also with the hydroxyl groups on the secondary rim. The less reactive reagents react at position 6 selectively. The best known examples of this feature are syntheses of 6^{A}–*O*-tosyl-cyclodextrin where multisubstituted derivatives are the only by-products (Řezanka 2016).

Bases first deprotonate the hydroxyl groups at position 2, because they are the most acidic, having $pK_a = 12.2$ (Sallas and Darcy 2008). The oxyanion formed is more nucleophilic than other non-deprotonated hydroxyl groups. Thus, the use of a corresponding amount of a strong base, e.g. NaH or LiH, often leads to the substitution at position 2 predominantly (Řezanka 2016).

Substitution at position 3 is the most difficult one for the abovementioned reasons. Fortunately, some reagents interfere with the cavity of cyclodextrins, making this process much easier. For example, cinnamyl bromide forms a complex with β-cyclodextrin and allows the substitution to be selective at position 3 with multisubstituted derivatives as the only by-products (Jindřich and Tišlerová 2005). On the other hand, such interference of a reagent with the cavity could be a complication in other modifications of cyclodextrins and should be always taken into account.

Solvents play another important role during the modification of a cyclodextrin. They can affect both nucleophilicity of oxyanions, as well as the strength of a complex with a substitution agent. If the complex is strong, the predominant product will be driven by the orientation of the substitution agent in the interior of cyclodextrin. Considering the abovementioned facts, the commonly achieved yields of substituted cyclodextrin derivatives are very low. Exceptions to this rule will be of interest in the next sections.

2.2.2 Persubstituted Cyclodextrin Derivatives

A great variety of persubstituted cyclodextrins, i. e. cyclodextrin derivatives, where every hydroxyl group is substituted by the same functional group, are available from common commercial sources. However, they are still synthesised by researches because of their simple synthesis and lower overall cost when compared to the commercial ones. The modification usually aims to increase the solubility of cyclodextrins – either in organic solvents or water – or to use them in deprotection reactions – see below.

As it was stated above, the persubstitutions proceed smoothly and they are usually carried out by the reaction of a cyclodextrin with an excess of the reagent, e.g. alkyl halogenide, in the presence of a base. Per-*O*-methylated cyclodextrins are obtained by the reaction of a corresponding cyclodextrin with NaH and methyl iodide in *N,N'*-dimethylformamide (Nakazono et al. 2010; Stefanache et al. 2014) or dimethyl sulfoxide (Szejtli et al. 1980). Similarly, per-*O*-benzylated cyclodextrins are prepared by the reaction of a cyclodextrin with benzyl halogenide (Bjerre et al. 2007; Normand et al. 2012); per-*O*-trimethylsilylated cyclodextrins with trimethylsilyl chloride (Kurochkina et al. 2014) or imidazole (Harabagiu et al. 2004) and per-*O*-(2-hydroxypropyl) cyclodextrins with propylene carbonate (Trotta et al. 2002).

Per-*O*-acetylated cyclodextrins are usually prepared by the reaction of a cyclodextrin with acetanhydride instead of acetyl halogenide – as the latter one is more expensive and less easy to handle. The reaction is promoted by acids (Zhang et al. 2011; Jicsinszky et al. 2015) or bases (Ghanem and Schurig 2001; Lian et al. 2014).

When a reagent for cyclodextrin persubstitution is chosen wisely, it allows a subsequent modification, which leads to other derivatives as it is illustrated by the example from Kraus et al. (Scheme 2.1) (Kraus et al. 2001). Firstly, per-*O*-allyl-β-cyclodextrin is prepared. The subsequent oxidation steps led to an unstable formylmethyl derivative and it was thus directly transformed into a hydroxyethyl derivative. Finally, the oxidation and methylation reactions were carried out to obtain per-*O*-methoxycarbonylmethyl-β-cyclodextrin.

Mini-Summary Persubstituted cyclodextrin derivatives are prepared by the reaction of native cyclodextrin with an excess of the reagent.

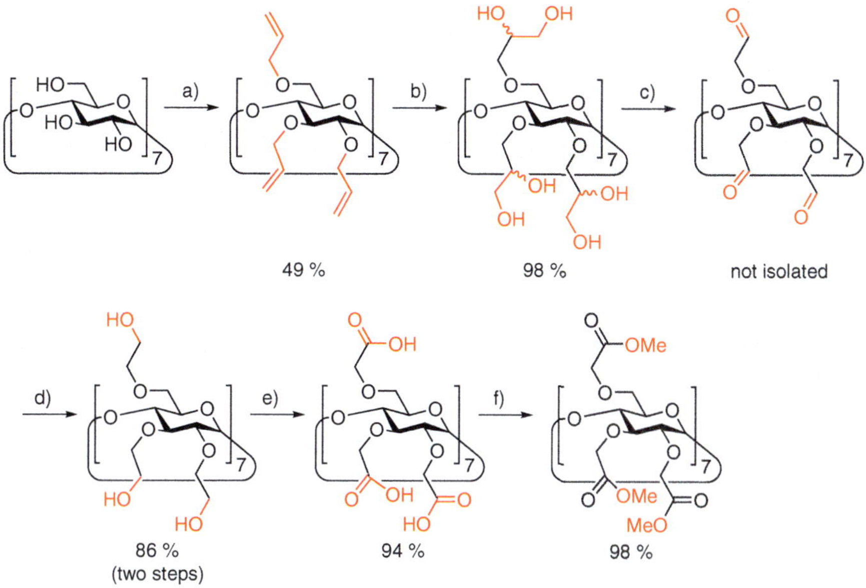

a) NaH, allyl bromide; b) OsO_4, *N*-methylmorpholine *N*-oxide; c) $NaIO_4$; d) $NaBH_4$; e) 1. (2,2,6,6-tetramethylpiperidin-1-yl)oxyl, NaClO, KBr, 2. Dowex; f) CH_2N_2 Me = methyl

Scheme 2.1 Synthesis of persubstituted cyclodextrin derivatives

2.2.3 *Randomly Substituted Cyclodextrin Derivatives*

Randomly substituted cyclodextrin derivatives are modified at various positions and they are usually characterised by degree of substitution. The exact structure and ratio of single derivatives forming the mixture of randomly substituted cyclodextrins is unknown. As well as in the case of persubstituted derivatives, randomly substituted cyclodextrin derivatives are largely available from common commercial sources and their synthesis is thus often claimed in patents (Wimmer 1995; Yanli 2005).

The purchase of randomly substituted cyclodextrin derivatives conceals many pitfalls. It was shown there could be a significant difference in the relative abundances of the isomers with the same degree of substitution between two batches (Estrada III and Vigh 2012). Moreover, the authors also found the information about degree of substitution from the supplier may be affected by an error.

Researches should be careful when preparing randomly substituted cyclodextrin derivatives as they do not know the exact composition of the single isomers in the product. As it has been shown before (Řezanka et al. 2016), diverse single cyclodextrin isomers have different properties and their ratio could therefore affect the properties of randomly substituted cyclodextrin derivatives. They could be synthesised using conditions similar to preparation of persubstituted derivatives. The only need is to add a lesser amount of a reagent that is needed for a fully

Table 2.2 Syntheses of common randomly substituted cyclodextrin derivatives

Randomly substituted cyclodextrin derivative	Reagent	Reference
Carboxymethylated β-cyclodextrin	Chloroacetic acid	(Lammers et al. 1971; Reuben et al. 1994)
Carboxymethylated β-cyclodextrin	Sodium chloroacetate	(Shirin et al. 2003)
2-hydroxybutylated β-cyclodextrin	Butylene oxide	(Ishiguro et al. 2011)
2-hydroxypropylated β-cyclodextrin	Propylene oxide	(Pitha et al. 1986; Rao et al. 1992; Yuan et al. 2015)
Methylated β-cyclodextrin	Dimethyl carbonate	(Gan et al. 2011)
Methylated β-cyclodextrin	Methyl chloride	(Yanli 2005; Cui et al. 2010)
Methylated β-cyclodextrin	Methyl iodide	(Wimmer 1995; Yanli 2005)
Methylated β-cyclodextrin	Dimethyl sulfate	(Bakó et al. 1994)
Phosphated β-cyclodextrin	Phosphoryl chloride	(Lee and Lim 1998)
Sulfated β-cyclodextrin	Sulfur trioxide pyridine complex	(Estrada III and Vigh 2012)
Sulfoethylated β-cyclodextrin	Sodium 2-bromoethanesulfonate	(Qu et al. 2002)
Sulfobutylated β-cyclodextrin	1,4-butanesultone	(Song et al. 2009a; Ma et al. 2016)
Sulfobutylated γ-cyclodextrin	1,4-butanesultone	(Jung and Francotte 1996; Tongiani et al. 2005)
Sulfopropylated β-cyclodextrin	1,3-propanesultone	(Lammers et al. 1971; Song et al. 2009a)

substituted derivative. Syntheses of the most common randomly substituted cyclodextrin derivatives are summarised in Table 2.2.

2.2.3.1 Random Cyclodextrin Polymers

The randomly substituted cyclodextrin derivatives also include a group of random cyclodextrin polymers, where cyclodextrins are interconnected to the other ones in a random way. Such polymers could be synthesised *via* three approaches: i) cyclodextrins or their derivatives are directly cross-linked by a suitable agent; ii) cyclodextrins are first randomly modified with reactive groups, which are subsequently used for the attachment onto a polymer backbone; iii) cyclodextrin is substituted with a functional group that is available for polymerisation, e.g. double bond. The first approach often gives branched polymers, whereas the second and the third result in linear polymers substituted with cyclodextrins. Prepared random cyclodextrin polymers could have various properties depending on the amount and nature of their cross-linking agent or the type of connection and the polymer used. However in the vast majority of cases, the cyclodextrin polymers retain their key property – the ability to include guests in their cavity.

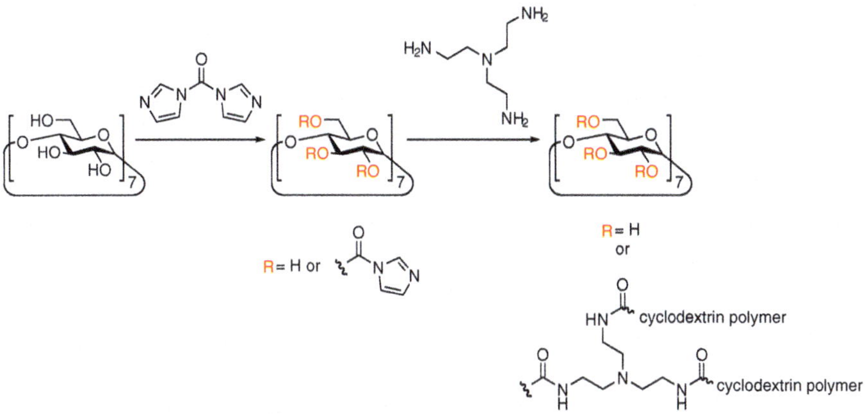

Scheme 2.2 Cross-linking of β-cyclodextrin using *N,N*-carbonyldiimidazole and tris(2-aminoethyl)amine

The best known example of the first method is the direct reaction of epichlorohydrin – 2-(chloromethyl)oxirane – with a cyclodextrin (van de Manakker et al. 2009; Morin-Crini and Crini 2013; Concheiro and Alvarez-Lorenzo 2013; Gidwani and Vyas 2014). If the degree of cross-linking is sufficiently high, the resulting polymer becomes insoluble in water. Epichlorohydrin could be used for cross-linking of randomly substituted cyclodextrin derivatives as well (Zhang et al. 2012).

Epichlorohydrin-cyclodextrin polymers are used in various applications (Morin-Crini and Crini 2013; Folch-Cano et al. 2014), for example in lithium rechargeable batteries, where cyclodextrin cross-linked hyperbranched network structure covers the electrode, which results in a great improvement in both capacity and cycle life (Jeong et al. 2014).

Another example of the first approach is the use of *N,N*-carbonyldiimidazole as a reagent allowing cyclodextrin to be reacted with a cross-linker. For example, *N,N*-carbonyldiimidazole was first reacted with β-cyclodextrin forming a reactive derivative, which was subsequently used for the cross-linking with tris(2-aminoethyl)amine (Scheme 2.2) (Wang et al. 2013).

Recently a review on cyclodextrin-polyurethane polymers for the removal of pollutants from waste water has been issued (Taka et al. 2017). Commonly used cross-linking agents for the synthesis of random cyclodextrin polymers are summarised in Fig. 2.3 (Mocanu et al. 2001; Concheiro and Alvarez-Lorenzo 2013; Karoyo and Wilson 2015).

Direct cross-linking of cyclodextrin has also recently been published in Nature (Alsbaiee et al. 2016). Cross-linking of β-cyclodextrin with tetrafluoroterephthalonitrile resulted in a high-surface-area mesoporous polymer capable of adsorption of organic micropollutants. Moreover, only a mild washing procedure was required for the polymer regeneration and the polymer exhibited no loss in performance after recycling.

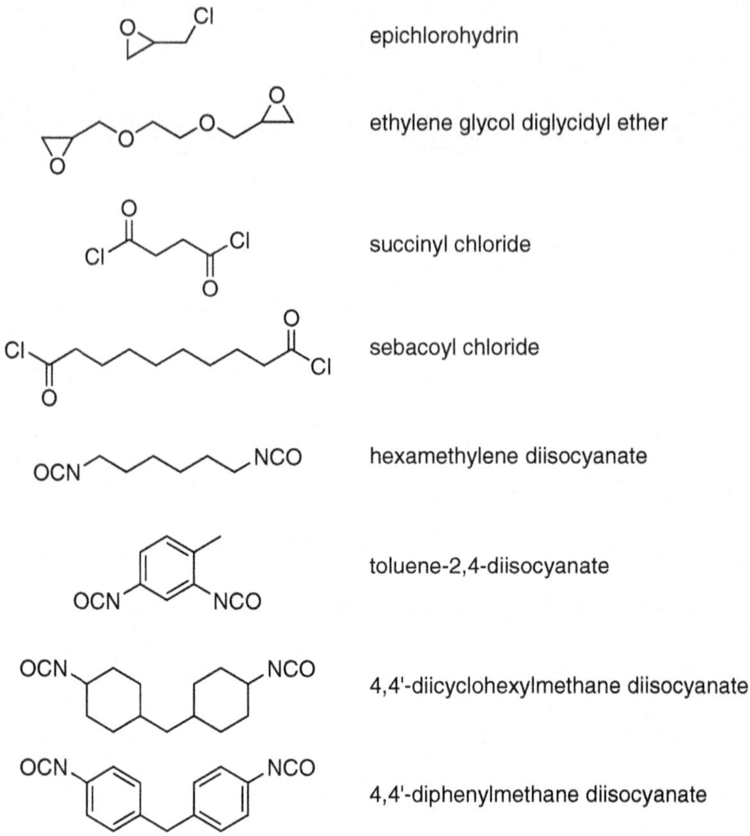

Fig. 2.3 Commonly used cross-linking agents for preparation of random cyclodextrin polymers

An example of the second approach, where cyclodextrins are first randomly modified by reactive groups and subsequently attached onto a polymer backbone, is the use of randomly carboxymethylated β-cyclodextrin for the attachment onto chitosan (Krauland and Alonso 2007; Prabaharan and Gong 2008). Opening of anhydride in poly[(methyl vinyl ether)-*alt*-(maleic anhydride)] by deprotonated β-cyclodextrin fits this approach as well (Renard et al. 2005).

The third approach requires the introduction of functional groups with the ability to be polymerised. This requirement is usually fulfilled by the synthesis of cyclodextrin ester of acrylic acid – e.g. by the reaction of cyclodextrin with *m*-nitrophenyl acrylate. Polymerisation of randomly acryloylated cyclodextrins is then initialised by potassium persulfate (Mocanu et al. 2001; Zhang et al. 2009).

Mini-Summary Randomly substituted cyclodextrin derivatives are prepared by the reaction of native cyclodextrins with fewer equivalents of the reagent than the number of hydroxyl groups is. Random cyclodextrin polymers are most often synthesised by cross-linking with epichlorohydrin.

2.2.4 *Cyclodextrins Persubstituted at Selected Positions*

Cyclodextrins persubstituted at selected positions include cyclodextrins persubstituted either at 2 or 3 or 6 positions or any combination of thereof (Fig. 2.2). Synthesis of these derivatives is based on the different reactivity of hydroxyl groups described above and often employs the protection/deprotection methodology to achieve a desired derivative. The most favourite reactions are those carried out at position 6, because it possesses primary hydroxyl groups. Cyclodextrins could be easily substituted at this position with *tert*-butyldimethylsilyl or halogens. Such derivatives are useful precursors for amphiphilic cyclodextrin derivatives (Sallas and Darcy 2008). Otherwise very popular *p*-toluenesulfonyl, also so-called "tosyl", group is not employed very often as it could spontaneously undergo the reaction to 3,6-anhydro form (Khan et al. 1998).

2.2.4.1 Syntheses Based on per-6-*O*-(*tert*-Butyldimethylsilyl) Cyclodextrins

Synthesis of per-6-*O*-(*tert*-butyldimethylsilyl)-α-, β- and γ-cyclodextrins is carried out by the reaction of native cyclodextrin with *tert*-butyldimethylsilyl chloride and BaO (Takeo et al. 1988; Takeo et al. 1989), pyridine (Fugedi 1989; Ashton et al. 1996) or imidazole (Vincent et al. 1997; Maynard and Vigh 2000) (Scheme 2.3). Among the bases, pyridine gave the best yields (Ashton et al. 1996).

Hexakis(6-*O*-*tert*-butyldimethylsilyl)-α-cyclodextrin could be protected at positions 2 and 3 by acetyl, methyl or benzyl groups. Subsequent deprotection of silyl groups by BF_3 with tetrahydrofurane, sodium methanolate or tetrabutylammonium fluoride results in useful derivatives, where hydroxyl groups at positions 6 are ready for any modification desired (Scheme 2.4) (Takeo et al. 1988). The reactions proceed similarly with β- (Jullien et al. 1994; Ashton et al. 1996; Vincent et al. 1997; Kirschner and Green 2005) or γ-cyclodextrin derivatives (Jullien et al. 1994). TBDMS group in heptakis(2,3-*O*-dimethyl-6-*O*-*tert*-butyldimethylsilyl)-β-cyclodextrin was also transformed by triphenylphosphine and bromine to heptakis(2,3-*O*-dimethyl-6-*O*-bromo)-β-cyclodextrin (Ashton et al. 1996).

Heptakis(6-*O*-*tert*-butyldimethylsilyl)-β-cyclodextrin was also used for the synthesis of carboxymethyl β-cyclodextrin derivatives (Scheme 2.5). Firstly, positions 2 and 3 were allylated and the silyl group deprotected by tetrabuty lammonium fluoride (Baer et al. 1992b). Subsequently, hydroxyl groups at position 6 were methylated and allyl groups oxidised in a similar way as described above (Kraus et al. 2001).

The β-cyclodextrin derivatives similar to those depicted in Scheme 2.4 were also used for syntheses of other useful precursors (Scheme 2.6 and 2.7). Heptakis(6-*O*-*tert*-butyldimethylsilyl)-β-cyclodextrin was tosylated and the subsequent desilylation yielded heptakis(2-*O*-tosyl)-β-cyclodextrin (Coleman et al. 1991). Although this derivative could also be obtained by the direct substitution reaction of

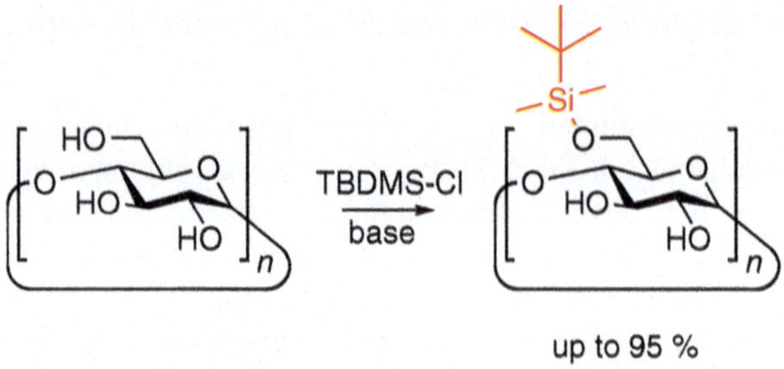

$n = 6$, α-cyclodextrin derivative
$n = 7$, β-cyclodextrin derivative
$n = 8$, γ-cyclodextrin derivative
TBDMS-Cl = *tert*-butyldimethylsilyl chloride

Scheme 2.3 Synthesis of per-6-*O*-(*tert*-butyldimethylsilyl)-α-, β- and γ-cyclodextrins

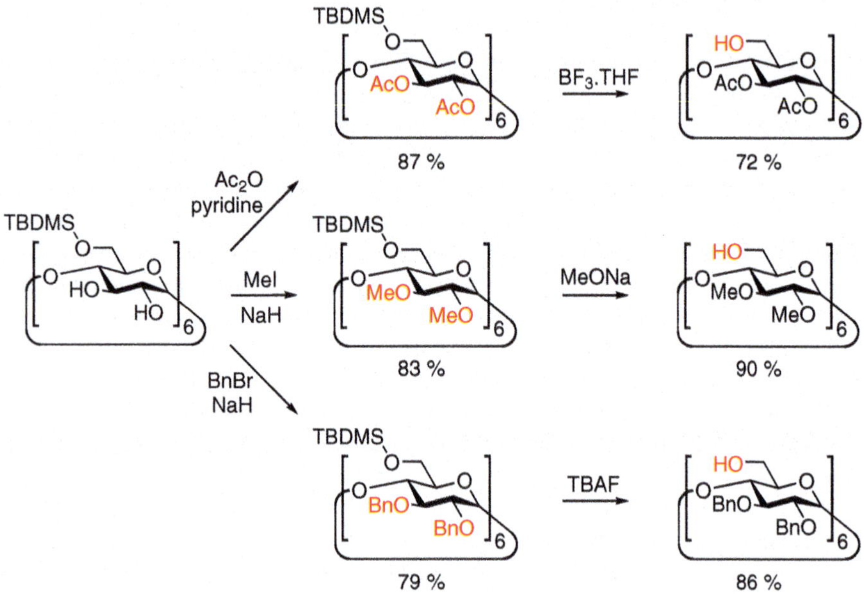

TBDMS = *tert*-butyldimethylsilyl
Ac = acetyl
THF = tetrahydrofurane
Me = methyl
Bn = benzyl
TBAF = tetrabutylammonium fluoride

Scheme 2.4 Synthesis of α-cyclodextrins protected at positions 2 and 3

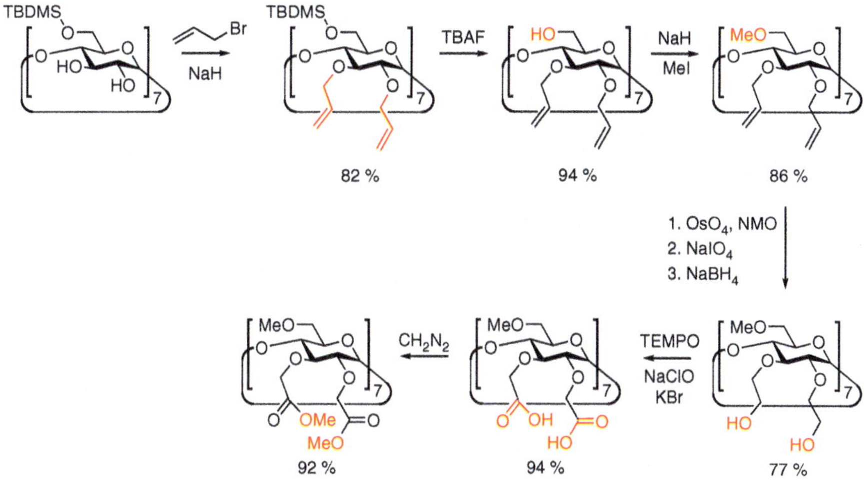

TBDMS = *tert*-butyldimethylsilyl
TBAF = tetrabutylammonium fluoride
Me = methyl
NMO = *N*-methylmorpholine N-oxide
TEMPO = (2,2,6,6-tetramethylpiperidin-1-yl)oxyl

Scheme 2.5 Synthesis of carboxymethyl β-cyclodextrin derivatives

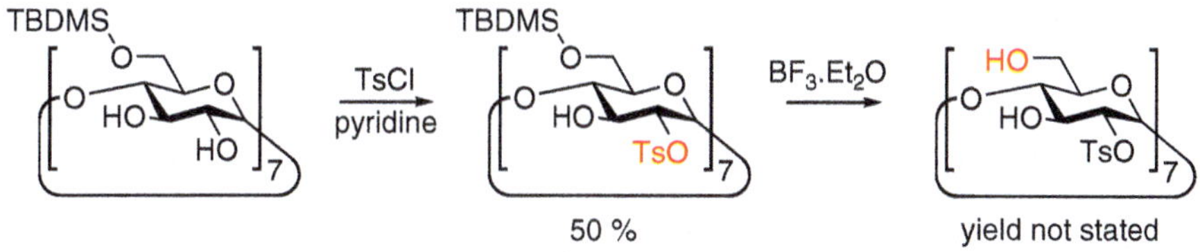

TBDMS = *tert*-butyldimethylsilyl
Ts = tosyl
Et = ethyl

Scheme 2.6 Synthesis of useful β-cyclodextrin precursors

β-cyclodextrin with tosylimidazole in the presence of Cs_2CO_3, the yield is only 5% (Yu et al. 2006). It is noteworthy the subsequent substitution of tosyl group leads (according to Walden inversion) to non-cyclodextrin derivatives with aldohexoses in *manno* configuration.

β-cyclodextrins protected at positions 2 and 3 with methyl or acetyl group could undergo oxidation by (2,2,6,6-tetramethylpiperidin-1-yl)oxyl forming 5-carboxy-5-dehydroxymethyl derivatives (Scheme 2.7). Moreover, the acetylated derivative enables deprotection at positions 2 and 3 (Kraus et al. 2000). Uccello-Barretta et al. also synthesised other derivatives persubstituted at position 6 by common reactions (Scheme 2.7) (Uccello-Barretta et al. 2005).

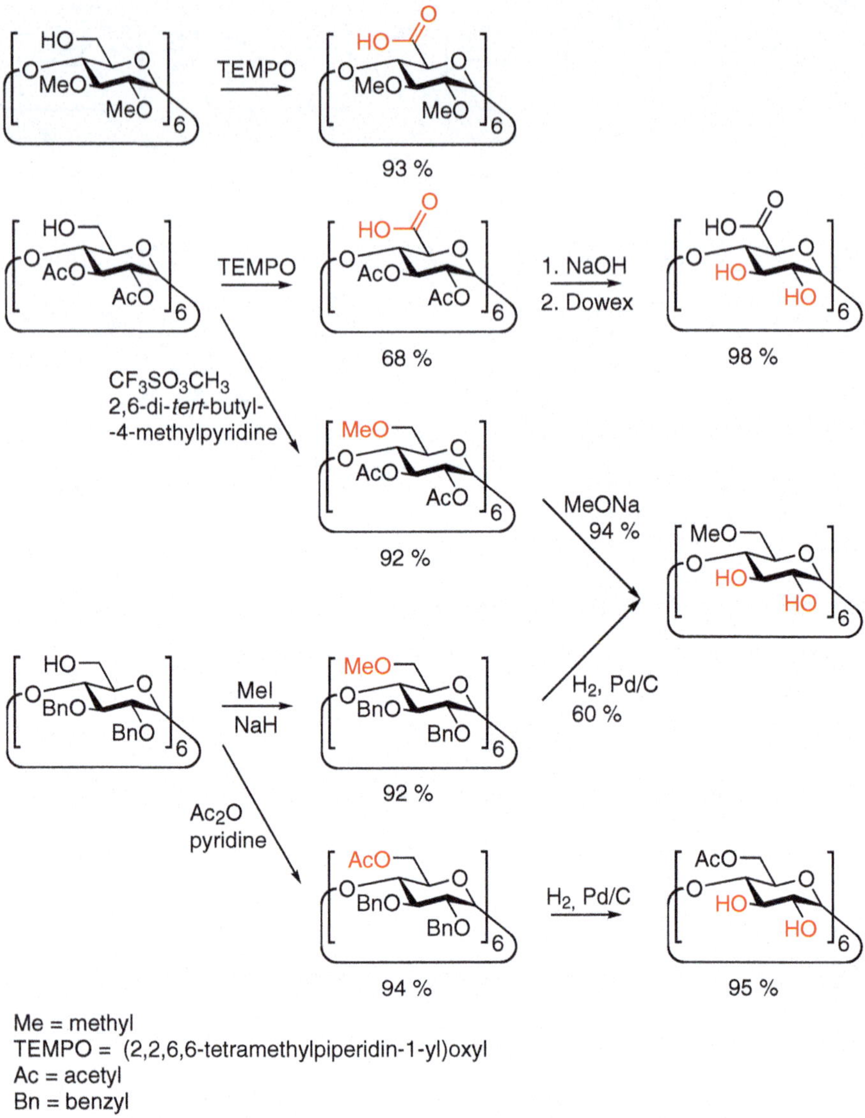

Scheme 2.7 Synthesis of β-cyclodextrins persubstituted at position 6

Vigh and his colleagues devoted a great effort to synthesis of cyclodextrin derivatives persubstituted at selected positions (Vincent et al. 1997; Maynard and Vigh 2000; Busby and Vigh 2005; Tutu and Vigh 2011). Their methodology is based on modifications of per-6-*O*-(*tert*-butyldimethylsilyl)-α-, β- and γ-cyclodextrins (Scheme 2.8, 2.9 and 2.10). Synthesis of heptakis(2-*O*-methyl)-β-cyclodextrin includes a crucial step that involves the migration of silyl groups to position 3 and

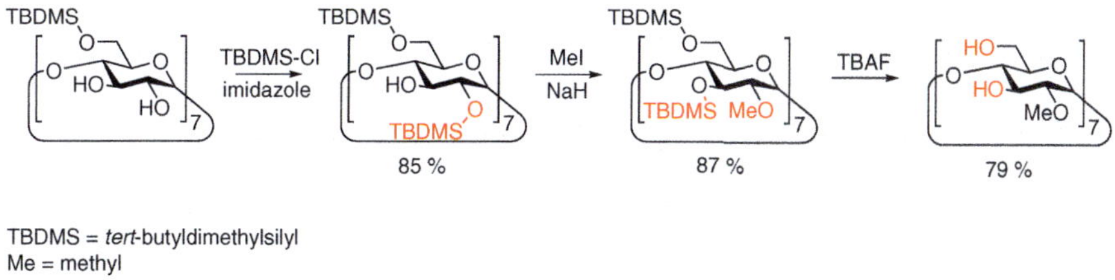

Scheme 2.8 Synthesis of heptakis(2-*O*-methyl)-β-cyclodextrin

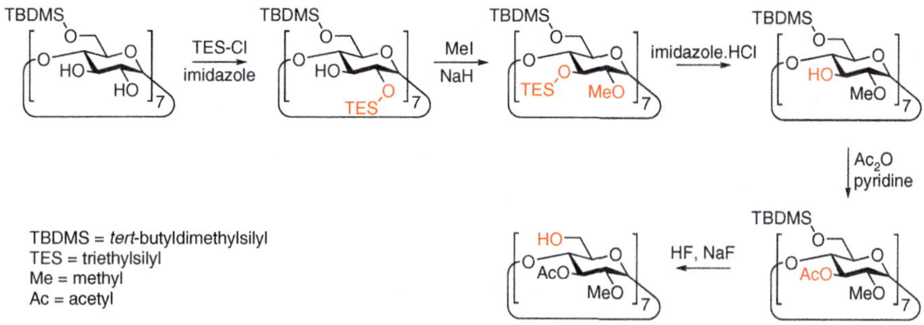

Scheme 2.9 Synthesis of heptakis(2-*O*-methyl-3-*O*-acetyl)-β-cyclodextrin. Yields are not stated

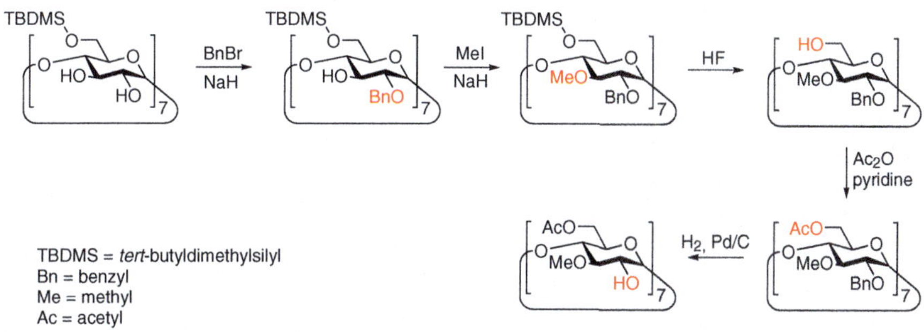

Scheme 2.10 Synthesis of heptakis(3-*O*-methyl-6-*O*-acetyl)-β-cyclodextrin. Yields are not stated

the simultaneous methylation of the hydroxyl group at position 2 (Scheme 2.8). The migration mechanism was suggested on the basis of NMR spectroscopy and subsequently confirmed by X-ray crystallography (Maynard and Vigh 2000).

The migration also allows synthesis of heptakis(2-*O*-methyl-3-*O*-acetyl)-β-cyclodextrin, where substituents at every position are different from each other (Scheme 2.9). However, in this case, protection and migration of triethylsilyl instead of *tert*-butyldimethylsilyl groups is taken into advantage. Triethylsilyl is subsequently selectively hydrolysed by imidazole hydrochloride and free hydroxyl groups at position 3 are acetylated. The final product is obtained by the removal of *tert*-butyldimethylsilyl groups (Busby and Vigh 2005).

Finally, Tutu and Vigh synthesised heptakis(3-*O*-methyl-6-*O*-acetyl)-β-cyclodextrin (Scheme 2.10). The product was prepared by the standard protection/deprotection methodology with the selective benzylation of per-6-*O*-(*tert*-butyldimethylsilyl)-β-cyclodextrin at position 2 as the key step (Tutu and Vigh 2011).

2.2.4.2 Syntheses Based on per-6-Halogeno-per-6-Deoxy Cyclodextrins

Another group of favourite starting materials for synthesis of cyclodextrins persubstituted at selected positions are cyclodextrins perhalogenated at position 6. They are synthesised from native cyclodextrins by the reaction with triphenylphosphine and bromine (Takeo et al. 1974) or iodine (Gadelle and Defaye 1991; Fernandez et al. 1995; Ashton et al. 1996; Benkhaled et al. 2008) in *N*,*N*′-dimethylformamide (Scheme 2.11). However, the synthesis of bromo derivatives has been declining due to more convenient handling of iodine.

These halogen derivatives are very useful precursors and could be easily transformed for example into azides and amines (Ashton et al. 1996; Gorin et al. 1996) or thiols (Rojas et al. 1995; Gorin et al. 1996) by standard procedures (Scheme 2.12). The remaining hydroxyl groups of these derivatives could be peracetylated (Boger et al. 1978; Baer et al. 1992a), methylated (Boger et al. 1978) or benzylated (Jullien et al. 1994) by the same methods described above.

Jicsinszky et al. have recently described the use of per-6-iodo-per-6-deoxy-β- and γ-cyclodextrins for the synthesis of azido or thio derivatives in a planetary ball mill under solvent-free conditions. The authors found out the mechanochemical synthesis not only simplified the isolation and purification processes, but also allowed easy scale-up (Jicsinszky et al. 2016a).

Per-6-azido-per-6-deoxy cyclodextrins are the perfect starting materials for nowadays favourite copper-catalysed azide-alkyne cycloaddition reactions (Faugeras et al. 2012; Letort et al. 2016) or, as described by Kraus et al., for modifications at positions 2 and 3 (Kraus et al. 2002) (Scheme 2.13). The key synthetic step is the use of $BaO/Ba(OH)_2$ for the introduction allyl groups at positon 2. The property of this base is not unique for azido derivatives and it is used as well for

Scheme 2.11 Synthesis of per-6-halogeno-per-6-deoxy-α-, β- and γ-cyclodextrins

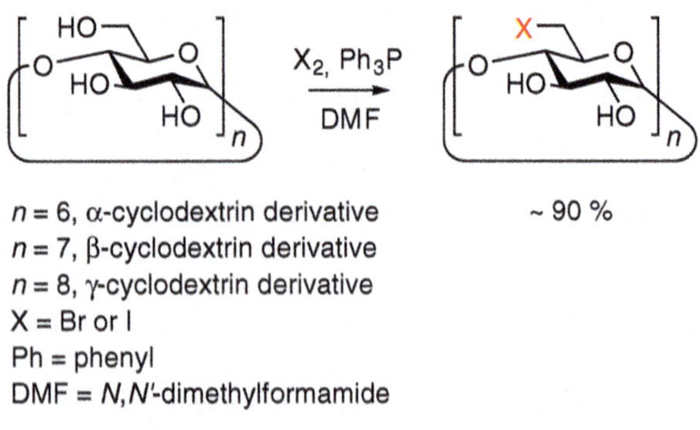

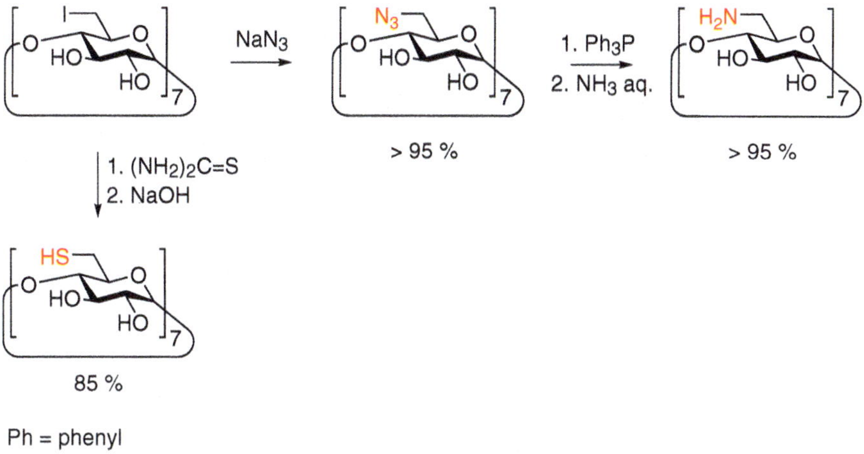

Scheme 2.12 Synthesis of azido, amino and thio derivatives of β-cyclodextrin

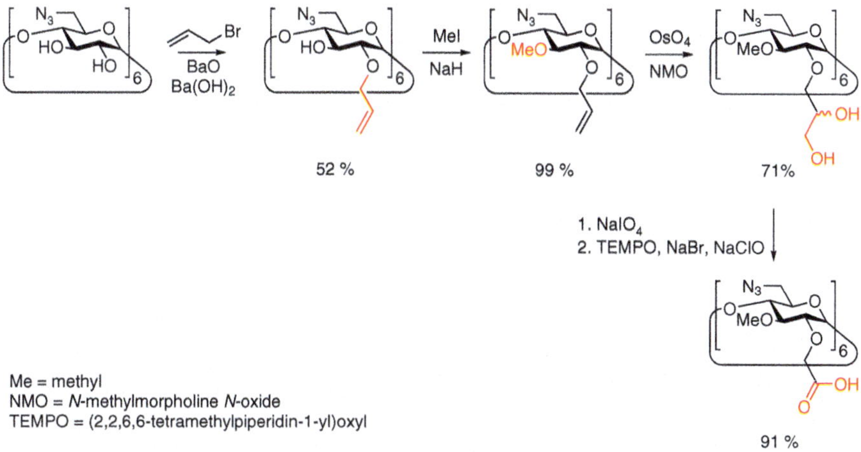

Scheme 2.13 Modifications of hexakis(6-azido-6-deoxy)-α-cyclodextrin

other syntheses, where the introduction of a substituent at position 2 is needed – see the beginning of the next section.

2.2.4.3 Other Syntheses

As it was described above, BaO/Ba$(OH)_2$ was successfully applied for the introduction of a substituent at position 2. Moreover, when the conditions are applied to a native cyclodextrin, the reaction proceeds to per-2,6-dialkyl cyclodextrin derivatives. This can be used e. g. in the synthesis of heptakis(2,6-*O*-dimethyl)-β-cyclodextrin (Szejtli

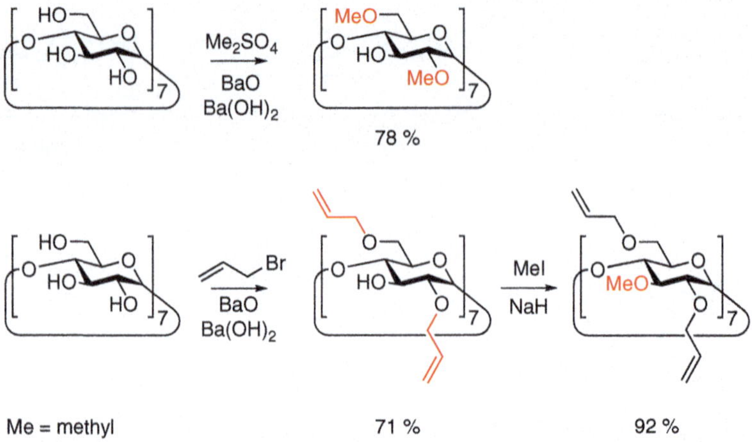

Scheme 2.14 Synthesis of heptakis(2,6-*O*-dialkyl) β-cyclodextrin derivatives

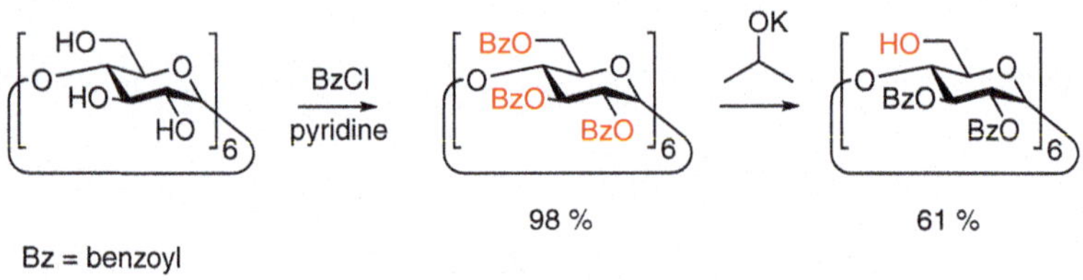

Scheme 2.15 Selective deprotection of per-*O*-benzoyl-α-cyclodextrin

et al. 1980) or heptakis(2,6-*O*-diallyl)-β-cyclodextrin (Bergeron et al. 1976) (Scheme 2.14). Allyl groups in heptakis(2-*O*-allyl-3-*O*-methyl-6-*O*-allyl)-β-cyclodextrin could be oxidised similarly to Scheme 2.5 to carboxymethyl derivatives (Kraus et al. 2001).

Boger et al. used the selective deprotection strategy as the key step for syntheses of various cyclodextrin derivatives (Boger et al. 1978). Firstly, per-*O*-benzoyl-α-cyclodextrin is prepared. Secondly, the deprotection step by potassium isopropylalcoholate is carried out. The base is selective due to its steric hindrance and thus is able to deprotect only the primary hydroxyl groups (Scheme 2.15). The resulting product was transformed to various 6-azido or 6-amino derivatives.

The selective deprotection strategy was also used for acetolysis of perbenzylated α-cyclodextrin (Angibeaud and Utille 1991). The reaction yielded hexakis(2,3-*O*-dibenzyl-6-*O*-acetyl)-α-cyclodextrin, in which either acetyl or benzyl groups could be selectively deprotected (Scheme 2.16). Moreover, perbenzylated α-cyclodextrin was also used for deprotection by triethylsilane and iodine (Guitet et al. 2012). The method was originally developed for debenzylation of multiple-*O*-benzylated mono- and disaccharides (Pastore et al. 2011) and in the case of α-cyclodextrin proceeds at position 3, which is normally the least accessible (Scheme 2.16).

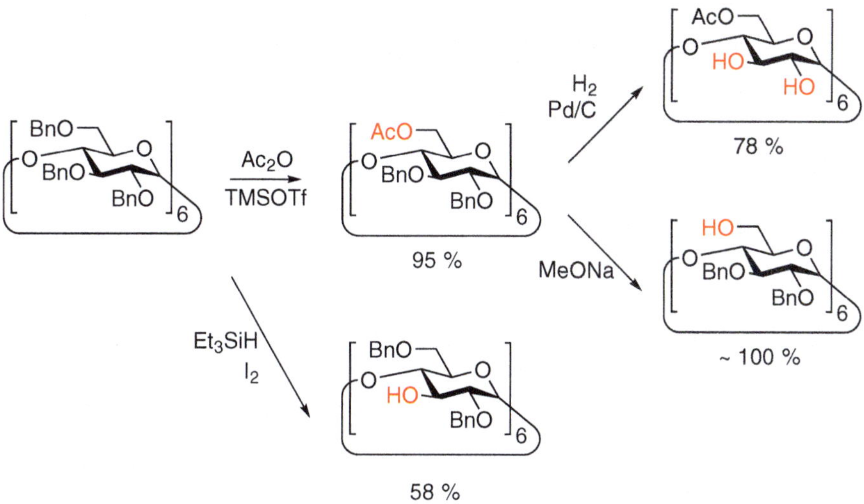

Bn = benzyl
Ac = acetyl
TMS = trimethylsilyl
Tf = trifluoromethanesulfonate
Me = methyl
Et = ethyl

Scheme 2.16 Selective deprotection of perbenzylated α-cyclodextrin

Mini-Summary Synthesis of cyclodextrins persubstituted at selected positions utilises the different reactivity of hydroxyl groups and employs protection/deprotection methodology to achieve a desired derivative. Most of the syntheses begin with the substitution at position 6 – either by *tert*-butyldimethylsilyl or a halogen. BaO/ $Ba(OH)_2$ direct the substitution to positions 2 and 6. When position 6 is blocked, the substitution proceeds selectively at position 2. The selective substitution at otherwise the least accessible position 3 could be achieved by deprotection of perbenzylated cyclodextrin by triethylsilane with iodine.

2.2.5 Selectively Substituted Cyclodextrins

Selectively substituted cyclodextrins fill the gap between all the above discussed cyclodextrin derivatives and monosubstituted cyclodextrins, i.e., they are single isomer compounds with a known structure, where two or more substituents are attached to the cyclodextrin skeleton. Their synthesis is the most challenging among all cyclodextrin derivatives, as the number of possible isomers starts at dozens and ends at millions for different substituents at different positions. For the purposes of this chapter, the term of selectively substituted cyclodextrins also refers to the derivatives that contain several free hydroxyl groups with the rest being protected by e. g. methyl, acetyl, benzyl or other groups.

2.2.5.1 Syntheses Based on Direct Substitution

One of the first attempts to synthesise selectively substituted cyclodextrins was the use of disulfonates for the selective modification of selected glucose units at position 6 (Scheme 2.17). $6^A,6^B$ derivative is formed when β-cyclodextrin is reacted with 4,6-dimethoxybenzene-1,3-disulfonyl chloride (Breslow et al. 1990). The use of benzophenone-3,3′-disulfonyl chloride led to $6^A,6^C$ derivative, while *trans*-stilbene-4,4′-disulfonyl chloride led to $6^A,6^D$ derivative (Tabushi et al. 1981). The disulfonates could be transformed into diiodo (Breslow et al. 1990), diazido, diamino (Tabushi et al. 1977; Di Blasio et al. 1996) or dithio derivatives (Tabushi et al. 1977) by common reactions.

Similarly, $2^A,2^B$ disulfonates were synthesised reacting benzophenone-3,3′-disulfonyl imidazole with α- or β-cyclodextrin in 30% yields (Teranishi 2000). However, a further substitution of these derivatives leads to non-glucose cyclodextrin derivatives due to the stereogenic centre inversion. This problem could be avoided using non-sulfonate derivatives.

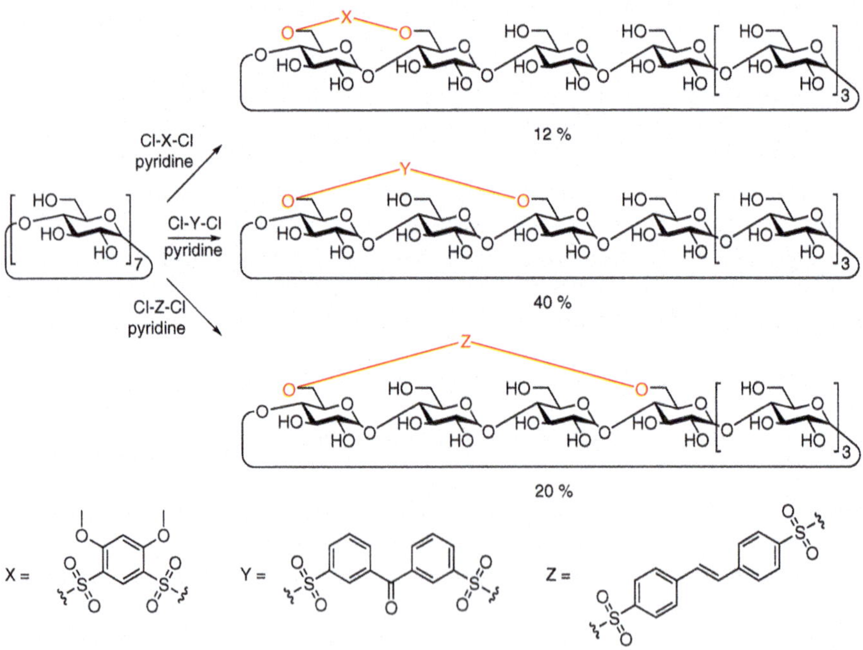

Scheme 2.17 Regioselective synthesis of disulfonate cyclodextrin derivatives

α,α-Dimethoxytoluene is able to protect two hydroxyl groups on a cyclodextrin skeleton, but the reaction is non-selective. Heptakis(6-*O*-pivaloyl)-β-cyclodextrin was therefore prepared to overcome this disadvantage (Scheme 2.18). Pivaloyl derivative is, unlike a native cyclodextrin, capable of furnishing $2^{A},3^{B}$ derivative in a good yield (Sakairi and Kuzuhara 1993). The subsequent benzylation of the remaining hydroxyl groups together with the exchange of pivaloyl residues to benzyl groups and deprotection of benzylidene led to per-*O*-benzyl-$2^{A},3^{B}$–dihydroxy-β-cyclodextrin. This approach was also applied for α- and γ-cyclodextrin as well (Sakairi et al. 1996b). Such derivatives are very useful precursors for further syntheses and are similar to the permethylated ones discussed below.

It is also possible to disubstitute just one glucose unit to obtain $2^{A},3^{A}$ derivative. Such regioselectivity was achieved for the reaction of β-cyclodextrin with α,α'-dibromo-*o*-xylene (Balbuena et al. 2007). The reaction proceeds with lithium diisopropylamide as a base in a 30% yield (Scheme 2.19). The subsequent methylation of all remaining hydroxyl groups together with hydrogenolysis of the xylene group gave per-*O*-methyl-$2^{A},3^{A}$–dihydroxy-β-cyclodextrin. This approach was also successfully applied on α- and γ-cyclodextrin (Balbuena et al. 2013).

Similarly, methallyl dichloride could be used instead of α,α'-dibromo-*o*-xylene (Fenger et al. 2011). The reaction proceeds on α- and β-cyclodextrin and in similar yields.

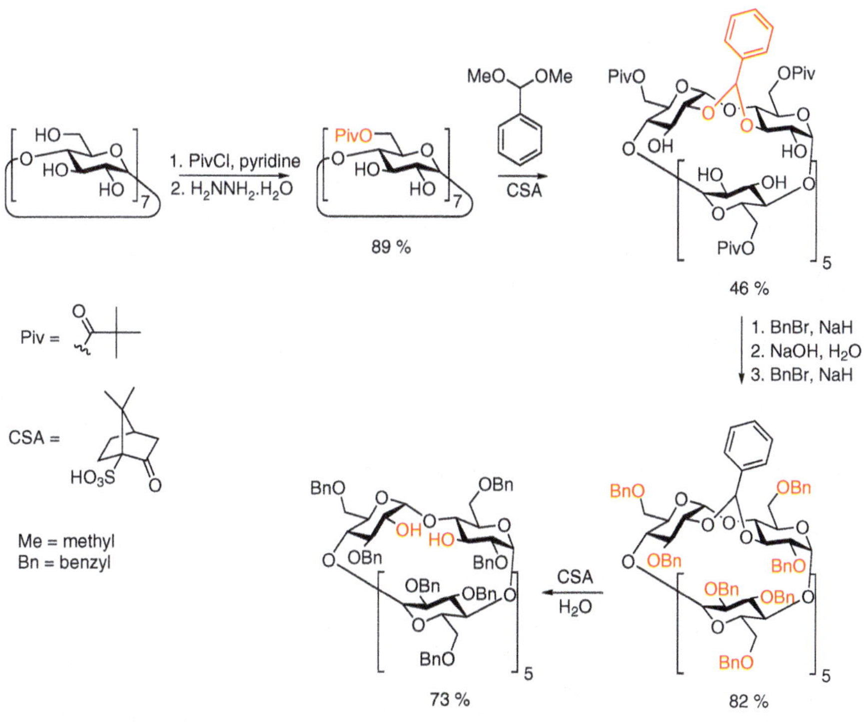

Scheme 2.18 Regioselective protection by α,α-dimethoxytoluene

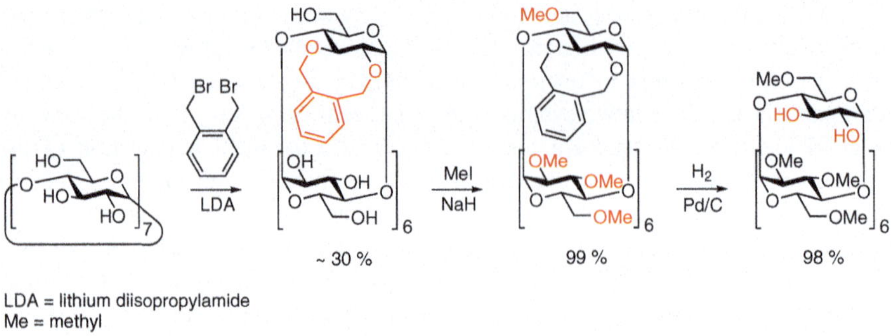

Scheme 2.19 Synthesis of per-*O*-methyl-2^{A},3^{A}–dihydroxy-β-cyclodextrin

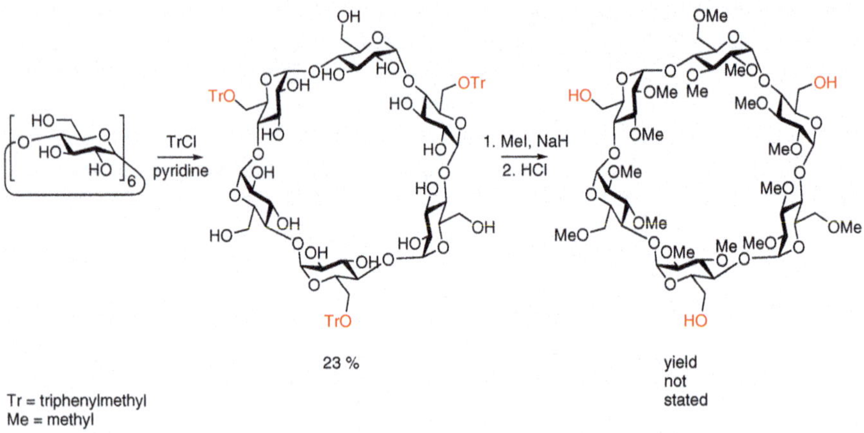

Scheme 2.20 Synthesis of symmetrically trisubstituted α-cyclodextrin

When a trisubstituted cyclodextrin derivative is needed, the reaction of α-cyclodextrin with triphenylmethyl chloride, so-called "trityl chloride", comes into play. The reaction yields symmetrically trisubstituted cyclodextrins due to sterical hindrance (Scheme 2.20). Protection of all the remaining hydroxyl groups by methyl iodide and hydrolysis of trityl groups proceeded to a useful trisubstituted precursor – per-*O*-methyl-6^{A},6^{C},6^{E}–trihydroxy-α-cyclodextrin (Boger et al. 1979). The quantitative analysis of α-cyclodextrin tritylation was studied later by ultra-fast liquid chromatography (Yoshikiyo et al. 2015).

2.2.5.2 Syntheses Based on Selective Debenzylation

Protection/deprotection strategies are vastly used for syntheses of selectively substituted cyclodextrins, where sterical properties of a reagent together with cyclodextrin reactivity play the key role. Selective deprotection of perbenzylated cyclodextrins

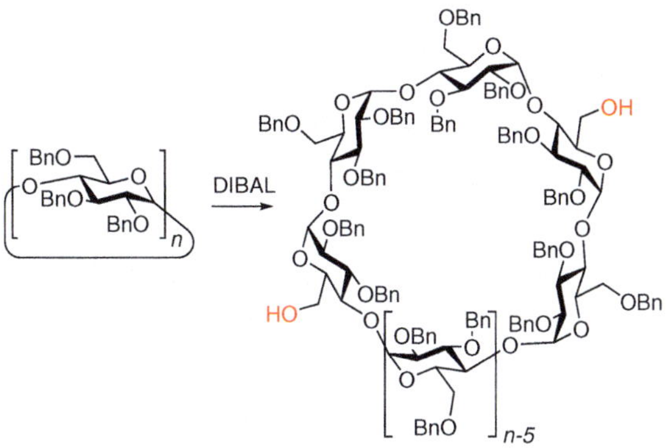

n = 6, α-cyclodextrin derivative, 82 %
n = 7, β-cyclodextrin derivative, 83 %
n = 8, γ-cyclodextrin derivative, 59 %
(together with $6^A,6^E$ derivative)

Bn = benzyl
DIBAL = diisobutylaluminium hydride

Scheme 2.21 Selective bis-*O*-debenzylation

by diisobutylaluminium hydride represents the most favourite method for synthesis of selectively substituted cyclodextrins, as well as monosubstituted cyclodextrins – see below. It was first described by Pearce and Sinaÿ (Pearce and Sinaÿ 2000). The method involves a selective deprotection of two opposite benzyl groups in high yields and is applicable to α-, β-, and γ-cyclodextrin (Scheme 2.21). Benzyl groups are removed from glucose units A and D at position 6, but in the case of γ-cyclodextrin also from glucose units A and E. Free hydroxyl groups could be subsequently used for organic chemistry transformations leading to the desired cyclodextrin derivative (Petrillo et al. 2009; Volkov et al. 2015).

It has been proposed the reaction involves at least two molecules of diisobutylaluminium hydride and the mechanism occurs by a stepwise process (Sollogoub 2013). This was proved by the synthesis of per-*O*-benzyl-6^A-hydroxy derivatives, see below. This derivative could be used for the synthesis of per-*O*- benzyl-6^A-deoxy-6^A-azido derivatives and for the second debenzylation (Scheme 2.22). Azido groups provide steric hindrance to direct the second deprotection step towards the opposed glucose unit (Guieu and Sollogoub 2008a) allowing to introduce two different functional groups.

Such step by step debenzylation and substitution reactions could lead even to far more complicated structures – cyclodextrins trisubstituted (Guieu and Sollogoub 2008b; Rawal et al. 2010), tetrasubstituted (Rawal et al. 2010; Sollogoub 2013), pentasubstituted (Guieu and Sollogoub 2008b) or heaxasubstituted at position 6, even with all the substituents different from each other (Wang et al. 2014). Moreover, together with Et_3SiH/I_2 hexakis-*O*-debenzylation at position 3 (Guitet et al. 2012) (see above) they allow a simultaneous selective deprotection on both the primary and secondary rim.

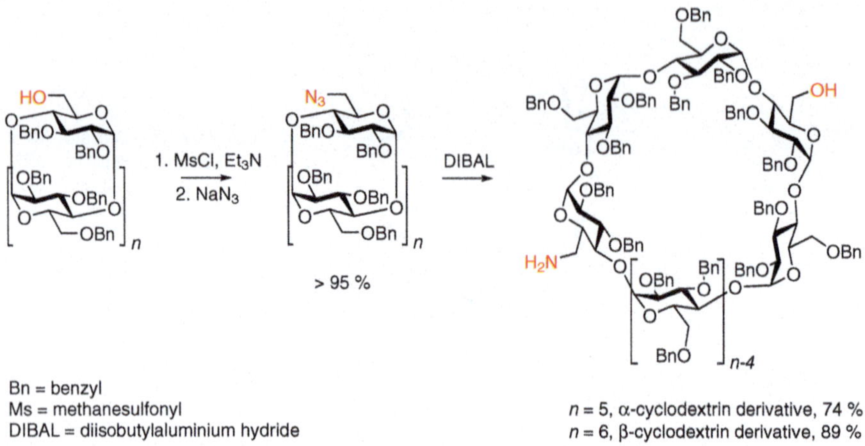

Scheme 2.22 Azide promoted debenzylation

2.2.5.3 Syntheses Based on Other Selective Deprotections

Shortly after the selective bis-*O*-debenzylation was discovered, another selective double diisobutylaluminium hydride deprotection from Sinaÿ saw the light of day (du Roizel et al. 2002). It was found out both per-*O*-methyl α- and β-cyclodextrin were able to undergo a regioselective bis-*O*-demethylation, but on the secondary rim (Scheme 2.23). This opened the way for a direct access to $2^A,3^B$ derivatives (Letort et al. 2015).

As it was shown by Xiao et al. the reaction requires 2^A methoxy group and an oxygen atom present at position 3^B (Xiao et al. 2013) to proceed. The authors also showed it was possible to carry out two or even three bis-*O*-demethylations on permethylated α- or β-cyclodextrins (Scheme 2.24). However, in contrast to the multiple bis-*O*-debenzylations described above, these bis-*O*-demethylations are carried out in one step.

The last from frequently used deprotections is selective bis-O-de(*tert*-butyldimethyl)silylation (Ghosh et al. 2012). It proceeds at $6^A,6^D$ positions in 70% yields and is carried out on hexakis(2,3-*O*-dimethyl-6-*O*-*tert*-butyldimethylsilyl)-α-cyclodextrin or heptakis(2,3-*O*-dimethyl-6-*O*-*tert*-butyldimethylsilyl)-β-cyclodextrin (Scheme 2.25). The desilylation proceeds in even better yields on substrates where secondary hydroxyl groups are protected by benzyl instead of methyl groups. The methylated precursors were used in further transformations (Ghosh et al. 2012; Tatar et al. 2017).

Mini-Summary The synthesis of selectively substituted cyclodextrins is quite a challenge due to the high number of theoretically possible isomers. However, selective bis-*O*-debenzylation (forming $6^A,6^D$ derivatives) and bis-*O*-demethylation (forming $2^A,3^B$ derivatives) make this process much easier. Such an approach allows to satisfy a sweet tooth of almost the majority of chemists. The deprotections could be used multiple times or combined with the other methods. Moreover, when

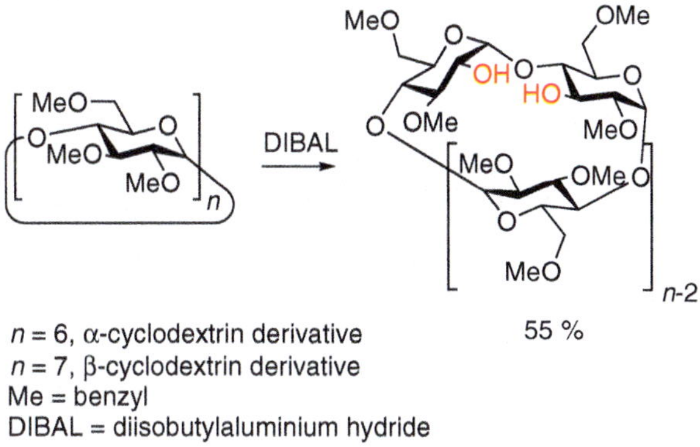

Scheme 2.23 Selective bis-*O*-demethylation

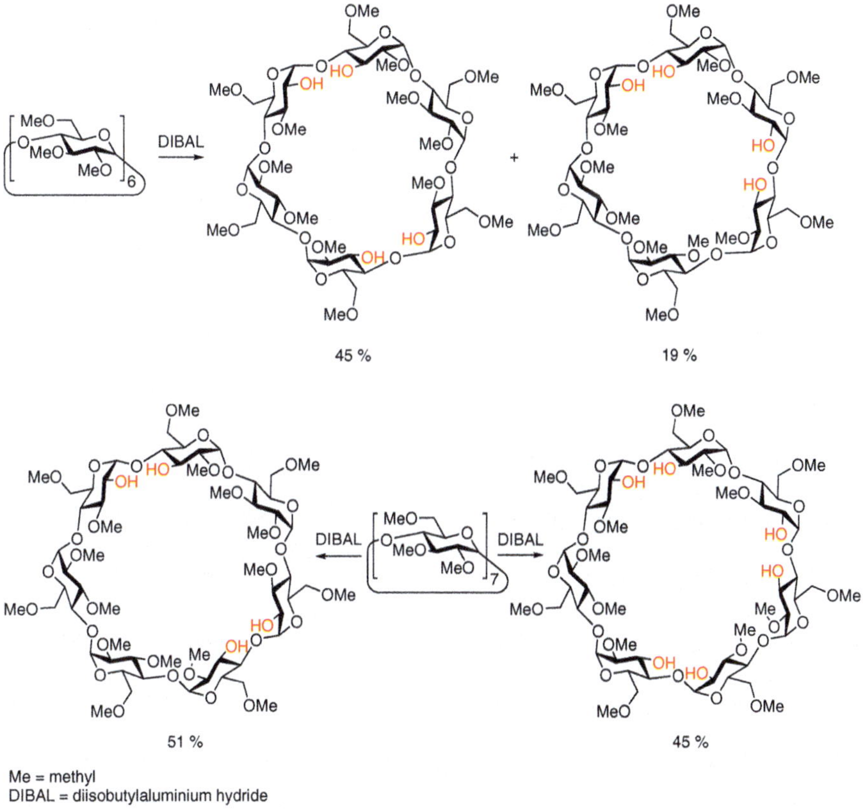

Scheme 2.24 Multiple bis-*O*-demethylations

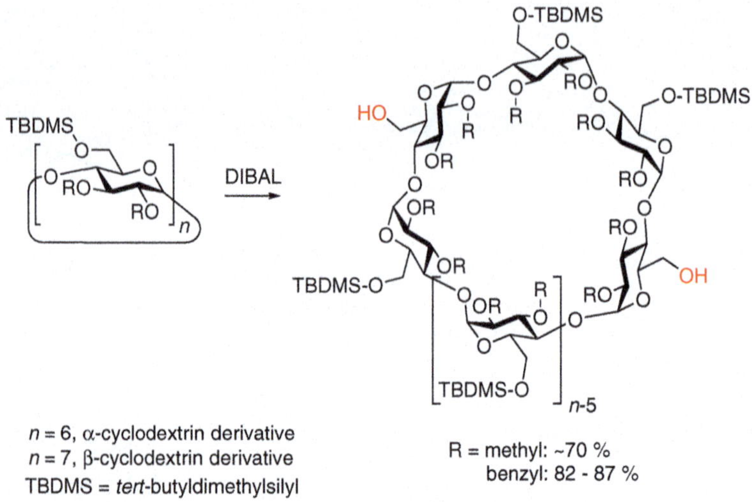

Scheme 2.25 Selective bis-O-*tert*-butyldimethylsilylation

chemists are still picky, selective 6^A,6^B-, 6^A,6^C-, 6^A,6^D–sulfonations, 2^A,3^A–xylylation or 6^A,6^C,6^E–tritylation could be offered.

2.2.6 *Monosubstituted Cyclodextrins*

Monosubstituted cyclodextrins could be either substituted at positions 2, 3 or 6. Nowadays, there are two main methods for their synthesis: direct and indirect (Řezanka 2016). When using the first method, a cyclodextrin is directly reacted with a substitution agent. A desired monosubstituted product is then purified from other isomers, unreacted cyclodextrins and multiple-substituted by-products. The location of a substituent on a glucose unit is driven by the cyclodextrin reactivity, which is described in Sect. 2.1. Yields usually strongly depend on solvent, base, type of cyclodextrin, and structure of the substitution agent and vary from units of percent up to a few dozens of percent. This dependency has been studied by several authors (Masurier et al. 2006; Martina et al. 2010; Řezanka 2016).

When reproducing synthesis according to the described procedure, please have in mind not all the authors use satisfactory purification steps and characterise the product sufficiently. Precipitation of a reaction mixture (e.g. by acetone) as the only purification method is inadequate. Without a proof of purity such as chromatogram from high-performance liquid chromatography, copy of ^{1}H nuclear magnetic resonance spectra etc. it is expected the product contains several impurities – e.g. unreacted cyclodextrin and multiple-substituted cyclodextrin derivatives.

The second – indirect – method is based on a high-yielding deprotection step of a persubstituted cyclodextrin, such as permethylated, perbenzylated etc. Persubstituted cyclodextrins are easily obtained in high yields and thus the selective deprotection furnishes a monodeprotected derivative, where the free hydroxyl group is ready for further transformations. This method is more universal than the first one, as the substitution of the free hydroxyl group usually results in high yields regardless the nature of the substitution agent used.

There are also other methods available for cyclodextrin monosubstitution. It is possible to synthesise a monosubstituted derivative *de novo*. However, the synthesis requires many steps and the overall yield is very low (Sakairi et al. 1991; Sakairi et al. 1995; Chaise et al. 2008). An interesting approach of monosubstituted cyclodextrin derivative synthesis is the use of proteases, which are able to direct the substituents regioselectively (Xiao et al. 2004; Wang et al. 2005). Unfortunately, this method has not been examined in detail yet. The last from the methods for monosubstituted cyclodextrin derivative synthesis is the use of several protection/deprotection steps to achieve a desired product. However, this method requires a lot of steps and its use is now being superseded with direct and indirect strategies.

The syntheses of monosubstituted allyl, cinnamyl, propargyl, formylmethyl, carboxymethyl, azido and amino cyclodextrin derivatives have recently been studied in detail (Řezanka 2016). The following sections thus summarise the main findings and add information about other derivatives, which could be used as precursors for further synthesis.

2.2.6.1 Monosubstitution at Position 2

The direct method takes advantage of the above mentioned fact the hydroxyl group at position 2 is the most acidic. The use of a strong base thus mostly leads to substitution predominantly at this position and the yields reach up to 40%. However, sometimes the desired 2^{A}-*O* substituted derivative is hard to separate from its isomer and the purification step is done after peracetylation (Scheme 2.26) (Řezanka and Jindřich 2011). Peracetylation has also other advantages such as easy distinguishing between 2, 3 and 6 isomers directly from ^{1}H nuclear magnetic resonance spectrum (Řezanka 2016) and protecting the rest of hydroxyl groups against side reactions, for example, when oxidation of double bond is needed. Peracetylated derivatives could be easily deprotected by Zemplén deacetylation (Řezanka et al. 2010).

The best results for alkylation are generally achieved in dimethyl sulfoxide with a strong base like LiH together with a catalytic amount of LiI (Hanessian et al. 1995; Řezanka and Jindřich 2011); sodium ethoxide (Masurier et al. 2006, 2009); lithium diisopropylamide (Masurier et al. 2006); or NaH (Kalakuntla et al. 2013).

It is also possible to use the indirect method for the synthesis of cyclodextrin monosubstituted at position 2. As it has been mentioned above, regioselective bis-*O*-demethylation yields permethylated $2^{A},3^{B}$-dihydroxy derivatives. The diol could be selectively alkylated at position 2 and subsequently methylated on the remaining

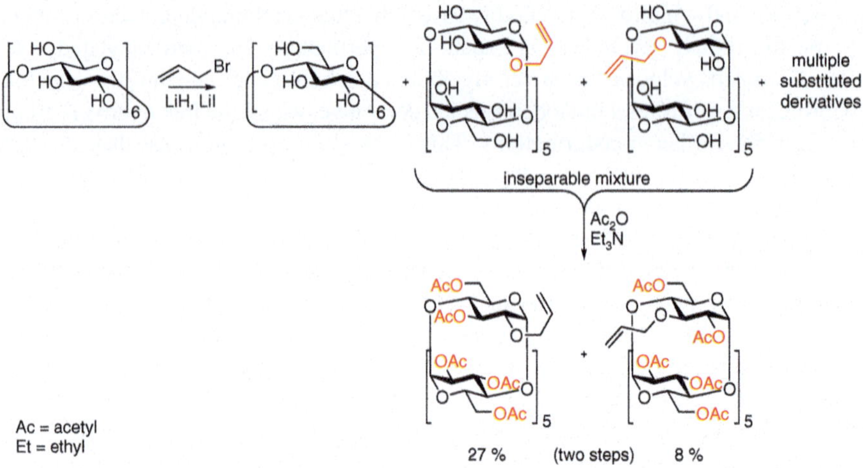

Scheme 2.26 Synthesis of allyl derivatives

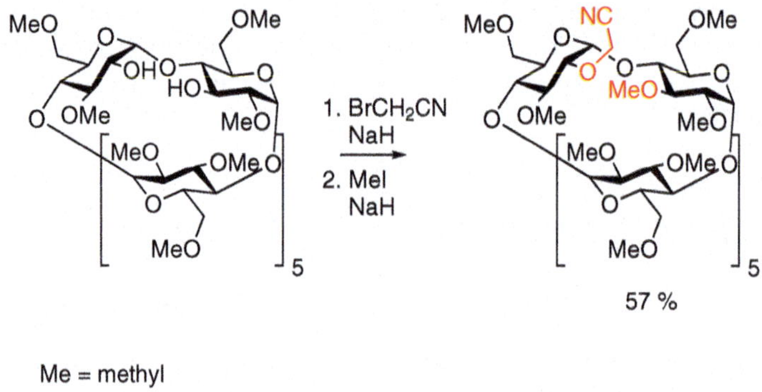

Scheme 2.27 Synthesis of permethylated cyclodextrin derivatives monosubstituted at position 2

hydroxyl group yielding permethylated 2-*O* derivative (Scheme 2.27) (Guan et al. 2009; Řezanka et al. 2015). A great variety of permethylated α-cyclodextrin derivatives was prepared using this methodology (Xiao et al. 2013).

The other option of a selective deprotection, which yields 2-*O* substituted derivative, is selective debenzoylation (Sakairi et al. 1996a). The deprotection is performed on perbenzoylated β-cyclodextrin by hydrazine (Scheme 2.28). Free benzoyl group is then ready for further transformations. However, be aware of Walden inversion at this position, e.g. when substituting tosyl group in this position. The product is a non-cyclodextrin derivative with one unit in *manno* configuration.

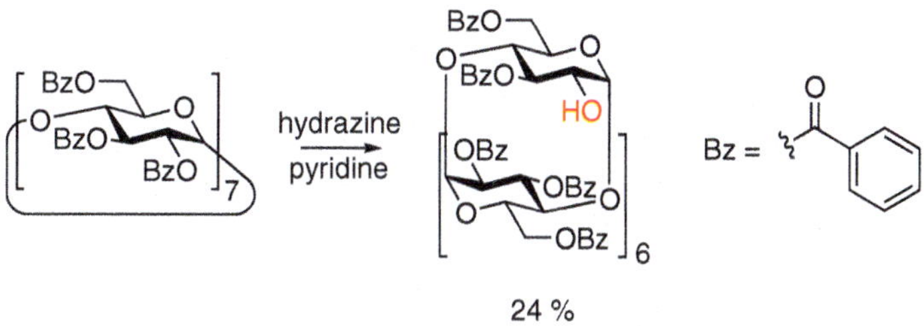

Scheme 2.28 Selective debenzoylation

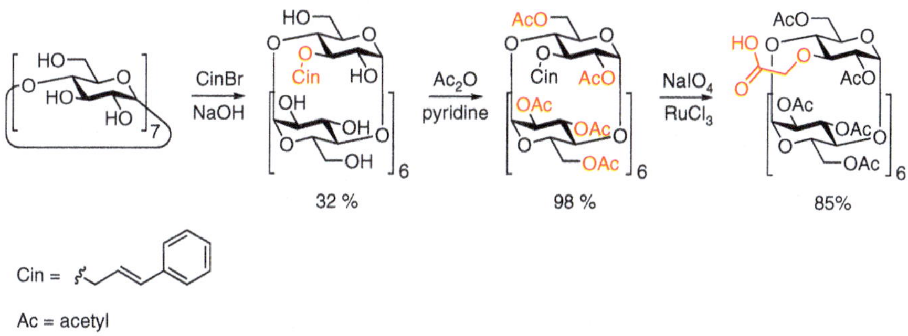

Scheme 2.29 Synthesis of 3^A^-*O*-cinnamyl-β-cyclodextrin

2.2.6.2 Monosubstitution at Position 3

As it has been mentioned above, position 3 is the least accessible, which results in difficulties during the synthesis of such derivatives. However, there are several methods allowing substitution at position 3. Jindřich and Tišlerová found the alkylation of β-cyclodextrin with cinnamyl bromide result selectively in 3^{A}-*O*-cinnamyl-β-cyclodextrin in a good yield (Jindřich and Tišlerová 2005). This behaviour is caused by the inclusion of cinnamyl bromide in the cavity. The resulting complex has the reactive centre of the alkylation agent oriented towards position 3. Cinnamyl group could be then oxidised after peracetylation and used for further synthesis (Scheme 2.29).

Another possibility to introduce a substituent at position 3 selectively is the use of copper(II) sulfate in water together with NaOH as a base (Masurier et al. 2009). Alkylation using these conditions for several allylic or benzylic reagents resulted in 3-substituted derivatives in 40% yields. Other alkylation methods using NaOH in water or water/acetonitrile result only in 10% yields for cyclodextrin derivatives monosubstituted at position 3. Moreover, they have to be separated from their 2^{A}-*O* and 6^{A}-*O* isomers formed as by-products during the reaction (Řezanka and Jindřich 2011; Zhou et al. 2012a, 2012b; Bláhová et al. 2013).

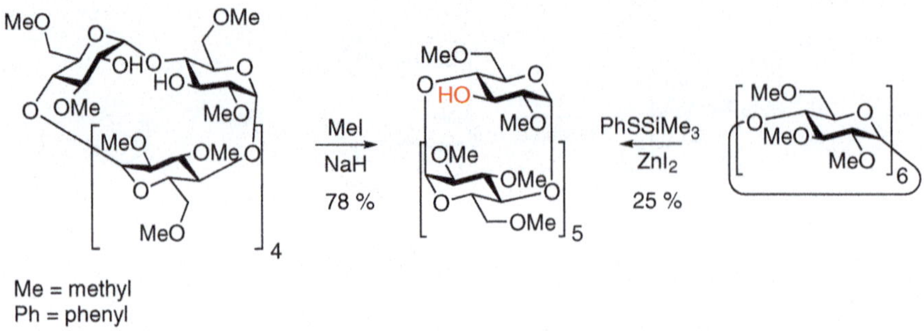

Scheme 2.30 Synthesis of per-*O*-methyl-3^{A}-hydroxy-α-cyclodextrin

The abovementioned regioselective bis-*O*-demethylation is also useful for the synthesis of permethylated cyclodextrin derivatives monosubstituted at position 3. The 2^{A},3^{B}-diol could be selectively methylated at position 2 (Scheme 2.30) and the free hydroxyl group at position 3 serves as the reaction centre for further modifications (Xiao et al. 2013). Alternatively, the same derivative was obtained by selective demethylation of permethylated α-cyclodextrin by phenylthiotrimethylsilane in the presence of zinc(II) iodide (Scheme 2.30) (Chaise et al. 2004).

2.2.6.3 Monosubstitution at Position 6

Cyclodextrins monosubstituted at position 6 represent unique precursors for the attachment to another molecule or to a surface. Connection through position 6 leaves the wider rim of a cyclodextrin open for interactions of guests with the cavity. Moreover, cyclodextrins monosubstituted at position 6 are easy to synthesise compared to the other derivatives. The reason has already been mentioned above – the hydroxyl group at position 6 is primary and the least acidic. The deprotonation of all hydroxyl groups thus predominantly leads to the substitution at this position, as there is the lowest steric hindrance. Although a lot of cyclodextrins monosubstituted at position 6 are commercially available, they are still synthesised by researchers due to their high price.

The most important cyclodextrins monosubstituted at position 6 are 6^{A}-deoxy-6^{A}-tosyl-α-, β- and γ-cyclodextrin. To synthesise them, cyclodextrin is reacted with tosyl chloride (Petter et al. 1990; Brown et al. 1993; Hamasaki et al. 1993; Gao et al. 1995; Liu et al. 1998; Tang and Ng 2008b; Trellenkamp and Ritter 2010; Xu et al. 2010), tosyl anhydride (Zhong et al. 1998) or 1-tosylimidazole (Tang and Ng 2007; Cao et al. 2009; Nielsen et al. 2010; Kulkarni et al. 2012). While some authors do not use purification methods, it is recommended to treat the product by chromatography (Brown et al. 1993; Hamasaki et al. 1993; Kulkarni et al. 2012) or to recrystallise it from water (Gao et al. 1995; Tang and Ng 2008b; Nielsen et al. 2010; Xu et al. 2010) or 50% MeOH in H_2O (Popr et al. 2014; Bednářová et al. 2016).

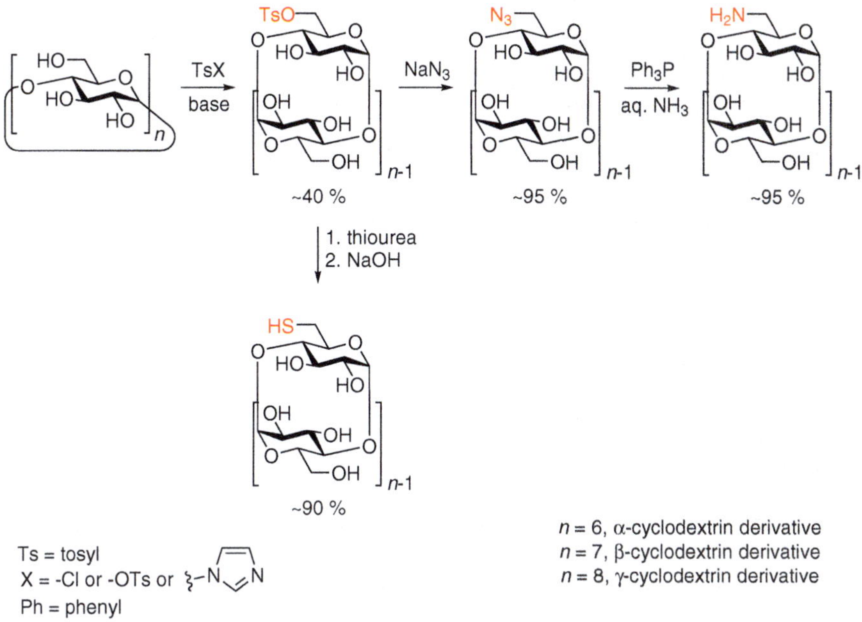

Scheme 2.31 Synthesis of cyclodextrins monosubstituted at position 6

Tosyl group is a useful precursor for azide, amino and thio groups (Scheme 2.31). Tosylated cyclodextrins could react with sodium azide in water (Hamasaki et al. 1993; Tang and Ng 2008b; Trellenkamp and Ritter 2010; Kulkarni et al. 2012) or *N,N′*-dimethylformamide (Petter et al. 1990; Nielsen et al. 2010; Xu et al. 2010). It is also possible to synthesise azide under solvent-free conditions in a ball mill (Jicsinszky et al. 2016b). The azide group could be subsequently transformed into amine by the reduction with triphenylphosphine and ammonia (Hamasaki et al. 1993; Xu et al. 2010; Kulkarni et al. 2012) or water (Tang and Ng 2008b). Both azido and amino derivatives are the favourite compounds for the synthesis of defined cyclodextrin polymers (Pun et al. 2004; Zhou and Ritter 2010).

In order to synthesise a thiol, tosyl group is treated with thiourea and subsequently with sodium hydroxide (Fujita et al. 1982; Fujita et al. 1985; Martinelli et al. 2014). Tosyl derivatives also serve as precursors for *N*-alkyl (Tang and Ng 2008a; Popr et al. 2014) or *S*-alkyl compounds (Bednářová et al. 2016).

When *O*-alkyl derivatives are needed, cyclodextrin could be directly alkylated in an excess of NaOH in water (Řezanka et al. 2010). It was found the method could be used for various alkylation agents on α-, β-, γ-cyclodextrins (Řezanka 2016). The yields typically reach values above 10% and no other monosubstituted isomers (2^{A}-*O* and 3^{A}-*O*) are formed.

The best known indirect method for the synthesis of cyclodextrin derivatives monosubstituted at position 6 is selective debenzylation of perbenzylated α-, β-, and γ-cyclodextrins (Scheme 2.32) (Pearce and Sinaÿ 2000). The reactions proceed in

Scheme 2.32 Selective debenzylation of cyclodextrins

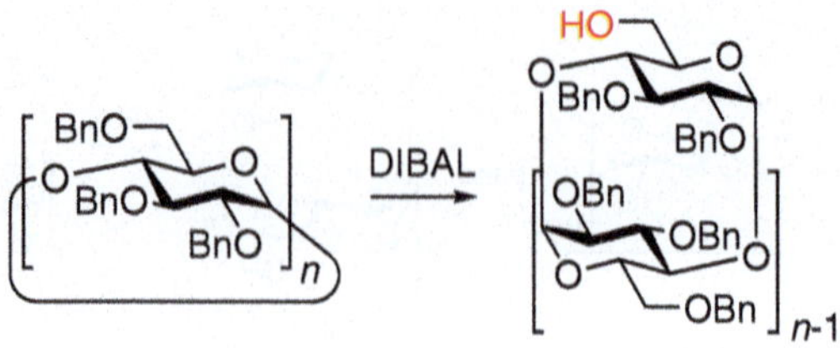

n = 6, α-cyclodextrin derivative, 64 %
n = 7, β-cyclodextrin derivative, 60 %
n = 8, γ-cyclodextrin derivative, 47 %
Bn = benzyl
DIBAL = diisobutylaluminium hydride

Scheme 2.33 Synthesis of per-*O*-methyl-6[A]-hydroxy-α- and β-cyclodextrins

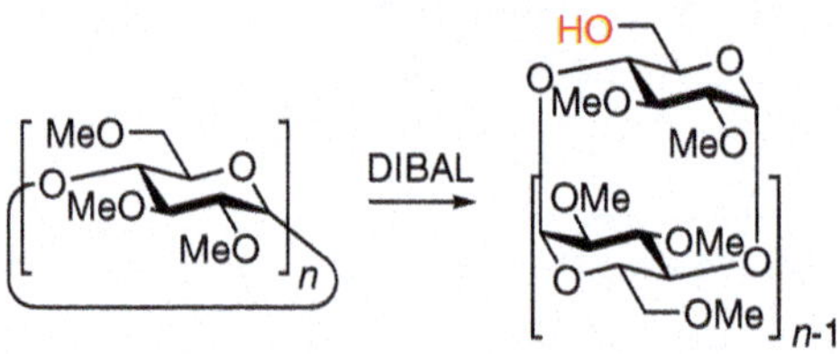

n = 6, α-cyclodextrin derivative, 20 %
n = 7, β-cyclodextrin derivative, 20 %

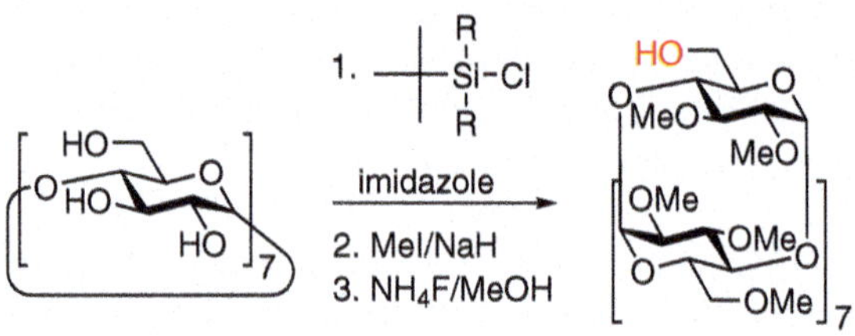

43 - 45 %

Me = benzyl
DIBAL = diisobutylaluminium hydride
R = methyl or phenyl

very good yields, unusual for monosubstituted cyclodextrins. The free hydroxyl group could be subsequently modified and benzyl groups removed by H_2 on Pd (Lindbäck et al. 2012).

The second indirect method is the abovementioned regioselective bis-*O*-demethylation. It provides per-*O*-methyl-6[A]-hydroxy-α- or β-cyclodextrin as by-products in 20% yield (Scheme 2.33) (du Roizel et al. 2002). It is also possible to use the protection/deprotection methodology for the synthesis of the latter one (Chen et al. 1996; Lupescu et al. 1999).

Mini-Summary The direct syntheses of monosubstituted cyclodextrin derivatives benefit from the different reactivity of hydroxyl groups. 2[A]-*O* substituted derivatives are obtained using a strong base in dimethyl sulfoxide. 3^A-*O* substituted derivatives are synthesised using either selective introduction of cinnamyl group or selected alkylation agents in the presence of copper(II) sulfate. Cyclodextrins monosubstituted at position 6 are the most favourite ones and 6^A-deoxy-6^A-tosyl-cyclodextrins overshadow all the other derivatives, as it is the most used precursor for further synthesis. Deprotection of methyl groups from permethylated cyclodextrins could lead selectively to either 2, 3, or 6 monosubstituted derivatives and debenzylation of perbenzylated cyclodextrins furnishes per-*O*-benzyl-6^A-hydroxy-cyclodextrins selectively in high yields.

2.3 Conclusion

A lot of cyclodextrin derivatives have become available on the market over the years. However, their price is in the range of fine chemicals and thus they are still often synthesised in laboratories. Randomly substituted cyclodextrin derivatives are the only exception. Synthesis of persubstituted cyclodextrin derivatives remains more or less the same and the methods for cyclodextrins persubstituted at selected positions are now very well examined.

The synthesis of selectively substituted and monosubstituted cyclodextrin derivatives has changed much over the years. Originally used direct methods subside and the indirect methods are now on the rise. The only exceptions are 6^A-deoxy-6^A-tosyl-α-, β- and γ-cyclodextrins. They are the most favourite precursors for further syntheses.

The synthesis of tosyl derivatives is quick, high yielding, using cheap chemicals and the tosyl group is suitable for further reactions. Moreover, the synthesis requires only an easy purification process, i.e. recrystallisation, which is the key step. Cyclodextrin derivatives aspiring to be similarly successful should fulfil these conditions.

Acknowledgements This work was supported by project LO1201 of the Ministry of Education, Youth and Sports in the framework of the targeted support of the "National Programme for Sustainability I"; and by the project 16-02316Y of the Czech Science Foundation (GA CR).

References

Adly FG, Antwi NY, Ghanem A (2016) Cyclodextrin-functionalized monolithic capillary columns: preparation and chiral applications. Chirality 28(2):97–109. https://doi.org/10.1002/chir.22550

Aghahosseini H, Ramazani A (2016) General overview on cyclodextrin-based artificial enzymes' activity. Curr Org Chem 20(26):2817–2836. https://doi.org/10.2174/1385272820666160328201207

Alsbaiee A, Smith BJ, Xiao L, Ling Y, Helbling DE, Dichtel WR (2016) Rapid removal of organic micropollutants from water by a porous beta-cyclodextrin polymer. Nature 529(7585):190–194. https://doi.org/10.1038/nature16185

Angibeaud P, Utille J-P (1991) Cyclodextrin chemistry; Part I. Application of a regioselective acetolysis method for benzyl ethers. Synthesis-Stuttgart 1991(9):737–738. https://doi.org/10.1055/s-1991-26560

Antoniuk I, Amiel C (2016) Cyclodextrin-mediated hierarchical self-assembly and its potential in drug delivery applications. J Pharm Sci 105(9):2570–2588. https://doi.org/10.1016/j.xphs.2016.05.010

Ashton PR, Königer R, Stoddart JF, Alker D, Harding VD (1996) Amino acid derivatives of β-cyclodextrin. J Org Chem 61(3):903–908. https://doi.org/10.1021/jo951396d

Astray G, Gonzalez-Barreiro C, Mejuto JC, Rial-Otero R, Simal-Gandara J (2009) A review on the use of cyclodextrins in foods. Food Hydrocoll 23(7):1631–1640. https://doi.org/10.1016/j.foodhyd.2009.01.001

Baer H, Berenguel A, Shu Y, Defaye J, Gadelle A, Gonzalez F (1992a) Improved preparation of Hexakis(6-deoxy)cyclomaltohexaose and Heptakis(6-deoxy)cyclomaltoheptaose. Carbohydr Res 228(1):307–314. https://doi.org/10.1016/S0008-6215(00)90568-8

Baer H, Shen Y, Gonzalez F, Berenguel A, Garcia J (1992b) Synthesis of a cycloheptaose consisting of (1-4)-linked 7-Amino-6,7-Dideoxy-alpha-D-Gluco-Heptopyranosyl units: a new analog of cyclomaltoheptaose. Carbohydr Res 235:129–139. https://doi.org/10.1016/0008-6215(92)80083-D

Bakó P, Fenichel L, Töke L, Szente L, Szejtli J (1994) Methylation of cyclodextrins by phase-transfer catalysis. J Incl Phenom Mol Recognit Chem 18(3):307–314. https://doi.org/10.1007/BF00708737

Balbuena P, Lesur D, Alvarez MJG, Mendicuti F, Mellet CO, Fernandez JMG (2007) One-pot regioselective synthesis of 2(I),3(I)-O-(o-xylylene)-capped cyclomaltooligosaccharides: tailoring the topology and supramolecular properties of cyclodextrins. Chem Commun 2007(31):3270–3272. https://doi.org/10.1039/b705644c

Balbuena P, Goncalves-Pereira R, Jimenez Blanco JL, Isabel Garcia-Moreno M, Lesur D, Ortiz Mellet C, Garcia Fernandez JM (2013) o-Xylylene protecting group in carbohydrate chemistry: application to the regioselective protection of a single vic-diol segment in cyclodextrins. J Org Chem 78(4):1390–1403. https://doi.org/10.1021/jo302178f

Bednářová E, Hybelbauerová S, Jindřich J (2016) Optimized methods for preparation of 6(I)-(omega-sulfanyl-alkylene-sulfanyl)-beta-cyclodextrin derivatives. Beilstein J Org Chem 12:349–352. https://doi.org/10.3762/bjoc.12.38

Bellia F, Mendola DL, Pedone C, Rizzarelli E, Saviano M, Vecchio G (2009) Selectively functionalized cyclodextrins and their metal complexes. Chem Soc Rev 38(9):2756–2781. https://doi.org/10.1039/B718436K

Benkhaled A, Cheradame H, Fichet O, Teyssie D, Buchmann W, Guegan P (2008) Synthesis and characterization of amphiphilic per-(6-thio-2,3-trimethylsilyl)cyclodextrin: application to Langmuir film formation. Carbohydr Polym 73(3):482–489. https://doi.org/10.1016/j.carbpol.2007.12.031

Bergeron R, Meeley M, Machida Y (1976) Selective alkylation of cycloheptaamylose. Bioorg Chem 5(1):121–126. https://doi.org/10.1016/0045-2068(76)90018-3

Bjerre J, Fenger TH, Marinescu LG, Bols M (2007) Synthesis of some trifluoromethylated cyclodextrin derivatives and analysis of their properties as artificial glycosidases and oxidases. Eur J Org Chem 2007(4):704–710. https://doi.org/10.1002/ejoc.200600762

Bláhová M, Bednářová E, Řezanka M, Jindřich J (2013) Complete sets of monosubstituted γ-cyclodextrins as precursors for further synthesis. J Org Chem 78(2):697–701. https://doi.org/10.1021/jo301656p

Boger J, Corcoran R, Lehn J (1978) Cyclodextrin chemistry - selective modification of all primary hydroxyl-groups of alpha-cyclodextrins and beta-cyclodextrins. Helv Chim Acta 61(6):2190–2218. https://doi.org/10.1002/hlca.19780610622

Boger J, Brenner D, Knowles J (1979) Symmetrical triamino-per-O-methyl-alpha-cyclodextrin - preparation and characterization of primary trisubstituted alpha-cyclodextrins. J Am Chem Soc 101(25):7630–7631. https://doi.org/10.1021/ja00519a035

Bom A, Bradley M, Cameron K, Clark JK, van Egmond J, Feilden H, MacLean EJ, Muir AW, Palin R, Rees DC, Zhang M-Q (2002) A novel concept of reversing neuromuscular block: chemical encapsulation of rocuronium bromide by a cyclodextrin-based synthetic host. Angew Chem Int Ed 41(2):265–270. https://doi.org/10.1002/1521-3773(20020118)41:2<265::AID-ANIE265>3.0.CO;2-Q

Bonnet V, Gervaise C, Djédaïni-Pilard F, Furlan A, Sarazin C (2015) Cyclodextrin nanoassemblies: a promising tool for drug delivery. Drug Discov Today 20(9):1120–1126. https://doi.org/10.1016/j.drudis.2015.05.008

Breslow R, Canary JW, Varney M, Waddell ST, Yang D (1990) Artificial transaminases linking pyridoxamine to binding cavities: controlling the geometry. J Am Chem Soc 112(13):5212–5219. https://doi.org/10.1021/ja00169a032

Brown SE, Coates JH, Coghlan DR, Easton CJ, van Eyk SJ, Janowski W, Lepore A, Lincoln SF, Luo Y, May BL, Schiesser DS, Wang P, Williams ML (1993) Synthesis and properties of 6A-Amino-6A-deoxy-α and β-cyclodextrin. Aust J Chem 46(6):953–958. https://doi.org/10.1071/CH9930953

Busby MB, Vigh G (2005) Synthesis of heptakis(2-O-methyl-3-O-acetyl-6-O-sulfo)-cyclomaltoheptaose, a single-isomer, sulfated β-cyclodextrin carrying nonidentical substitutents at all the C2, C3, and C6 positions and its use for the capillary electrophoretic separation of enantiomers in acidic aqueous and methanolic background electrolytes. Electrophoresis 26(10):1978–1987. https://doi.org/10.1002/elps.200500057

di Cagno MP (2017) The potential of cyclodextrins as novel active pharmaceutical ingredients: a short overview. Molecules 22(1):1. https://doi.org/10.3390/molecules22010001

Cameron KS, Fletcher D, Fielding L (2002) An NMR study of cyclodextrin complexes of the steroidal neuromuscular blocker drug Rocuronium Bromide. Magn Reson Chem 40(4):251–260. https://doi.org/10.1002/mrc.1008

Cao H, He J, Deng L, Gao X (2009) Fabrication of cyclodextrin-functionalized superparamagnetic Fe3O4/amino-silane core–shell nanoparticles via layer-by-layer method. Appl Surf Sci 255(18):7974–7980. https://doi.org/10.1016/j.apsusc.2009.04.199

Chaise T, Bourgeaux E, Cardinael P, Combret JC (2004) Direct and convenient access to mono 3-hydroxy per-O-methylated alpha-cyclodextrin. Tetrahedron Lett 45(30):5853–5856. https://doi.org/10.1016/j.tetlet.2004.06.003

Chaise T, Cardinael P, Tisse S, Combret J-C, Bouillon J-P (2008) Indirect and direct approaches in the synthesis of a new mono-6-O-benzyl methylated gamma-cyclodextrin as chiral selector for enantioselective gas chromatography. Tetrahedron Asymmetry 19(3):348–357. https://doi.org/10.1016/j.tetasy.2008.01.015

Chen Z, Bradshaw JS, Lee ML (1996) A convenient synthesis of mono-6-hydroxy permethylated beta-cyclodextrin via tert-butyldimethylsilylation. Tetrahedron Lett 37(38):6831–6834. https://doi.org/10.1016/0040-4039(96)01545-6

Coisne C, Tilloy S, Monflier E, Wils D, Fenart L, Gosselet F (2016) Cyclodextrins as emerging therapeutic tools in the treatment of cholesterol-associated vascular and neurodegenerative diseases. Molecules 21(12):1748. https://doi.org/10.3390/molecules21121748

Coleman A, Zhang P, Parrotlopez H, Ling C, Miocque M, Mascrier L (1991) The first selective per-tosylation of the secondary OH-2 of beta-cyclodextrin. Tetrahedron Lett 32(32):3997–3998. https://doi.org/10.1016/0040-4039(91)80609-A

Concheiro A, Alvarez-Lorenzo C (2013) Chemically cross-linked and grafted cyclodextrin hydrogels: from nanostructures to drug-eluting medical devices. Adv Drug Deliv Rev 65(9):1188–1203. https://doi.org/10.1016/j.addr.2013.04.015

Crini G (2014) Review: a history of cyclodextrins. Chem Rev 114(21):10940–10975. https://doi.org/10.1021/cr500081p

Croft AP, Bartsch RA (1983) Synthesis of chemically modified cyclodextrins. Tetrahedron 39(9):1417–1474. https://doi.org/10.1016/S0040-4020(01)88551-3

Cui Y, Wang C, Mao J, Yu Y (2010) A facile and practical approach to randomly methylated beta-cyclodextrin. J Chem Technol Biotechnol 85(2):248–251. https://doi.org/10.1002/jctb.2295

Di Blasio B, Galdiero S, Saviano M, Pedone C, Benedetti E, Rizzarelli E, Pedotti S, Vecchio G, Gibbons WA (1996) Synthesis and structural characterization of 6(I),6(II)-diamino-6(I),6(II)-dideoxy-cyclomaltoheptaose, a difunctionalized beta-cyclodextrin. Carbohydr Res 282(1):41–52. https://doi.org/10.1016/0008-6215(95)00357-6

Donati F (2008) Sugammadex: a cyclodextrin to reverse neuromuscular blockade in anaesthesia. Expert Opin Pharmacother 9(8):1375–1386. https://doi.org/10.1517/14656560802046211

Endo T (2011) Large-ring cyclodextrins. Trends Glycosci Glycotechnol 23(130):79–92. https://doi.org/10.4052/tigg.23.79
Engeldinger E, Armspach D, Matt D (2003) Capped cyclodextrins. Chem Rev 103(11):4147–4174. https://doi.org/10.1021/cr030670y
Estrada R III, Vigh G (2012) Comparison of charge state distribution in commercially available sulfated cyclodextrins used as chiral resolving agents in capillary electrophoresis. J Chromatogr A 1226:24–30. https://doi.org/10.1016/j.chroma.2011.08.040
Faugeras P-A, Boens B, Elchinger P-H, Brouillette F, Montplaisir D, Zerrouki R, Lucas R (2012) When cyclodextrins meet click chemistry. Eur J Org Chem 2012(22):4087–4105. https://doi.org/10.1002/ejoc.201200013
Fenger TH, Marinescu LG, Bols M (2011) Cyclodextrin ketones with the catalytic group at the secondary rim and their effectiveness in enzyme-like epoxidation of stilbenes. Eur J Org Chem 2011(12):2339–2345. https://doi.org/10.1002/ejoc.201001696
Fenyvesi E, Vikmon M, Szente L (2016) Cyclodextrins in food technology and human nutrition: benefits and limitations. Crit Rev Food Sci Nutr 56(12):1981–2004. https://doi.org/10.1080/10408398.2013.809513
Fernandez J, Mellet C, Blanco J, Mota J, Gadelle A, Costesarguet A, Defaye J (1995) Isothio cyanoates and cyclic thiocarbamates of alpha,alpha'-trehalose sucrose, and cyclomaltooligosaccharides. Carbohydr Res 268(1):57–71. https://doi.org/10.1016/0008-6215(94)00312-4
Folch-Cano C, Yazdani-Pedram M, Olea-Azar C (2014) Inclusion and functionalization of polymers with cyclodextrins: current applications and future prospects. Molecules 19(9):14066–14079. https://doi.org/10.3390/molecules190914066
Fugedi P (1989) Synthesis of heptakis(6-O-tert-butyldimethylsilyl)cyclomaltoheptaose and octakis(6-O-tert-butyldimethylsilyl)cyclomalto-octaose. Carbohydr Res 192:366–369. https://doi.org/10.1016/0008-6215(89)85197-3
Fujita K, Ueda T, Imoto T, Tabushi I, Toh N, Koga T (1982) Guest-induced conformational change of beta-cyclodextrin capped with an environmentally sensitive chromophore. Bioorg Chem 11(1):72–84. https://doi.org/10.1016/0045-2068(82)90049-9
Fujita K, Ejima S, Imoto T (1985) 1:2 Host-guest binding by double gamma-cyclodextrin. Chem Lett 14(1):11–12. https://doi.org/10.1246/cl.1985.11
Gadelle A, Defaye J (1991) Selective halogenation at primary positions of cyclomaltooligosaccharides and a synthesis of per-3,6-anhydro cyclomaltooligosaccharides. Angew Chem Int Ed Engl 30(1):78–80. https://doi.org/10.1002/anie.199100781
Gan Y, Zhang Y, Xiao C, Zhou C, Zhao Y (2011) A novel preparation of methyl-beta-cyclodextrin from dimethyl carbonate and beta-cyclodextrin. Carbohydr Res 346(3):389–392. https://doi.org/10.1016/j.carres.2010.05.028
Gao X, Tong L, Inoue Y, Tai A (1995) Synthesis and characterization of novel multifunctional host compounds. 4. Cyclodextrin derivatives rearing chromophores. Synth Commun 25(5): 703–710. https://doi.org/10.1080/00397919508011407
Garcia-Rio L, Otero-Espinar FJ, Luzardo-Alvarez A, Blanco-Mendez J (2014) Cyclodextrin based rotaxanes, polyrotaxanes and polypseudorotaxanes and their biomedical applications. Curr Top Med Chem 14(4):478–493. https://doi.org/10.2174/1568026613666131219123910
Ghanem A, Schurig V (2001) Peracetylated beta-cyclodextrin as additive in enzymatic reactions: enhanced reaction rate and enantiomeric ratio in lipase-catalyzed transesterifications in organic solvents. Tetrahedron-Asymmetry 12(19):2761–2766. https://doi.org/10.1016/S0957-4166(01)00482-7
Ghosh R, Zhang P, Wang A, Ling C-C (2012) Diisobutylaluminum hydride mediated regioselective o desilylations: access to multisubstituted cyclodextrins. Angew Chem Int Ed 51(7):1548–1552. https://doi.org/10.1002/anie.201105737
Gidwani B, Vyas A (2014) Synthesis, characterization and application of epichlorohydrin-beta-cyclodextrin polymer. Colloids Surf B Biointerfaces 114:130–137. https://doi.org/10.1016/j.colsurfb.2013.09.035

Gorin BI, Riopelle RJ, Thatcher GRJ (1996) Efficient perfacial derivatization of cyclodextrins at the primary face. Tetrahedron Lett 37(27):4647–4650. https://doi.org/10.1016/0040-4039(96)00916-1

Gould S, Scott RC (2005) 2-hydroxypropyl-beta-cyclodextrin (HP-beta-CD): a toxicology review. Food Chem Toxicol 43(10):1451–1459. https://doi.org/10.1016/j.fct.2005.03.007

Grachev MK (2013) Phosphorus-containing cyclodextrins. Characteristics of the synthesis and chemical behaviour. Russ Chem Rev 82(11):1034–1046. https://doi.org/10.1070/RC2013v082n11ABEH004381

Guan Z, Wang Y, Chen Y, Zhang L, Zhang Y (2009) Novel approach for synthesis of 2:1 permethylated β-cyclodextrin–C60 conjugate. Tetrahedron 65(6):1125–1129. https://doi.org/10.1016/j.tet.2008.12.001

Guieu S, Sollogoub M (2008a) Regiospecific tandem azide-reduction/deprotection to afford versatile amino alcohol-functionalized α- and β-cyclodextrins. Angew Chem Int Ed 47(37):7060–7063. https://doi.org/10.1002/anie.200801573

Guieu S, Sollogoub M (2008b) Multiple homo- and hetero-functionalizations of alpha-cyclodextrin through oriented deprotections. J Org Chem 73(7):2819–2828. https://doi.org/10.1021/jo7027085

Guitet M, de Beaumais SA, Bleriot Y, Vauzeilles B, Zhang Y, Menand M, Sollogoub M (2012) Cyclodextrins selectively modified on both rims using an O-3-debenzylative post-functionalisation, a consequence of the sorrento meeting. Carbohydr Res 356:278–281. https://doi.org/10.1016/j.carres.2011.12.002

Hamasaki K, Ikeda H, Nakamura A, Ueno A, Toda F, Suzuki I, Osa T (1993) Fluorescent sensors of molecular recognition. Modified cyclodextrins capable of exhibiting guest-responsive twisted intramolecular charge transfer fluorescence. J Am Chem Soc 115(12):5035–5040. https://doi.org/10.1021/ja00065a012

Hanessian S, Benalil A, Laferriere C (1995) The synthesis of functionalized cyclodextrins as scaffolds and templates for molecular diversity, catalysis, and inclusion phenomena. J Org Chem 60(15):4786–4797. https://doi.org/10.1021/jo00120a023

Harabagiu V, Simionescu BC, Pinteala M, Merrienne C, Mahuteau J, Guegan P, Cheradame H (2004) Synthesis and characterization of persilylated cyclodextrins. Carbohydr Polym 56(3):301–311. https://doi.org/10.1016/j.carbpol.2003.12.007

Harada A, Hashidzume A, Yamaguchi H, Takashima Y (2009) Polymeric rotaxanes. Chem Rev 109(11):5974–6023. https://doi.org/10.1021/cr9000622

Iacovino R, Caso JV, Di Donato C, Malgieri G, Palmieri M, Russo L, Isernia C (2017) Cyclodextrins as complexing agents: preparation and applications. Curr Org Chem 21(2):162–176. https://doi.org/10.2174/1385272820666160909111842

Ishiguro T, Morishita E, Iohara D, Hirayama F, Wada K, Motoyama K, Arima H, Uekama K (2011) Some pharmaceutical and inclusion properties of 2-hydroxybutyl-beta-cyclodextrin derivative. Int J Pharm 419(1–2):161–169. https://doi.org/10.1016/j.ijpharm.2011.07.044

Jambhekar SS, Breen P (2016) Cyclodextrins in pharmaceutical formulations I: structure and physicochemical properties, formation of complexes, and types of complex. Drug Discov Today 21(2):356–362. https://doi.org/10.1016/j.drudis.2015.11.017

Jeong YK, Kwon T, Lee I, Kim T-S, Coskun A, Choi JW (2014) Hyperbranched beta-cyclodextrin polymer as an effective multidimensional binder for silicon anodes in lithium rechargeable batteries. Nano Lett 14(2):864–870. https://doi.org/10.1021/nl404237j

Jicsinszky L, Martina K, Caporaso M, Cintas P, Zanichelli A, Cravotto G (2015) Complexes of peracetylated cyclodextrin in a non-aqueous aprotic medium: the role of residual water. Phys Chem Chem Phys 17(26):17380–17390. https://doi.org/10.1039/c5cp02379c

Jicsinszky L, Caporaso M, Martina K, Gaudino EC, Cravotto G (2016a) Efficient mechanochemical synthesis of regioselective persubstituted cyclodextrins. Beilstein J Org Chem 12:2364–2371. https://doi.org/10.3762/bjoc.12.230

Jicsinszky L, Caporaso M, Tuza K, Martina K, Gaudino EC, Cravotto G (2016b) Nucleophilic substitutions of 6I-O-monotosyl-beta-cyclodextrin in a planetary ball mill. ACS Sustain Chem Eng 4(3):919–929. https://doi.org/10.1021/acssuschemeng.5b01006

Jindřich J, Tišlerová I (2005) Simple preparation of 3I-O-substituted β-cyclodextrin derivatives using cinnamyl bromide. J Org Chem 70(22):9054–9055. https://doi.org/10.1021/jo051339c

Jullien L, Canceill J, Lacombe L, Lehn J (1994) Analysis of the conformational behavior of perfunctionalized beta-cyclodextrins. Part 1. Evidence for insertion of one of the rim substituents into the cyclodextrin cavity in organic-solvents. J Chem Soc Perkin Trans 2(5):989–1002. https://doi.org/10.1039/p29940000989

Jung M, Francotte E (1996) Comparison of γ-cyclodextrin sulfobutyl ether and unmodified γ-cyclodextrin as chiral selectors in capillary electrophoresis. J Chromatogr A 755(1):81–88. https://doi.org/10.1016/S0021-9673(96)00589-4

Kalakuntla RK, Wille T, Le Provost R, Letort S, Reiter G, Mueller S, Thiermann H, Worek F, Gouhier G, Lafont O, Estour F (2013) New modified beta-cyclodextrin derivatives as detoxifying agents of chemical warfare agents (I). Synthesis and preliminary screening: evaluation of the detoxification using a half-quantitative enzymatic assay. Toxicol Lett 216(2–3):200–205. https://doi.org/10.1016/j.toxlet.2012.11.020

Karoyo AH, Wilson LD (2015) Nano-sized cyclodextrin-based molecularly imprinted polymer adsorbents for perfluorinated compounds-a mini-review. Nanomaterials 5(2):981–1003. https://doi.org/10.3390/nano5020981

Khan AR, Forgo P, Stine KJ, D'Souza VT (1998) Methods for selective modifications of cyclodextrins. Chem Rev 98(5):1977–1996. https://doi.org/10.1021/cr970012b

Kirschner DL, Green TK (2005) Nonaqueous synthesis of a selectively modified, highly anionic sulfopropyl ether derivative of cyclomaltoheptaose (beta-cyclodextrin) in the presence of 18-crown-6. Carbohydr Res 340(11):1773–1779. https://doi.org/10.1016/j.carres.2005.04.012

Kiss T, Fenyvesi F, Bacskay I, Varadi J, Fenyvesi E, Ivanyi R, Szente L, Tosaki A, Vecsernyes M (2010) Evaluation of the cytotoxicity of beta-cyclodextrin derivatives: evidence for the role of cholesterol extraction. Eur J Pharm Sci 40(4):376–380. https://doi.org/10.1016/j.ejps.2010.04.014

Krauland AH, Alonso MJ (2007) Chitosan/cyclodextrin nanoparticles as macromolecular drug delivery system. Int J Pharm 340(1–2):134–142. https://doi.org/10.1016/j.ijpharm.2007.03.005

Kraus T, Budesinsky M, Zavada J (2000) Synthesis of per(5-carboxy-5-dehydroxymethyl)-alpha-cyclodextrin and beta-cyclodextrin - self-assembly of the per(2,3-di-O-methyl)-protected homologues into highly stable dimers, driven by multiple hydrogen bonds. Eur J Org Chem 2000(18):3133–3137. https://doi.org/10.1002/1099-0690(200009)2000:18<3133::AID-EJOC3133>3.0.CO;2-P

Kraus T, Buděšínský M, Závada J (2001) General approach to the synthesis of persubstituted hydrophilic and amphiphilic β-cyclodextrin derivatives. J Org Chem 66(13):4595–4600. https://doi.org/10.1021/jo010046q

Kraus T, Budesinsky M, Cisarova I, Zavada J (2002) Per(6-amino-2-O-carboxymethyl-6-deoxy-3-O-methyl)-alpha-cyclodextrin: helical self-assembly of a polyionic amino acid into nanotubes. Angew Chem Int Ed 41(10):1715–1717. https://doi.org/10.1002/1521-3773(20020517)41:10<1715::AID-ANIE1715>3.0.CO;2-K

Kryjewski M, Goslinski T, Mielcarek J (2015) Functionality stored in the structures of cyclodextrin-porphyrinoid systems. Coord Chem Rev 300:101–120. https://doi.org/10.1016/j.ccr.2015.04.009

Kulkarni A, DeFrees K, Hyun S-H, Thompson DH (2012) Pendant polymer: amino-β-cyclodextrin:siRNA guest:host nanoparticles as efficient vectors for gene silencing. J Am Chem Soc 134(18):7596–7599. https://doi.org/10.1021/ja300690j

Kurochkina GI, Grachev MK, Levina II, Nifant'ev EE (2014) (29)Si NMR spectroscopy as a reliable method for the determination of silyl protective groups positions at the cyclodextrin frame. Phosphorus Sulfur Silicon Relat Elem 189(1):33–39. https://doi.org/10.1080/10426507.2013.788006

Lammers JNJJ, Koole JL, Hurkmans J (1971) Properties of cyclodextrins. Part VI. Water-soluble cyclodextrin-derivatives. Preparation and analysis. Starch - Stärke 23(5):167–171. https://doi.org/10.1002/star.19710230504

Leclercq L (2016) Interactions between cyclodextrins and cellular components: towards greener medical applications? Beilstein J Org Chem 12:2644–2662. https://doi.org/10.3762/bjoc.12.261

Lee SA, Lim ST (1998) Preparation and solubility of phosphorylated beta-cyclodextrins. Cereal Chem 75(5):690–694. https://doi.org/10.1094/CCHEM.1998.75.5.690

Letort S, Mathiron D, Grel T, Albaret C, Daulon S, Djédaïni-Pilard F, Gouhier G, Estour F (2015) The first 2(IB), 3(IA)-heterodifunctionalized beta-cyclodextrin derivatives as artificial enzymes. Chem Commun 51(13):2601–2604. https://doi.org/10.1039/c4cc09189b

Letort S, Balieu S, Erb W, Gouhier G, Estour F (2016) Interactions of cyclodextrins and their derivatives with toxic organophosphorus compounds. Beilstein J Org Chem 12:204–228. https://doi.org/10.3762/bjoc.12.23

Li P, Song J, Ni X, Guo Q, Wen H, Zhou Q, Shen Y, Huang Y, Qiu P, Lin S, Hu H (2016) Comparison in toxicity and solubilizing capacity of hydroxypropyl-β-cyclodextrin with different degree of substitution. Int J Pharm 513(1–2):347–356. https://doi.org/10.1016/j.ijpharm.2016.09.036

Lian C-M, Jiang L-P, Liu D-L (2014) Synthesis and characterization of acetylated sept-D-glucopyranose carbamate as an oligosaccharide donor. Chin Chem Lett 25(1):134–136. https://doi.org/10.1016/j.cclet.2013.10.025

Lima PSS, Lucchese AM, Araujo-Filho HG, Menezes PP, Araujo AAS, Quintans-Junior LJ, Quintans JSS (2016) Inclusion of terpenes in cyclodextrins: preparation, characterization and pharmacological approaches. Carbohydr Polym 151:965–987. https://doi.org/10.1016/j.carbpol.2016.06.040

Lindbäck E, Zhou Y, Pedersen CM, Bols M (2012) Two diastereomeric artificial enzymes with different catalytic activity. Eur J Org Chem 2012(27):5366–5372. https://doi.org/10.1002/ejoc.201200699

Liu L, Guo Q-X (2002) The driving forces in the inclusion complexation of cyclodextrins. J Incl Phenom Macrocycl Chem 42(1–2):1–14. https://doi.org/10.1023/A:1014520830813

Liu Y, Han B-H, Li B, Zhang Y-M, Zhao P, Chen Y-T, Wada T, Inoue Y (1998) Molecular recognition study on supramolecular system. 14. Synthesis of modified cyclodextrins and their inclusion complexation thermodynamics with l-tryptophan and some naphthalene derivatives. J Org Chem 63(5):1444–1454. https://doi.org/10.1021/jo971466b

Luke DR, Tomaszewski K, Damle B, Schlamm HT (2010) Review of the basic and clinical pharmacology of sulfobutylether-beta-cyclodextrin (SBECD). J Pharm Sci 99(8):3291–3301. https://doi.org/10.1002/jps.22109

Lupescu N, Ho CKY, Jia GC, Krepinsky JJ (1999) A convenient synthesis of per-O-methylated 6-O-monosubstituted beta-cyclodextrins. J Carbohydr Chem 18(1):99–104. https://doi.org/10.1080/07328309908543982

Ma D-Y, Zhang Y-M, Xu J-N (2016) The synthesis and process optimization of sulfobutyl ether beta-cyclodextrin derivatives. Tetrahedron 72(22):3105–3112. https://doi.org/10.1016/j.tet.2016.04.039

Macaev F, Boldescu V (2015) Cyclodextrins in asymmetric and stereospecific synthesis. Symmetry-Basel 7(4):1699–1720. https://doi.org/10.3390/sym7041699

van de Manakker F, Vermonden T, van Nostrum CF, Hennink WE (2009) Cyclodextrin-based polymeric materials: synthesis, properties, and pharmaceutical/biomedical applications. Biomacromolecules 10(12):3157–3175. https://doi.org/10.1021/bm901065f

Martina K, Puntambekar DS, Barge A, Gallarate M, Chirio D, Cravotto G (2010) Synthesis, characterization and potential application of monoacyl-cyclodextrins. Carbohydr Res 345(2):191–198. https://doi.org/10.1016/j.carres.2009.11.009

Martinelli J, Thangavel K, Tei L, Botta M (2014) Dendrimeric beta-cyclodextrin/Gd-III chelate supramolecular host-guest adducts as high-relaxivity MRI probes. Chem Eur J 20(35):10944–10952. https://doi.org/10.1002/chem.201402418

Masurier N, Estour F, Lefèvre B, Brasme B, Masson P, Lafont O (2006) Improved access to 2-O-monobenzyl ethers of beta-cyclodextrin as precursors of catalysts for organophosphoryl esters hydrolysis. Carbohydr Res 341(7):935–940. https://doi.org/10.1016/j.carres.2006.02.012

Masurier N, Lafont O, Le Provost R, Lesur D, Masson P, Djédaïni-Pilard F, Estour F (2009) Regioselective access to 3(I)-O-substituted-beta-cyclodextrin derivatives. Chem Commun 5:589–591. https://doi.org/10.1039/b812325j

Maynard DK, Vigh G (2000) Synthesis and analytical characterization of the sodium salt of heptakis(2-O-methyl-3,6-di-O-sulfo)cyclomalto-heptaose, a chiral resolving agent candidate for capillary electrophoresis. Carbohydr Res 328(3):277–285. https://doi.org/10.1016/S0008-6215(00)00114-2

Mejia-Ariza R, Grana-Suarez L, Verboom W, Huskens J (2017) Cyclodextrin-based supramolecular nanoparticles for biomedical applications. J Mater Chem B 5(1):36–52. https://doi.org/10.1039/c6tb02776h

Mocanu G, Vizitiu D, Carpov A (2001) Cyclodextrin polymers. J Bioact Compat Polym 16(4):315–342. https://doi.org/10.1106/JJUV-8F2K-JGYF-HNGF

Morin-Crini N, Crini G (2013) Environmental applications of water-insoluble beta-cyclodextrin-epichlorohydrin polymers. Prog Polym Sci 38(2):344–368. https://doi.org/10.1016/j.progpolymsci.2012.06.005

Nakagawa T, Ueno K, Kashiwa M, Watanabe J (1994) The stereoselective synthesis of cyclomaltopentaose. A novel cyclodextrin homologue with D.P. five. Tetrahedron Lett 35(12):1921–1924. https://doi.org/10.1016/S0040-4039(00)73196-0

Nakazono K, Takashima T, Arai T, Koyama Y, Takata T (2010) High-yield one-pot synthesis of permethylated alpha-cyclodextrin-based polyrotaxane in hydrocarbon solvent through an efficient heterogeneous reaction. Macromolecules 43(2):691–696. https://doi.org/10.1021/ma902161d

Nielsen TT, Wintgens V, Amiel C, Wimmer R, Larsen KL (2010) Facile synthesis of β-cyclodextrin-dextran polymers by "Click" chemistry. Biomacromolecules 11(7):1710–1715. https://doi.org/10.1021/bm9013233

Normand M, Kirillov E, Carpentier J-F, Guillaume SM (2012) Cyclodextrin-centered polyesters: controlled ring-opening polymerization of cyclic esters from beta-cyclodextrin-diol. Macromolecules 45(2):1122–1130. https://doi.org/10.1021/ma202400e

Oliveri V, Vecchio G (2016) Cyclodextrins as protective agents of protein aggregation: an overview. Chem Asian J 11(11):1648–1657. https://doi.org/10.1002/asia.201600259

Pastore A, Valerio S, Adinolfi M, Iadonisi A (2011) An easy and versatile approach for the regioselective de-O-benzylation of protected sugars based on the I2/Et3SiH combined system. Chem-Eur J 17(21):5881–5889. https://doi.org/10.1002/chem.201003332

Pearce AJ, Sinaÿ P (2000) Diisobutylaluminum-promoted regioselective de-O-benzylation of perbenzylated cyclodextrins: a powerful new strategy for the preparation of selectively modified cyclodextrins. Angew Chem Int Ed 39(20):3610–3612. https://doi.org/10.1002/1521-3773(20001016)39:20<3610::AID-ANIE3610>3.0.CO;2-V

Petrillo M, Marinescu L, Rousseau C, Bols M (2009) Selective discrimination of cyclodextrin diols using cyclic sulfates. Org Lett 11(9):1983–1985. https://doi.org/10.1021/ol900413s

Petter RC, Salek JS, Sikorski CT, Kumaravel G, Lin FT (1990) Cooperative binding by aggregated mono-6-(alkylamino)-beta-cyclodextrins. J Am Chem Soc 112(10):3860–3868. https://doi.org/10.1021/ja00166a021

Pitha J, Milecki J, Fales H, Pannell L, Uekama K (1986) Hydroxypropyl-beta-cyclodextrin - preparation and characterization - effects on solubility of drugs. Int J Pharm 29(1):73–82. https://doi.org/10.1016/0378-5173(86)90201-2

Popr M, Hybelbauerova S, Jindrich J (2014) A complete series of 6-deoxy-monosubstituted tetraalkylammonium derivatives of alpha-, beta-, and gamma-cyclodextrin with 1, 2, and 3 permanent positive charges. Beilstein J Org Chem 10:1390–1396. https://doi.org/10.3762/bjoc.10.142

Prabaharan M, Gong S (2008) Novel thiolated carboxymethyl chitosan-g-beta-cyclodextrin as mucoadhesive hydrophobic drug delivery carriers. Carbohydr Polym 73(1):117–125. https://doi.org/10.1016/j.carbpol.2007.11.005

Pun SH, Bellocq NC, Liu AJ, Jensen G, Machemer T, Quijano E, Schluep T, Wen SF, Engler H, Heidel J, Davis ME (2004) Cyclodextrin-modified polyethylenimine polymers for gene delivery. Bioconjug Chem 15(4):831–840. https://doi.org/10.1021/bc049891g

Qu QI, Tucker E, Christian SD (2002) Sulfoalkyl ether β-cyclodextrin derivatives: synthesis and characterizations. J Incl Phenom Macrocycl Chem 43(3–4):213–222. https://doi.org/10.1023/A:1021255314835

Radu C-D, Parteni O, Ochiuz L (2016) Applications of cyclodextrins in medical textiles - review. J Control Release 224:146–157. https://doi.org/10.1016/j.jconrel.2015.12.046

Rao CT, Pitha J, Lindberg B, Lindberg J (1992) Distribution of substituents in O-(2-hydroxypropyl) derivatives of cyclomalto-oligosaccharides (cyclodextrins): influence of increasing substitution, of the base used in the preparation, and of macrocyclic size. Carbohydr Res 223:99–107. https://doi.org/10.1016/0008-6215(92)80009-P

Rawal GK, Rani S, Ward S, Ling C-C (2010) DIBAL-H mediated triple and quadruple debenzylations of perbenzylated cyclodextrins. Org Biomol Chem 8(1):171–180. https://doi.org/10.1039/b915450g

Rekharsky MV, Inoue Y (1998) Complexation thermodynamics of cyclodextrins. Chem Rev 98(5):1875–1918. https://doi.org/10.1021/cr970015o

Renard E, Volet G, Amiel C (2005) Synthesis of a novel linear water-soluble beta-cyclodextrin polymer. Polym Int 54(3):594–599. https://doi.org/10.1002/pi.1742

Reuben J, Trinadha Rao C, Pitha J (1994) Distribution of substituents in carboxymethyl ethers of cyclomaltoheptaose. Carbohydr Res 258:281–285. https://doi.org/10.1016/0008-6215(94)84094-6

Řezanka M (2016) Monosubstituted cyclodextrins as precursors for further use. Eur J Org Chem 2016(32):5322–5334. https://doi.org/10.1002/ejoc.201600693

Řezanka M, Jindřich J (2011) Complete sets of monosubstituted cyclomaltohexaoses (alpha-cyclodextrins) as precursors for further synthesis. Carbohydr Res 346(15):2374–2379. https://doi.org/10.1016/j.carres.2011.08.011

Řezanka M, Eignerová B, Jindřich J, Kotora M (2010) Synthesis of mono(perfluoroalkyl) cyclodextrins via cross metathesis. Eur J Org Chem 2010(32):6256–6262. https://doi.org/10.1002/ejoc.201000807

Řezanka P, Navrátilová K, Řezanka M, Král V, Sýkora D (2014) Application of cyclodextrins in chiral capillary electrophoresis. Electrophoresis 35(19):2701–2721. https://doi.org/10.1002/elps.201400145

Řezanka M, Langton MJ, Beer PD (2015) Anion recognition in water by a rotaxane containing a secondary rim functionalised cyclodextrin stoppered axle. Chem Commun 51(21):4499–4502. https://doi.org/10.1039/C5CC00171D

Řezanka P, Řezanková K, Sedláčková H, Mašek J, Rokosová L, Bláhová M, Řezanka M, Jindřich J, Sýkora D, Král V (2016) Influence of substituent position and cavity size of the regioisomers of monocarboxymethyl-α-, β-, and γ-cyclodextrins on the apparent stability constants of their complexes with both enantiomers of Tröger's base. J Sep Sci 39(5):980–985. https://doi.org/10.1002/jssc.201500845

du Roizel B, Baltaze J-P, Sinaÿ P (2002) Diisobutylaluminum-promoted secondary rim selective de-O-methylation of permethylated cyclodextrins. Tetrahedron Lett 43(13):2371–2373. https://doi.org/10.1016/S0040-4039(02)00274-5

Rojas MT, Koeniger R, Stoddart JF, Kaifer AE (1995) Supported monolayers containing preformed binding sites. Synthesis and interfacial binding properties of a thiolated beta-cyclodextrin derivative. J Am Chem Soc 117(1):336–343. https://doi.org/10.1021/ja00106a036

Sakairi N, Kuzuhara H (1993) Efficient and regioselective preparation of an eight-membered interglycosidic benzylidene derivative of β-cyclodextrin. Chem Lett 22(12):2077–2080. https://doi.org/10.1246/cl.1993.2077

Sakairi N, Wang L-X, Kuzuhara H (1991) Insertion of a D-glucosamine residue into the α-cyclodextrin skeleton; a model synthesis of 'chimera cyclodextrins'. J Chem Soc Chem Commun (5):289–290. https://doi.org/10.1039/C39910000289

Sakairi N, Wang L-X, Kuzuhara H (1995) Modification of cyclodextrins by insertion of a heterogeneous sugar unit into their skeletons. Synthesis of 2-amino-2-deoxy-β-cyclodextrin from α-cyclodextrin. J Chem Soc Perkin 1(4):437–443. https://doi.org/10.1039/P19950000437

Sakairi N, Kuzuhara H, Okamoto T, Yajima M (1996a) Synthesis and biological evaluation of 2-amino-2-deoxy- and 6-amino-6-deoxy-cyclomaltoheptaose polysulfates as synergists for angiogenesis inhibition. Bioorg Med Chem 4(12):2187–2192. https://doi.org/10.1016/S0968-0896(96)00222-2

Sakairi N, Nishi N, Tokura S, Kuzuhara H (1996b) Interglycosidic acetals. Part 3. Synthesis and structure determination of cyclic monobenzylidene acetals of cyclodextrin derivatives bridging between two contiguous D-glucopyranosyl residues. Carbohydr Res 291:53–62. https://doi.org/10.1016/S0008-6215(96)00192-9

Sallas F, Darcy R (2008) Amphiphilic cyclodextrins – advances in synthesis and supramolecular chemistry. Eur J Org Chem 2008(6):957–969. https://doi.org/10.1002/ejoc.200700933

Saokham P, Loftsson T (2017) Gamma-cyclodextrin. Int J Pharm 516(1–2):278–292. https://doi.org/10.1016/j.ijpharm.2016.10.062

Saz JM, Marina ML (2016) Recent advances on the use of cyclodextrins in the chiral analysis of drugs by capillary electrophoresis. J Chromatogr A 1467:79–94. https://doi.org/10.1016/j.chroma.2016.08.029

Sharma N, Baldi A (2016) Exploring versatile applications of cyclodextrins: an overview. Drug Deliv 23(3):739–757. https://doi.org/10.3109/10717544.2014.938839

Shirin S, Buncel E, vanLoon GW (2003) The use of β-cyclodextrins to enhance the aqueous solubility of trichloroethylene and perchloroethylene and their removal from soil organic matter: effect of substituents. Can J Chem 81(1):45–52. https://doi.org/10.1139/v02-205

Sollogoub M (2013) Site-selective heterofunctionalization of cyclodextrins: discovery, development, and use in catalysis. Synlett 24(20):2629–2640. https://doi.org/10.1055/s-0033-1339877

Song A, Wang J, Liu C, Deng L (2009a) Sulfoalkyl ether β-cyclodextrin derivatives synthesized by a single step method as pharmaceutical biomaterials. Chin Sci Bull 54(18):3187–3199. https://doi.org/10.1007/s11434-009-0262-8

Song LX, Bai L, Xu XM, He J, Pan SZ (2009b) Inclusion complexation, encapsulation interaction and inclusion number in cyclodextrin chemistry. Coord Chem Rev 253(9–10):1276–1284. https://doi.org/10.1016/j.ccr.2008.08.011

Stefanache A, Silion M, Stoica I, Fifere A, Harabagiu V, Farcas A (2014) Poly[2,7-(9,9-dioctylfluorene)-alt-(5,5′-bithiophene/permethylated beta-cyclodextrin)] main-chain polyrotaxane: synthesis, characterization and surface morphology. Eur Polym J 50:223–234. https://doi.org/10.1016/j.eurpolymj.2013.11.001

Swaminathan S, Cavalli R, Trotta F (2016) Cyclodextrin-based nanosponges: a versatile platform for cancer nanotherapeutics development. Wiley Interdiscip Rev Nanomed Nanobiotechnol 8(4):579–601. https://doi.org/10.1002/wnan.1384

Szejtli J (1998) Introduction and general overview of cyclodextrin chemistry. Chem Rev 98(5):1743–1754. https://doi.org/10.1021/cr970022c

Szejtli J, Lipták A, Jodál I, Fügedi P, Nánási P, Neszmélyi A (1980) Synthesis and 13C-NMR spectroscopy of methylated beta-cyclodextrins. Starch - Stärke 32(5):165–169. https://doi.org/10.1002/star.19800320506

Szente L, Fenyvesi E (2017) Cyclodextrin-lipid complexes: cavity size matters. Struct Chem 28(2):479–492. https://doi.org/10.1007/s11224-016-0884-9

Szente L, Szejtli J (1999) Highly soluble cyclodextrin derivatives: chemistry, properties, and trends in development. Adv Drug Deliv Rev 36(1):17–28. https://doi.org/10.1016/S0169-409X(98)00092-1

Tabushi I, Shimokawa K, Fujita K (1977) Specific bifunctionalization on cyclodextrin. Tetrahedron Lett 18(18):1527–1530. https://doi.org/10.1016/S0040-4039(01)93093-X

Tabushi I, Kuroda Y, Yokota K, Yuan L (1981) Regiospecific A,C- and A,D-disulfonate capping of beta-cyclodextrin. J Am Chem Soc 103(3):711–712. https://doi.org/10.1021/ja00393a056

Taka AL, Pillay K, Mbianda XY (2017) Nanosponge cyclodextrin polyurethanes and their modification with nanomaterials for the removal of pollutants from waste water: a review. Carbohydr Polym 159:94–107. https://doi.org/10.1016/j.carbpol.2016.12.027

Takeo K, Sumimoto T, Kuge T (1974) An improved synthesis of 6-deoxy-analogues of cyclodextrins and amylose - further interpretations of the proton magnetic resonance spectra of the peracetates of cyclodextrins and amylose. Starch 26(4):111–118. https://doi.org/10.1002/star.19740260403

Takeo K, Uemura K, Mitoh H (1988) Derivatives of alpha-cyclodextrin and the synthesis of 6-O-alpha-D-glucopyranosyl-alpha-cyclodextrin. J Carbohydr Chem 7(2):293–308. https://doi.org/10.1080/07328308808058926

Takeo K, Mitoh H, Uemura K (1989) Selective chemical modification of cyclomalto- oligosaccharides via tert-butyldimethylsilylation. Carbohydr Res 187(2):203–221. https://doi.org/10.1016/0008-6215(89)80004-7

Tang W, Ng S-C (2007) Synthesis of cationic single-isomer cyclodextrins for the chiral separation of amino acids and anionic pharmaceuticals. Nat Protoc 2(12):3195–3200. https://doi.org/10.1038/nprot.2007.479

Tang W, Ng SC (2008a) Monosubstituted positively charged cyclodextrins: synthesis and applications in chiral separation. J Sep Sci 31(18):3246–3256. https://doi.org/10.1002/jssc.200800357

Tang W, Ng S-C (2008b) Facile synthesis of mono-6-amino-6-deoxy-α-, β-, γ-cyclodextrin hydrochlorides for molecular recognition, chiral separation and drug delivery. Nat Protoc 3(4):691–697. https://doi.org/10.1038/nprot.2008.37

Tatar A, Grishina A, Budesinsky M, Kraus T (2017) Per-2,3-O-alkylated beta-cyclodextrin duplexes connected with disulfide bonds. Supramol Chem 29(1):40–48. https://doi.org/10.1080/10610278.2016.1164860

Teranishi K (2000) Practicable regiospecific bifunctionalization on the secondary face of alpha- and beta-cyclodextrins. Chem Commun (14):1255–1256. https://doi.org/10.1039/b002445g

Tongiani S, Velde DV, Ozeki T, Stella VJ (2005) Sulfoalkyl ether-alkyl ether cyclodextrin derivatives, their synthesis, NMR characterization, and binding of 6 alpha-methylprednisolone. J Pharm Sci 94(11):2380–2392. https://doi.org/10.1002/jps.20367

Trellenkamp T, Ritter H (2010) Poly(N-vinylpyrrolidone) bearing covalently attached cyclodextrin via click-chemistry: synthesis, characterization, and complexation behavior with phenolphthalein. Macromolecules 43(13):5538–5543. https://doi.org/10.1021/ma100812q

Trotta F, Costa L, Giancarlo C (2002) A new easy access to enantiomerically pure 2,2′-dihydroxy-1,1′-binaphthyl. J Incl Phenom Macrocycl Chem 44(1–4):341–344. https://doi.org/10.1023/A:1023069324382

Tutu E, Vigh G (2011) Synthesis, analytical characterization and initial capillary electrophoretic use in an acidic background electrolyte of a new, single-isomer chiral resolving agent: heptakis(2- O- sulfo-3-O-methyl-6-O-acetyl)-β-cyclodextrin. Electrophoresis 32(19):2655–2662. https://doi.org/10.1002/elps.201100104

Uccello-Barretta G, Sicoll G, Balzano F, Salvadori P (2005) NMR spectroscopy: a powerful tool for detecting the conformational features of symmetrical persubstituted mixed cyclomaltoheptaoses (beta-cyclodextrins). Carbohydr Res 340(2):271–281. https://doi.org/10.1016/j.carres.2004.11.022

Venuti V, Rossi B, Mele A, Melone L, Punta C, Majolino D, Masciovecchio C, Caldera F, Trotta F (2017) Tuning structural parameters for the optimization of drug delivery performance of cyclodextrin-based nanosponges. Expert Opin Drug Deliv 14(3):331–340. https://doi.org/10.1080/17425247.2016.1215301

Vincent JB, Kirby DM, Nguyen TV, Vigh G (1997) A family of single-isomer chiral resolving agents for capillary electrophoresis. 2. Hepta-6-sulfato-β-cyclodextrin. Anal Chem 69(21):4419–4428. https://doi.org/10.1021/ac970418o

Volkov S, Kumprecht L, Budesinsky M, Lepsik M, Dusek M, Kraus T (2015) A gamma- cyclodextrin duplex connected with two disulfide bonds: synthesis, structure and inclusion complexes. Org Biomol Chem 13(10):2980–2985. https://doi.org/10.1039/c4ob02464h

Wang N, Wu Q, Xiao YM, Chen CX, Lin MF (2005) Regioselective synthesis of cyclodextrin mono-substituted conjugates of non-steroidal anti-inflammatory drugs at C-2 secondary hydroxyl by protease in non-aqueous media. Bioorg Med Chem 13(11):3667–3671. https://doi.org/10.1016/j.bmc.2005.03.031

Wang Y, Zhao Y, Han B (2013) Preparation of tris(2-aminoethyl)amine-cross-linked cyclodextrin-based porous nanospheres and their application as drug delivery systems. Chin J Chem 31(5):657–662. https://doi.org/10.1002/cjoc.201300296

Wang B, Zaborova E, Guieu S, Petrillo M, Guitet M, Blériot Y, Ménand M, Zhang Y, Sollogoub M (2014) Site-selective hexa-hetero-functionalization of α-cyclodextrin an archetypical C6-symmetric concave cycle. Nat Commun 5:5354. https://doi.org/10.1038/ncomms6354

Wang J, Qiu Z, Wang Y, Li L, Guo X, Pham D-T, Lincoln SF, Prud'homme RK (2016) Supramolecular polymer assembly in aqueous solution arising from cyclodextrin host-guest complexation. Beilstein J Org Chem 12:50–72. https://doi.org/10.3762/bjoc.12.7

Wenz G (1994) Cyclodextrins as building blocks for supramolecular structures and functional units. Angew Chem Int Ed 33(8):803–822. https://doi.org/10.1002/anie.199408031

Wimmer T (1995) Process for the preparation of alkylated cyclodextrin derivatives, the methylated cyclodextrin derivatives obtainable from this process and the use of the products, Patent DE4333598 (A1)

Xiao Y, Wu Q, Wang N, Lin X (2004) Regioselective monoacylation of cyclomaltoheptaose at the C-2 secondary hydroxyl groups by the alkaline protease from Bacillus subtilis in nonaqueous media. Carbohydr Res 339(7):1279–1283. https://doi.org/10.1016/j.carres.2004.02.024

Xiao S, Yang M, Yu F, Zhang L, Zhou D, Sinaÿ P, Zhang Y (2013) Synthesis of four mono-functionalized α-cyclodextrin derivatives for further confirming DIBAL-H-promoted bis- de-O-methylation mechanism. Tetrahedron 69(20):4053–4060. https://doi.org/10.1016/j.tet.2013.03.070

Xu M, Wu S, Zeng F, Yu C (2010) Cyclodextrin supramolecular complex as a water-soluble ratiometric sensor for ferric ion sensing. Langmuir 26(6):4529–4534. https://doi.org/10.1021/la9033244

Yanli M (2005) Process for synthesizing random methylated beta-cyclodextrin, Patent CN1709918 (A)

Yoshikiyo K, Matsui Y, Yamamoto T (2015) Qualitative evaluation of regioselectivity in the formation of di- and tri-6-O-tritylates of alpha-cyclodextrin. Beilstein J Org Chem 11:1530–1540. https://doi.org/10.3762/bjoc.11.168

Yu H, Teramoto A, Fukudome M, Xie R-G, Yuan D-Q, Fujita K (2006) A facile sulfonylation method enabling direct syntheses of per(2-O-sulfonyl)-beta-cyclodextrins. Tetrahedron Lett 47(50):8837–8840. https://doi.org/10.1016/j.tetlet.2006.10.061

Yuan C, Liu B, Liu H (2015) Characterization of hydroxypropyl-beta-cyclodextrins with different substitution patterns via FTIR, GC-MS, and TG-DTA. Carbohydr Polym 118:36–40. https://doi.org/10.1016/j.carbpol.2014.10.070

Zhang W, Qin L, He X-W, Li W-Y, Zhang Y-K (2009) Novel surface modified molecularly imprinted polymer using acryloyl-beta-cyclodextrin and acrylamide as monomers for selective recognition of lysozyme in aqueous solution. J Chromatogr A 1216(21):4560–4567. https://doi.org/10.1016/j.chroma.2009.03.056

Zhang J, Zhang B, Zhou J, Li J, Shi C, Huang T, Wang Z, Tang J (2011) H_2SO_4-SiO_2: highly efficient and reusable catalyst for per-O-acetylation of carbohydrates under solvent-free conditions. J Carbohydr Chem 30(3):165–177. https://doi.org/10.1080/07328303.2011.621042

Zhang XM, Peng CS, Xu GC (2012) Synthesis of modified beta-cyclodextrin polymers and characterization of their fuchsin adsorption. J Incl Phenom Macrocycl Chem 72(1–2):165–171. https://doi.org/10.1007/s10847-011-9956-z

Zhong N, Byun H-S, Bittman R (1998) An improved synthesis of 6-O-monotosyl-6-deoxy-beta-cyclodextrin. Tetrahedron Lett 39(19):2919–2920. https://doi.org/10.1016/S0040-4039(98)00417-1

Zhou J, Ritter H (2010) Cyclodextrin functionalized polymers as drug delivery systems. Polym Chem 1(10):1552–1559. https://doi.org/10.1039/c0py00219d

Zhou Y, Lindbäck E, Marinescu LG, Pedersen CM, Bols M (2012a) Synthesis of cyclodextrins with carboxylic acids at the secondary rim and an evaluation of their properties as chemzymes for glycoside hydrolysis. Eur J Org Chem 2012(21):4063–4070. https://doi.org/10.1002/ejoc.201200404

Zhou Y, Marinescu L, Pedersen CM, Bols M (2012b) Synthesis of tin-containing cyclodextrins as potential enzyme models. Eur J Org Chem 2012(32):6383–6389. https://doi.org/10.1002/ejoc.201200756

Zhou J, Tang J, Tang W (2015) Recent development of cationic cyclodextrins for chiral separation trac. Trends Anal Chem 65:22–29. https://doi.org/10.1016/j.trac.2014.10.009

Zhu Q, Scriba GKE (2016) Advances in the use of cyclodextrins as chiral selectors in capillary electrokinetic chromatography: fundamentals and applications. Chromatographia 79(21–22):1403–1435. https://doi.org/10.1007/s10337-016-3167-0

Chapter 3
Cyclodextrins and Nanostructured Porous Inorganic Materials

Rudina Bleta, Eric Monflier, and Anne Ponchel

Contents

Abstract The discovery of ordered mesoporous materials in 1992 by Mobil Oil Corporation (Mobils) scientists has opened great opportunities for new applications in many emerging fields such as heterogeneous catalysis, biocatalysis, energy conversion, biosensors, photocatalytic devices and environmental technologies. Porous materials are grouped by the International Union of Pure and Applied Chemistry (IUPAC) into three classes according to their pore diameter: microporous (< 2 nm), mesoporous (2–50 nm) and macroporous (> 50 nm). One of the most versatile methods for the preparation of these materials is the soft-template approach which combines the sol-gel process with the molecular self-assembly. Materials with monodisperse particle sizes, well-defined architectures and tunable porosity can be built by this approach using micelles formed by ionic or non-ionic surfactants, as well as amphiphilic polymers, as templates. On the other hand, the supramolecular

R. Bleta (✉) · E. Monflier · A. Ponchel
Univ. Artois, CNRS, Centrale Lille, ENSCL, Univ. Lille, UMR 8181,
Unité de Catalyse et Chimie du Solide (UCCS), F-62300 Lens, France
e-mail: rudina.bleta@univ-artois.fr

S. Fourmentin et al. (eds.), *Cyclodextrin Fundamentals, Reactivity and Analysis*,
Environmental Chemistry for a Sustainable World,
https://doi.org/10.1007/978-3-319-76159-6_3

assemblies formed between cyclodextrins and block copolymers have been less investigated in the literature for the preparation of inorganic materials or nanocomposites, despite their large chemical and structural diversity.

This chapter focuses mainly on nanostructured porous inorganic materials derived from cyclodextrins or cyclodextrin-based assemblies. Examples from literature (more than 100 references) are described and discussed, in which we look both at the synthesis and characterization of those materials, as well as their applications in photocatalysis and heterogeneous catalysis. We first give a brief introduction to mesoporous silica prepared from conventional surfactants and polymers, and then look at attempts to the development of new synthetic strategies using cyclodextrin-based assemblies specifically designed to meet specific applications. Thus, cyclodextrin-based poly*pseudo*rotaxanes have been successfully used as templates for the preparation of mesoporous SiO_2 with tunable porosity. Moreover, multiple levels of porosity have been also integrated within the same structure, yielding silica materials with accessible porosities and hierarchical order at multiple length scales. On the other hand, for transition metal oxides, which are more difficult to prepare *via* the direct templating approach due to the very fast hydrolysis and condensation of the corresponding alkoxides in aqueous phase, the template-directed colloidal self-assembly approach is preferred and described. This approach involves the use of pre-synthesized colloidal particles which have the ability to self-assemble around a soft template (e.g., block copolymer or biopolymer). After drying and calcination, the recovered materials are usually robust and present high surface areas, large pore volumes and tunable porosities. For instance, supramolecular assemblies prepared from the randomly methylated β-cyclodextrin (RaMeβ-CD) have been successfully used as soft templates for the preparation of metal oxide nanostructures, such as TiO_2 and Al_2O with tunable properties.

From the perspective of applications, this book chapter presents recent developments involving the use of this new class of cyclodextrin-derived porous materials and nanocomposites in the emerging fields of heterogeneous catalysis and photocatalysis. A special attention is paid to the evaluation of the critical parameters that need to be controlled for improving their (photo) catalytic performances.

3.1 Introduction

Nanostructured porous materials have attracted tremendous research interest during the past two decades due to their unique structural and functional properties (Davis 2002; Lu and Schüth 2006; Imhof and Pine 1997; Velev and Kaler 2000). As reported in a large number of well-documented reviews, these materials offer potential applications in many emerging fields such as heterogeneous catalysis, energy conversion and storage devices, electrochemistry, optics, membranes and separation, drug delivery, biocatalysis, biosensors, photocatalytic devices and environmental technologies (Corma 1997;Sun et al. 2016; Zhu et al. 2015; Walcarius 2013).

Although a wide range of synthesis and assembly strategies have been developed so far for the fabrication of materials with tunable properties, the template synthesis remains one of the most flexible and important approaches to generate self-assembled nanostructures with well-defined architectures and hierarchical order at multiple length scales (Velev and Kaler 2000; Bleta et al. 2006; Blin et al. 2006; Liu et al. 2013; Yang et al. 2017). As both the structure and the composition of the template play a crucial role in the properties of the resulting material, there is a need to develop new supramolecular templates with tailored properties. In this context, cyclodextrins are of special interest owing to their unusual structural polymorphism as well as their ability to generate, in association with polymers, a wide range of fascinating supramolecular architectures (Wenz 1994; Born and Ritter 1995; Harada 1996; Herrmann et al. 1997; Szejtli 1998; Breslow and Dong 1998; Harada 2001).

In this book chapter, we aim to highlight the recent developments on the use of the native and modified cyclodextrins, as well as the supramolecular assemblies that they form by host-guest interactions with polymers, as excellent candidates for fabricating inorganic materials with precise control over the pore size and structure, the particle shape, the crystal phase composition and the spatial arrangement of nanoparticles. The three main topics that we aim to highlight are: (1) the use of nanocasting as a flexible and versatile strategy for replicating the architecture of cyclodextrin-based aggregates, poly*pseudo*rotaxanes and three-dimensional hydrogels to create hierarchically structured porous silica materials with well-defined spatial arrangement of the mico-, meso- and macropores, (2) elaboration of nanostructured metal oxide materials, other than silica, by template-directed colloidal self-assembly using cyclodextrin-based supramolecular assemblies as soft template and sol-gel synthesized colloidal particles as building blocks; (3) some applications of these materials in the emerging fields of heterogeneous catalysis and photocatalytic degradation of pesticides from wastewater.

3.2 Natural Porous Materials and Biomimetic Design

Porous materials occur widely in nature. Diatoms and radiolarians are most famous examples of natural porous materials living in most aquatic environments and using photosynthesis as a source of energy. One of the most remarkable features of these unicellular microalgae is their cell wall which is build-up of a silicified shell, also known as frustule. The diatom frustule displays a highly porous hierarchical structure and exhibits an extremely high mechanical stability which can be retained over geological timescales (Kooistra et al. 2007). Moreover, diatoms also play a major role in the carbon cycle of our Earth since they produce through photosynthesis about one-fifth of the oxygen we breathe (Kröger and Poulsen 2008). They act therefore as major contributors to global carbon dioxide fixation (Armbrust 2009).

Diatoms can be recognized in a microscope by their highly ornamented frustule (Fig. 3.1) (Hildebrand 2008). The diatom frustule is partly organic (proteins and

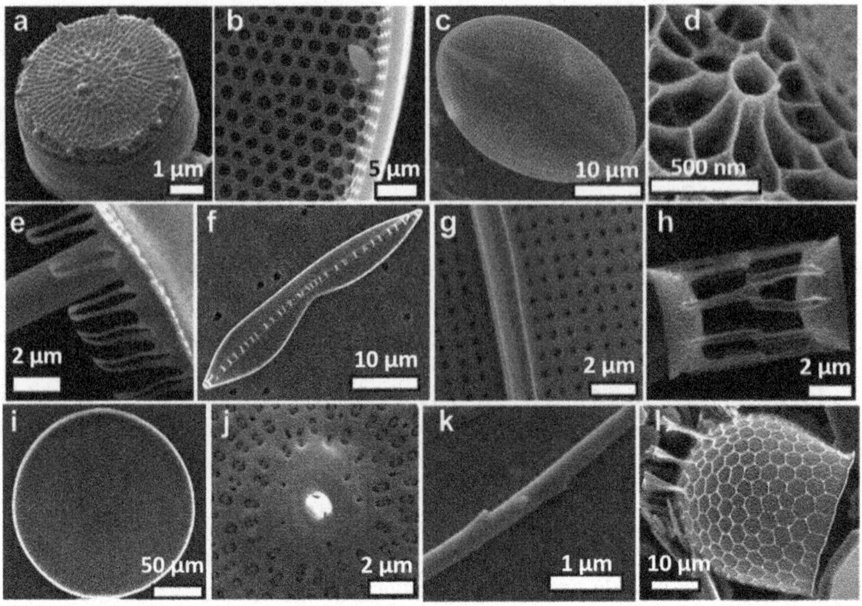

Fig. 3.1 Diversity of diatom silica structures. Acid-cleaned material from (**a**) *Thalassiosira pseudonana,* (**b**) close up of *Coscinodiscus wailesii,* (**c**) *Cocconeis sp.*, (**d**) rimoportula from *Thalassiosira weissflogii*, (**e**) corona structure *of Ditylum brightwellii*, (**f**) *Bacilaria paxillifer*, (**g**) close up of pores in *Gyrosigma balticum*, (**h**) *Skeletonema costatum*, (**i**) valve of *C. wailesii*, (**j**) close up of pores in *D. brightwellii*, (**k**) seta of *Chaetoceros gracilis*, and (**l**) *Stephanopyxis turris*. Reprinted with permission from (Hildebrand 2008). Copyright 2008 American Chemical Society

polysaccharides) and partly bioinorganic (hydrated silicon, $SiO_2{\cdot}[H_2O]_n$) and is formed through a remarkably rapid biomineralization process that is accomplished under mild physiological condition (Sumper et al. 2007). The frustule presents a well-defined hierarchical structure where several nanoscale elements (pores, channels) are regularly arranged in a complex 3D architecture with high level of precision. This porous structure is incredibly robust and confers to diatoms remarkable properties, such as mechanical strength, density, permeability, color and hydrophobicity, which protect them against desiccation, grazers and parasites (Hamm et al. 2003).

Pioneering biochemical studies of frustule composition have been performed by Kröger *et al.* in the diatom *Cylindrotheca fusiformis* and have led to the identification of three groups of biopolymers, *i.e.* the silaffins, the silacidines and the long chain polyamines, acting as major contributors in the biosilification process (Kröger and Poulsen 2008; Kröger et al. 1999). Silaffins and silacidins are peptides or proteins that carry many phosphate residues attached to amino acids (serine and threonine), whereas long chain polyamines are essentially linear non-protein components consisting of oligo-propyleneimine chains. Through self-assembly processes, essentially under electrostatic control, these biomolecules form bio-aggregates. Both the size of the aggregates and the quantity of the silicic oligomers that bind them determine the final size of the silica spheres in the architecture of the diatom frustule.

The recent sequencing of the genome in diatom *Thalassiosira pseudonana* greatly simplified the identification of silaffin genes and qualified these species as model organisms in future research on silica biomineralization (Armbrust et al. 2004; Poulsen and Kröger 2004). It was found that some rules exist during the construction of the diatom frustule implicating the action of an enzymatic machinery which transforms the amino acid sequence information within the silaffin polypeptide into the observed modified pattern. The structure of these silaffin polypeptides was found to influence the self-assembly process which guides the silica formation. Understanding the mechanism of fabrication of the diatom frustule is an inspiration for developing new synthetic routes towards novel structures that mimic natural systems.

3.3 Main Strategies toward the Synthesis of Nanostructured Silica Materials

3.3.1 The Concept of Nanocasting

Over the past 25 years, since the discovery of mesoporous materials by the group of Mobil Oil Corporation (Kresge et al. 1992; Beck et al. 1992), many research groups have proposed different biomimetic approaches to build nanostructured materials with multi-scale porous structures, similar to those of diatoms. One of the most well-known approaches is the nanocasting route. This approach has proved to be straightforward and highly effective for the synthesis of nanosystems that are able to mimic the original structure of a template (Wan and Zhao 2007). Generally, two kinds of templates, defined as hard- and soft-templates, are described in the literature. Hard templates are solid-state materials with particular structure and morphology, such as mesoporous silica, usually used for the synthesis of ordered carbon materials. On the other hand, soft templates are usually in a fluid-like state and have attracted more attention for the synthesis of mesoporous solids with tailorable pore structures.

Various supramolecular assemblies prepared from ionic or non-ionic surfactants or water-soluble polymers can be used as soft templates to prepare well-ordered mesoporous silica with monodisperse pore sizes (Lu and Schüth 2006). Most common mesoporous silicas are the small pore hexagonal Mobil Composition of Matter-41 (MCM-41), cubic MCM-48, lamellar MCM-50, the large pore hexagonal Santa Barbara Amorphous-15 (SBA-15) and cubic SBA-16, as well as Hexagonal Mesoporous Silica (HMS) and Michigan State University (MSU) materials with wormlike pore structure (Fig. 3.2).

Most of them are prepared quite easily, under mild conditions, using the sol-gel process. The sol-gel synthesis involves the hydrolysis and catalytic polycondensation

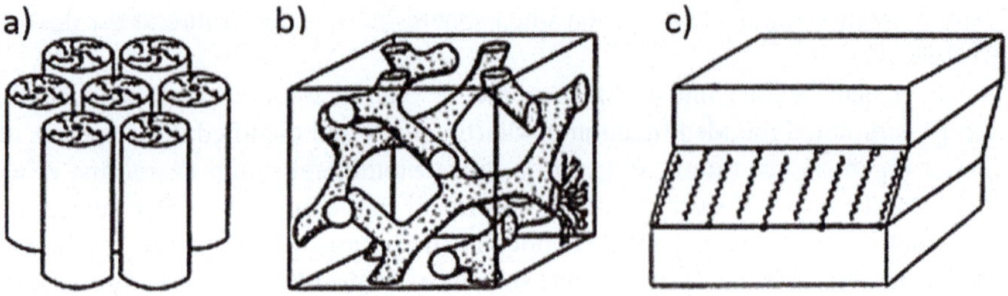

Fig. 3.2 Different types of mesoporous materials. (**a**) Hexagonal arrangement of pores: type Mobil Composition of Matter-41 (MCM-41); (**b**) Cubic arrangement of pores: type MCM-48; (**c**) Lamellar arrangement of pores: type MCM-50. Reprinted with permission of the Royal Society of Chemistry from (Corma et al. 2008)

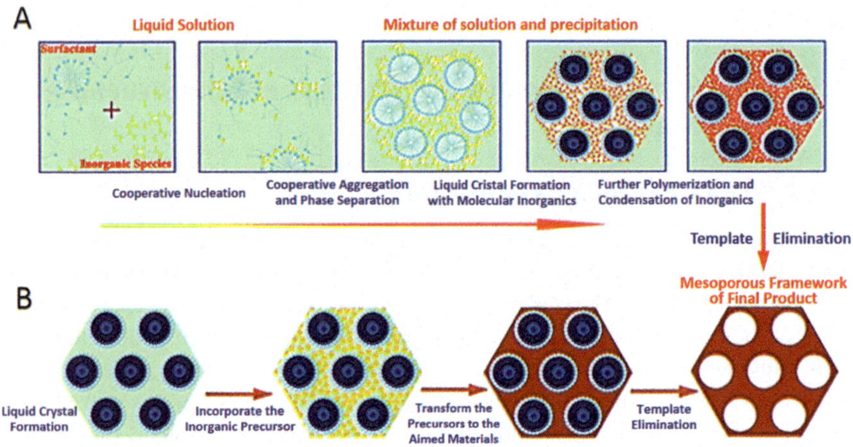

Fig. 3.3 Schematic representation of the soft-templating method *via* two synthetic strategies: (**A**) cooperative self-assembly and (**B**) "true liquid-crystal templating" (TLCT) process for the synthesis of ordered mesoporous materials. Reprinted with permission from (Wan and Zhao 2007). Copyright 2007 American Chemical Society

of a silicon alkoxide precursor (e.g., tetramethoxysilane or tetraethoxysilane) in the presence of a supramolecular template (typically micelles formed by amphiphilic molecules). The macromolecular network of siloxane bonds can be obtained by two different ways, *i.e. via* a cooperative assembly process taking place in-situ between the structure directing agent and the silica precursor in solution, or through a true liquid-crystal-phase templating (TLCT) mechanism, by formation of a silica framework around the pre-formed liquid crystal mesophase (Fig. 3.3). It is worth emphasizing that the cooperative assembly process is not a nanocasting route since it does not replicate a preformed surfactant structure. For instance, in the synthesis of mesoporous silica, concentrations of surfactant even below the critical micelle concentration (CMC) can yield highly ordered mesoporous structures. These materials are usually highly porous (pore volume higher than 0.7 $cm^3\ g^{-1}$), possess

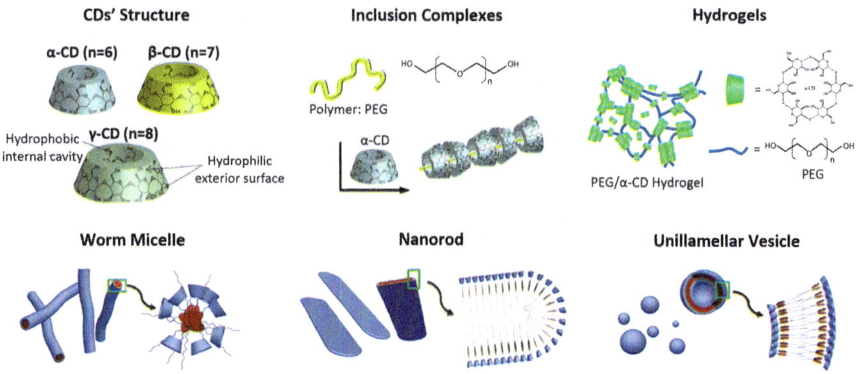

Fig. 3.4 Cyclodextrins structure and cyclodextrin-based nanosystems formed by their self-assembly with polymers. Adapted with permission from (Harada et al. 2009) and (Jiang et al. 2010). Copyright 2009 and 2010 American Chemical Society

large specific surface areas (up to 1500 $m^2\ g^{-1}$), well-defined multiscale porous networks, tunable pore sizes and interconnectivity. These properties make them viable for applications in different emerging fields ranging from biotechnology, biomedicine, drug delivery, catalysis, energy storage, optics, separation processes to immobilization of biomolecules and bio-organisms as well as bone regeneration (Su et al. 2012).

3.3.2 Using Cyclodextrin-Based Supramolecular Assemblies as Templates

The structure and properties of the template play a crucial role with respect to the properties of the replicated porous materials. In this context, cyclodextrins (CDs) offer attractive alternative to surfactants owing to the rich structural diversity of the supramolecular assemblies that they form in aqueous phase in association with polymers. Cyclodextrins are water-soluble cyclic oligosaccharides composed of a hydrophobic internal cavity and a hydrophilic exterior surface due to the presence of a large number of hydroxyl groups. Most common cyclodextrins contain six (α-CD), seven (β-CD) or eight (γ-CD) glucopyranose units in the ring (Fig. 3.4). These molecules demonstrate multifunctional properties, such as the ability to form supramolecular assemblies with a large variety of intriguing structures with amphiphilic surfactants of appropriate size and shape (Szejtli 1998; Breslow and Dong 1998; Wenz 1994; Born and Ritter 1995; Harada 2001; Harada 1996; Herrmann et al. 1997) (Fig. 3.4). Since the pioneering works of Harada *et al.* (Harada and Kamachi 1990), who showed that α-CD can form inclusion complexes with poly(ethylene glycol) (PEG) in aqueous solution to give poly*pseudo*rotaxanes with a necklace-like structure, a range of polymeric guests have been investigated

(Harada et al. 2009). A particular attention has been paid to the interactions occurring between the native β-CD and nonionic triblock copolymers of the poly(ethylene oxide) (PEO)-*b*-poly(propylene oxide) (PPO)-*b*-poly(ethylene oxide) (PEO) family, also known as Pluronics. Thus, it has been shown that the native β-CD can slide along the hydrophilic extremity PEO blocks of the Pluronic P84 ($PEO_{19}PPO_{43}PEO_{19}$) to selectively thread the middle hydrophobic PPO blocks and form poly*pseudo*rotaxanes with a well-ordered channel structure.

Despite their rich structural polymorphism, the supramolecular assemblies formed between block copolymers and CD-derivatives have been very little used as soft templates. In the field of materials science, micelles and lyotropic phases are the most commonly used templates for the synthesis of porous materials with controlled architectures. On the other hand, the possibility of using CDs or CD/polymer assemblies as supramolecular templates have been reported only in few studies and most of them have been devoted to the synthesis of mesoporous silica by sol-gel process (Polarz et al. 2001; Han and Antonietti 2002; Han et al. 2003).

3.3.3 Microporous Silica from Cyclodextrins

The possibility to extend the principle of nanocasting from micelles and lyotropic phases to cyclodextrins was first demonstrated by Antonietti *et al.* (Polarz et al. 2001; Polarz and Antonietti 2002; Antonietti 2006) who employed the aggregates formed by native and modified cyclodextrins in water as templates in the nanocasting process. Cyclodextrins have the ability to self-assemble in water into assemblies with a "molecular barrel" structure (Saenger 1980). These "barrels" have an exterior surface covered by -OH groups, while the interior cavity is quite hydrophobic, characterized by exposure of -CH_2- groups. Owing to the rich OH-functionality on their outside surface, cyclodextrin assemblies in water can be nanocasted. Interestingly, the silica materials prepared by nanocasting were "worm-type" and presented a very similar structure to those obtained with classical amphiphiles (Fig. 3.5). The pore morphology did not agreed with the structure of a molecularly dispersed system, thus confirming the formation of cyclodextrin aggregates that can be replicated by nanocasting. The pore size corresponded exactly to the diameter of the cyclodextrin, while the length was significantly larger. The pore cross-section of these materials, as determined by porosimetry, was directly related to the diameter of the single circular dichroism (CD)-units. Thus, using the hydroxypropylated α-CD (HPα-CD) assemblies, 1.4 nm pores were obtained, while the replication of the HPγ-CD assemblies led to 1.8 nm pores. Such a direct correlation is an important indication that the pore systems results really from the "molecular barrel" structure of the CD assemblies. The difference between the hydrophilic exterior and the hydrophobic interior of the CD-molecule was supposed to be the driving force for this assembly.

This result was particularly interesting since it showed for the first time that, besides offering the possibility to synthesize nanostructured silica materials,

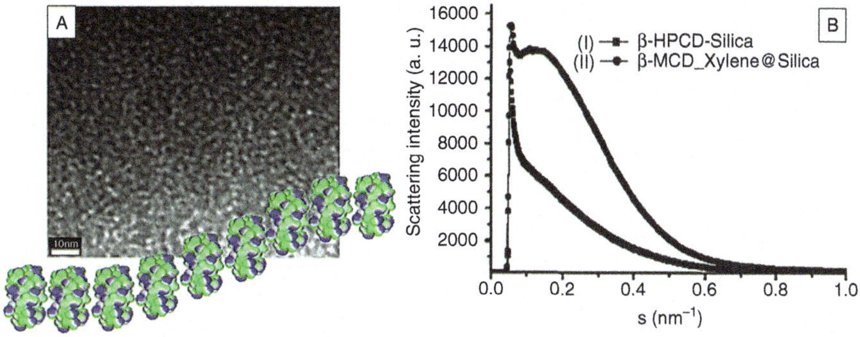

Fig. 3.5 (**A**) Transmission electron microscopy (TEM) image of a representative CD-based silica demonstrating the 'worm-type' architecture of the pores and a schematic image of the CD-alignment indicating the template structure; (**B**) small angle X-ray scattering (SAXS) diffractograms of two CD-based porous materials: (i) one was prepared using the hydrophilic β-HPCD showing a rather disordered pore-system, (ii) the other one was prepared with xylene@β-MCD possessing strong quadrupolar amphiphilic character. The latter pore-system is much more ordered and uniform. Reprinted with permission of the Royal Society of Chemistry from (Polarz and Antonietti 2002)

nanocasting can also be used as "analytical tool" to examine the soft and delicate structure of these assemblies. Thus, instead of investigating the soft-matter structure in its dispersion medium itself, it is possible to examine its "hardcopy" derived by nanocasting. This route has several potential advantages in comparison with the direct examination due to the higher electron contrast and higher stability of the solid matter against most experimental conditions, such as electron microscopy or scattering techniques.

3.3.4 *Mesoporous Silica from Cyclodextrin-Based poly*pseudo*rotaxanes*

Besides cyclodextrins, poly*pseudo*rotaxanes can also be replicated into mesoporous silica (Han and Antonietti 2002). It has been reported that in the case of triblock copolymers, the native α-CD preferentially binds the PEO units, whereas the native β-CD can slide along the hydrophilic extremity PEO blocks to selectively thread the middle hydrophobic PPO blocks (Harada 1996; Harada et al. 2009). The resulting poly*pseudo*rotaxanes have "string-of-pearl" morphologies. The structural models proposed by Harada (Harada 1996) for these channel-like inclusion complexes are depicted in Fig. 3.6. Remarkably, these poly*pseudo*rotaxanes were shown to spontaneously self-assemble in water to generate other types of supramolecular assemblies with a well-defined crystalline structure. Thus, by using small-angle neutron scattering (SANS) and atomic force microscopy (AFM), it was demonstrated that the self-assembly of β-CD and Pluronic F68 ($PEO_{80}PPO_{27}PEO_{80}$) micelles may lead to the formation of cylindrical bundles, which can then act as building blocks

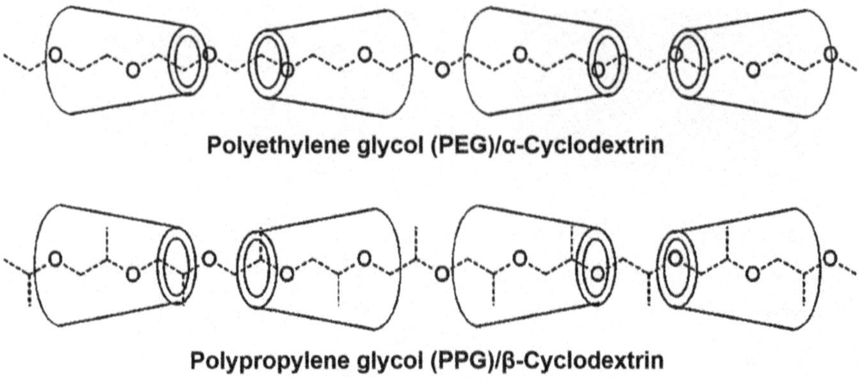

Fig. 3.6 Poly*pseudo*rotaxanes formed by threading of α-cyclodextrin along polyethylene glycol and β-cyclodextrin along polypropylene glycol. Reprinted with permission from (Han and Antonietti 2002). Copyright 2002 American Chemical Society

for the formation of flat and rigid platelets with well-defined angles at 40 °C (Travelet et al. 2008; 2009; 2010). When these *in-situ* formed supramolecular assemblies were used as templates for the synthesis of silica materials, the pore diameter was found to depend on the pH of the solution (Han and Antonietti 2002). Direct replication was obtained at pH 2.0, whereas larger pores were formed at higher pH values between pH 3 and pH 4 (Fig. 3.7). Usually, the structure of the formed silica does not change strongly in the pH range between 2.0 and 4.0. Therefore, the increase in the pore size was directly related to a pH-dependent behavior of the template. Accordingly, poly*pseudo*rotaxanes aggregate at pH 4.0 into arrays or bundle structures that can generate larger pores by nanocasting.

The nanocasting procedure was also extended to stable isolated poly*pseudo*rotaxanes obtained from α-CD and polyamines (Han et al. 2003). The obtained silica materials possessed elongated mesopores and high porosity. An approximate replica of the original rod-like rotaxanes was produced with a homogeneous distribution of the pores in the silica matrix.

3.3.4.1 Hierarchically Porous Silica from Polyethylene Glycol/α--Cyclodextrin Hydrogels

Not only cyclodextrin barrels or poly*pseudo*rotaxanes, but also highly reticulated hydrogels and their self-assembly motifs, can be replicated. This was clearly illustrated by silicification by sol-gel process of the polyethylene glycol/α--cyclodextrin (PEG/α-CD) hydrogels that yielded nanostructured silica materials with bimodal or trimodal pore size distributions (Fig. 3.8) (Bleta et al. 2014c).

Hydrogels can be defined as water-swollen hydrophilic polymers, formed by covalent bonds or physical cohesion forces between the polymer segments, such as van der Waals forces, hydrophobic interactions or hydrogen bonds. The pioneering

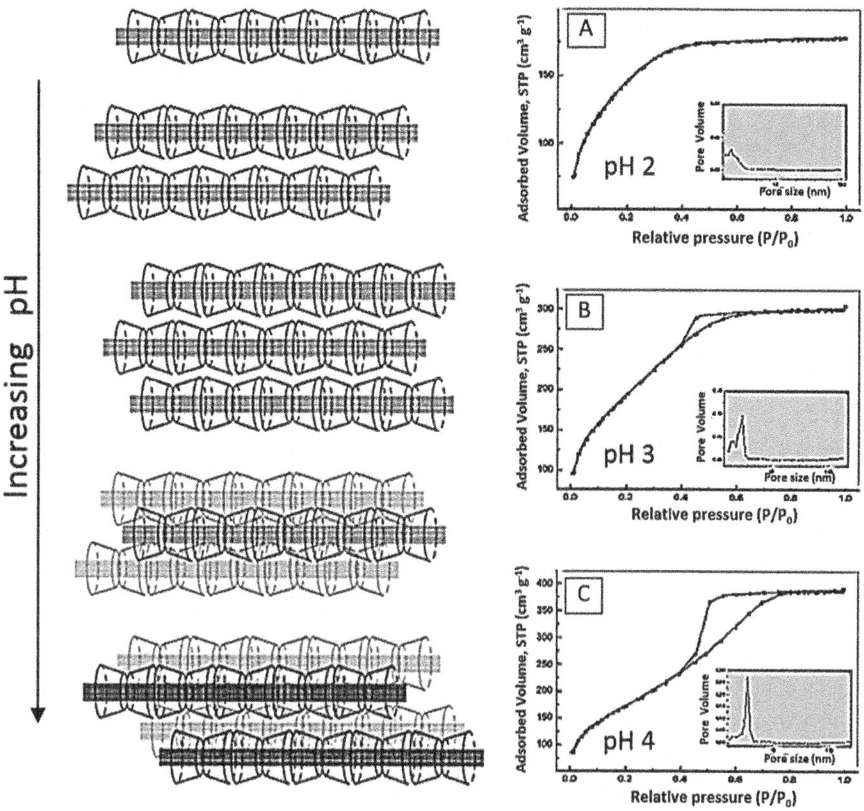

Fig. 3.7 Schematics of the pH-dependent aggregation of poly*pseudo*rotaxanes of polyethers with cyclodextrins. At pH 2.0, a single poly*pseudo*rotaxane domain predominates. With increasing pH values (pH 3.0 and 4.0), the aggregation of poly*pseudo*rotaxanes is enhanced giving bundle domains. N_2-sorption data of the three silicas prepared from α-cyclodextrin and polyethylene glycol (PEG) 2000 condensed at pH 2.0 (**A**), 3.0 (**B**) and 4.0 (**C**). Insets: pore size distribution as calculated by the Barrett-Joyner-Halenda (BJH) theory from the desorption branch. Reprinted with permission from (Han and Antonietti 2002). Copyright 2002 American Chemical Society

work of Li et al. (Li et al. 1994) showed that the native α-CD and a high molecular weight PEG (higher than 10,000 g mol^{-1}) can spontaneously form a three-dimensional physically cross-linked hydrogel by host-gust interactions. A large amount of water can be entrapped within the macroscopic voids. Gelation was proposed to occur *via* the cross-linking of the partially formed PEG/α-CD inclusion complex, which further self-assemble into water insoluble poly*pseudo*rotaxanes with a channel-like structure (Li et al. 1994). Subsequently, the individual poly*pseudo*rotaxanes further grow in size, aggregate and finally phase-separate. As stated by Weickenmeier and Wenz (Weickenmeier and Wenz 1997), the phase-separation is followed by the threading of additional α-CD onto the polymer chains, leading to the aggregation and growth of columnar poly*pseudo*rotaxane-based

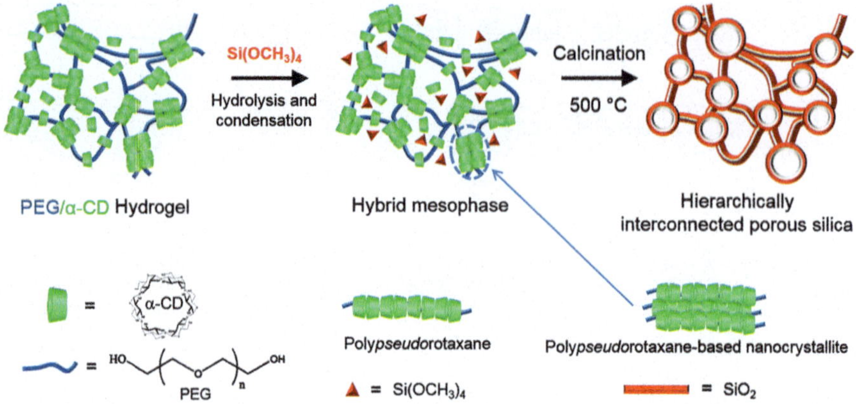

Fig. 3.8 Schematic illustration of the template-directed synthesis of silica materials with hierarchical pore architectures where the large mesopores are interconnected by small mesopores in a three-dimensional framework. The crosslinking of these poly*pseudo*rotaxanes into columnar nanocrystallites with a channel-like architecture directs the condensation of silica in a bi- or tri-modal porous network. Reprinted with permission of the Royal Society of Chemistry from (Bleta et al. 2014c)

nanocrystallites (Travelet et al. 2009, 2010). This nanoscale arrangement of poly*pseudo*rotaxanes into nanocrystallites is assumed to play a crucial role in maintaining the hydrogel in a water-swollen state.

To investigate in detail how the PEG concentration affects the gelation behavior, a series of silica materials were prepared by templating the hydrogels prepared with increasing amounts of PEG (1–30 mg mL^{-1}) in a saturated α-CD solution (100 mg mL^{-1}) at pH 4 and 2 (Bleta et al. 2014c). The chemistry of silica does not change significantly in this pH range, which is close to the isoelectric point of silicic acid, where the condensation rate is the slowest. So, the variations monitored in the porosity of the materials can be mainly attributed to the structural changes occurring within the hydrogel template in response to pH.

The adsorption isotherms and corresponding pore size distributions of the silica materials prepared at pH 4 are shown in Fig. 3.9A. In contrast to the materials prepared from PEG alone (average pore size 2.0 nm), those prepared from α-CD alone showed a well-defined type IV isotherm with a pronounced hysteresis loop typical of a mesoporous material. Mesopores have an average diameter of 6.8 nm which can be associated with the formation of large α-CD aggregates in water. Interestingly, when a small amount of PEG (1 mg mL^{-1}) was added to the saturated α-CD solution (100 mg mL^{-1}), two populations of small mesopores were observed at 1.8 nm and 2.0 nm attributed to the naked polymer chains and poly*pseudo*rotaxanes made from α-CD threaded onto PEG in a dynamic threading-dethreading equilibrium. For higher PEG concentrations (4–30 mg mL^{-1}), the examination of the pore structure of the resulting materials gave evidence of a hierarchical pore structure with three types of pores: (i) micropores (1.8 nm) associated with naked polymer chains, (ii) small mesopores (2.0 and 2.3 nm) associated with individual

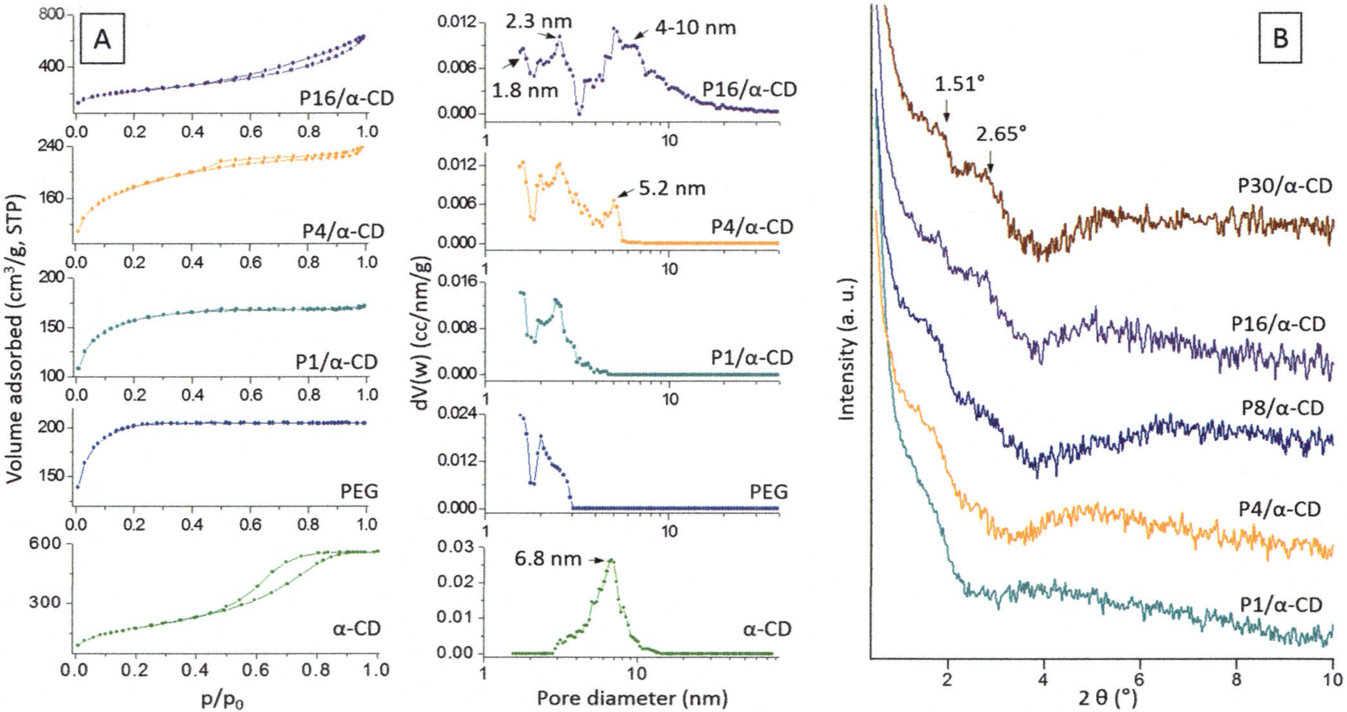

Fig. 3.9 (A) N_2-adsorption isotherms and corresponding pore size distributions determined by non-local density functional theory (NLDFT) for silica prepared at pH 4 from a saturated α-CD solution (100 mg mL^{-1}), from PEG (16 mg mL^{-1}) and from PEG/α-CD mixtures prepared with increasing amounts of polymer, from 1 mg mL^{-1} (P1) to 16 mg mL^{-1} (P16). With increasing PEG concentration, the aggregation of poly*pseudo*rotaxanes is enhanced giving rise to three types of pores: micropores (1.8 nm) associated with naked polymer chains, small mesopores (2.0 and 2.3 nm) associated with individual poly*pseudo*rotaxanes and larger mesopores (5.8 nm) associated with the columnar α-CD crystallites formed by the self-assembly of poly*pseudo*rotaxanes. (B) Corresponding small angle X-ray scattering (SAXS) plots. All samples were calcined at 500 °C for 16 h. Adapted with permission of the Royal Society of Chemistry from (Bleta et al. 2014c)

Table 3.1 Textural characteristics of the silicified α-CD after calcination at 500 °C

Sample	S_{BET} (m^2 g^{-1})	V_{cum} (cm^3 g^{-1})	PS (nm)	V_{micro} (cm^3 g^{-1})	% micro
CD100-pH 4	634	0.839	6.8	0.001	0.2
Bare PEG-pH 4	688	0.309	2.0	0.195	63
P1CD100-pH 4	639	0.251	2.0 and 2.3	0.140	56
P2CD100-pH 4	607	0.242	2.0 and 2.3 and 5.2	0.111	46
P4CD100-pH 4	618	0.338	2.0 and 2.3 and 5.2	0.086	26
P8CD100-pH 4	663	0.755	2.0 and 4–10	0.062	8.2
P16CD100-pH 4	777	0.912	2.0 and 4–10	0.075	8.2
P30CD100-pH 4	760	0.979	2.0 and 4–10	0.026	2.7
Bare PEG-pH 2	545	0.249	2.0	0.184	74
P4CD100-pH 2	635	0.373	2.0 and 2.3	0.179	48
P8CD100-pH 2	616	0.394	2.0 and 2.3 and 5.2	0.175	44
P16CD100-pH 2	698	0.446	2.0 and 2.3 and 5.2	0.188	42
P30CD100-pH 2	656	0.454	2.0 and 2.3 and 5.2	0.145	32

S_{BET} = BET specific surface area determined in the relative pressure range 0.1–0.25, V_{cum}, PS = cumulative pore volume and pore size resulting from non-local density functional theory (NLDFT) calculations, V_{micro} = micropore volume determined from t-plot, % micro = (V_{micro}/V_{cum})*100. Reprinted with permission of the Royal Society of Chemistry from (Bleta et al. 2014c)

poly*pseudo*rotaxanes, as well as (iii) larger mesopores (5.8 nm) attributed to the columnar α-CD crystallites formed by the self-assembly of poly*pseudo*rotaxanes at their CD-rich segments. Therefore, various self-assembled motifs exist within the hydrogel network including naked PEG chains, poly*pseudo*rotaxanes and nanocrystallites. Moreover, nanocrystallites were found to gradually grow in size as the solution was loaded with more PEG and as the hydrogel network became denser. On the other hand, all materials prepared at pH 2 contained a high portion of micropores (1.8 nm) and small mesopores (2 nm), even at a high PEG concentration (30 mg mL^{-1}) (Table 3.1) indicating that, under more acidic conditions, a higher portion of unthreaded α-CD and uncovered PEG segments is available, thus leading to lower crosslinking densities within the hydrogel network.

Interestingly, the small angle X-ray scattering (SAXS) data (Fig. 3.9B) indicated the presence of two reflection peaks at $2\theta = 1.51°$ and $2.65°$ (q ratios $1:\sqrt{3}$), corresponding to d-spacing values of 5.84 nm and 3.33 nm, respectively, which can be indexed to the (100) and (110) Bragg reflections of a hexagonal mesostructured material. This means that some local order exists within these silica replicas, due to the self-assembly of poly*pseudo*rotaxanes in rather regular columnar bundles within the crystallites.

Representative TEM images of the silica replicas prepared at pH 4 with 30 mg mL^{-1} PEG showed the presence long tubular structures with 30–170 nm diameters (Fig. 3.10 a–c) made-up of several linear strings associated with the bundles of poly*pseudo*rotaxanes. Moreover, some local order can be noticed in the mesoporous network (Fig. 3.10 d), in agreement with the SAXS pattern. On the other hand, sharper striations were formed in the silicified hydrogel at pH 2 (Fig. 3.10 e and f). The mesoporous network was characterized by the presence of regular lattice fringes

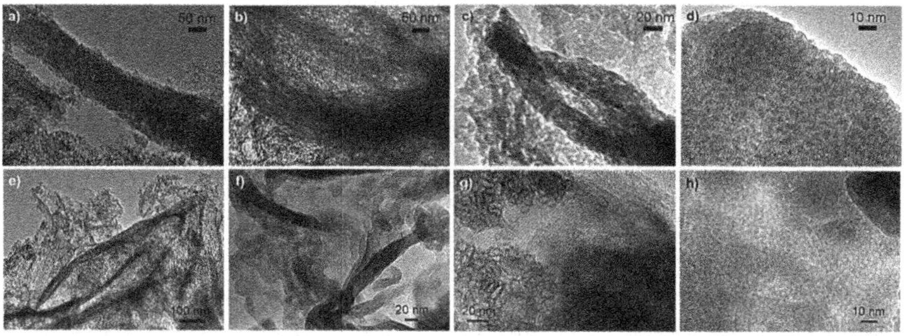

Fig. 3.10 Transmission electron microscopy (TEM) images of the calcined silica materials prepared from PEG (30 mg mL^{-1})/ α-CD (100 mg mL^{-1}) hydrogels at pH 4 (**a-d**) and pH 2 (**e-h**). Note the presence of long tubular structures with 30–170 nm diameters associated with the bundles of poly*pseudo*rotaxanes. Reprinted with permission of the Royal Society of Chemistry from (Bleta et al. 2014c)

(Fig. 3.10 g) consistent with a higher degree of structuration, as evidenced by SAXS. Worm-like pores with diameters of approximately 5 nm, represented by the bright points, were also observed (Fig. 3.10 h) in a relatively good agreement with the pore sizes previously determined by N_2-adsorption measurements.

These results demonstrated that the threading of α-CD onto PEG chains and the resulting nanocrystallites are affected by the pH of the aqueous solution. A schematic comparison of the replication process that occurs in PEG and PEG/α-CD systems at pH 4 and 2 is shown in Fig. 3.11. Depending on the pH of the hydrogel, several domains with various crosslinking densities can be formed within the organic templates giving rise to a wide range of pore sizes within the silica scaffolds. Therefore, in the case of the hydrogel prepared at pH 2, the presence of an excess of H_3O^+ ions may induce a protonation of the polymer -OH groups resulting in a lower degree of α-CD threading and a higher portion of naked polymer chains. The low level of α-CD threading onto PEG can explain the formation of less dense nanocrystallites within the hydrogel network. On the contrary, at pH 4, the higher concentration of threaded α-CD resulted in the formation of larger and denser nanocrystallites. Such differences in the supramolecular organization of PEG/α-CD assemblies may explain the higher degree of microporosity observed for materials prepared at pH 2 and the higher amount of large mesopores with a reduced degree of microporosity generated at pH 4. Thus, for instance, the silica material prepared with 30 mg mL^{-1} PEG and 100 mg mL^{-1} α-CD at pH 4 had a pore volume as high as 0.979 cm^3 g^{-1} and contained only 2.7% micropores, whereas the one prepared with the same amount of organics at pH 2 had a pore volume of 0.454 cm^3 g^{-1}, less than half that obtained at pH 4, and contains as much as 32% micropores (Table 3.1).

The explanation for the observed difference in porosity relies on the pH-induced structural changes that occur in the crosslinking densities of the nanocrystallites within the hydrogel network. Considering the low level threading of α-CD at pH 2 and the small size of the resulting nanocrystallites, materials with well-defined

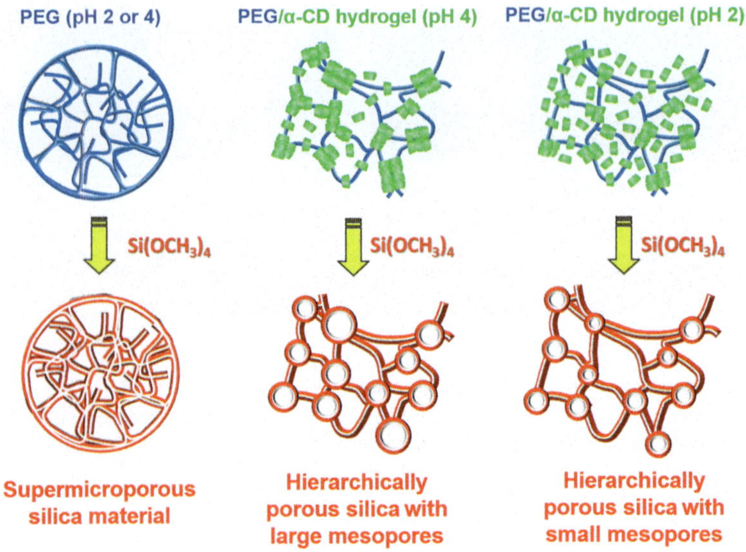

Fig. 3.11 Schematic illustration of the formation of hierarchical porous silica materials from PEG and PEG/α-CD hydrogels at pH 4 and pH 2. The crosslinking density of the nanocrystallites is enhanced with increasing pH from 2 to 4 yielding hierarchically porous silica with large mesopores (4–10 nm). Reprinted with permission of the Royal Society of Chemistry from (Bleta et al. 2014c)

small mesopores can be produced due to the lower, but better-controlled dispersion of the nanocrystallites within the hydrogel. On the other hand, the formation of large mesopores appears to be more favorable under mild acidic conditions (pH 4) due to the higher overall crystallinity of these hydrogels.

3.4 Main Strategies Toward the Synthesis Non-siliceous Mesoporous Oxides

3.4.1 The Template-Directed Colloidal Self-Assembly Approach

Non siliceous mesoporous metal oxides such as TiO_2 and Al_2O_3 are very difficult to prepare *via* a direct templating approach (*i.e.* using the cooperative self-assembly or the true liquid-crystal-phase templating (TLCT) processes, Fig. 3.3). Indeed, compared to silicon alkoxides, the hydrolysis and condensation of transition metal alkoxides are not easy to control precisely in aqueous phase. The resulting materials are generally not robust enough to maintain the mesostructure after the formation of the oxide framework, and they usually exhibit very poor crystallinity and low thermal stability after the template removal (Van Der Voort et al. 2008).

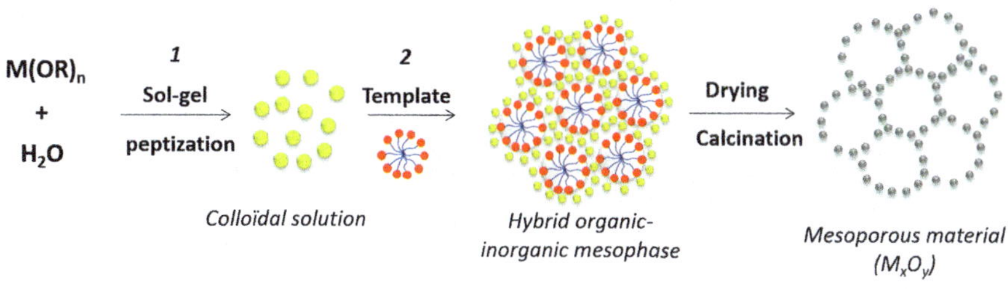

Fig. 3.12 Schematic illustration of the template-directed colloidal self-assembly approach were the colloidal nanoparticles act as building blocks for the construction of the inorganic network around the supramolecular template. Reprinted with permission from (Bleta et al. 2013). Copyright 2013 American Chemical Society

A versatile strategy to overcome the limitations of the cooperative self-assembly or the TLCT processes, and extend the scope of nanostructured porous materials beyond silica, is the so-called template-directed colloidal self-assembly approach (or the nanoparticle route) (Bleta et al. 2010). This approach involves the use of pre-synthesized colloidal particles which have the ability to self-assemble around a soft template (e.g., block copolymer or biopolymer). After drying and calcination, the recovered mesoporous materials are usually robust and present high surface areas, large pore volumes and tunable porosities (Bleta et al. 2010, 2012) (Fig. 3.12). Nanoparticles can be held together either by weak noncovalent forces, such as hydrogen bonding (Boal et al. 2000; Hao and Lian 2000) or by strong covalent bonds *via* different functional groups that can be fixed onto the nanoparticle surface (Patolsky et al. 2002; Lu et al. 2003). So far, the template-directed colloidal self-assembly has been successfully employed for the preparation of a variety of inorganic materials, especially semiconductor metal oxide nanoparticles, with various morphologies that depend on the characteristics of both the template and the colloidal nanocrystals (Yu and Peng 2002; Rajh et al. 1993).

3.4.2 Mesoporous Transition Metal Oxides from RaMeβ-CD/Pluronic P123 Assemblies

3.4.2.1 Interactions between RaMeβ-CD and Pluronic P123

Before using the cyclodextrin/polymer assemblies as template, it is necessary to study their interactions in aqueous phase. Since the pioneering work of Harada *et al.* (Harada and Kamachi 1990) who showed that α-CD can form inclusion complexes with poly(ethylene glycol) (PEG) in aqueous solution to give poly*pseudo*rotaxanes with a necklace-like structure, a range of polymeric guests have been found to form inclusion complexes with cyclodextrins (Harada 1996). However, compared to the native CDs, the interactions between the modified CDs and polymers have been less investigated in the literature. The results reported by

Gaitano et al. (1997), and later supported by other authors (Joseph et al. 2007; Lazzara and Milioto 2008; Nogueiras-Nieto et al. 2009; Dreiss et al. 2009; Tsai et al. 2010), showed that, similarly to native β-CD, the macrocycle of a dimethylated β-CD is able to complex the hydrophobic poly(propylene oxide) blocks of different nonionic triblock copolymers of the poly(ethylene oxide) (PEO)-b-poly(propylene oxide) (PPO)-b-poly(ethylene oxide) (PEO) (*i.e.* Pluronics). On the other hand, spectroscopic and time-resolved small angle neutron scattering measurements performed on mixtures of heptakis (2,6-di-o-methyl)-β-CD and three Pluronics [P123 ($PEO_{20}PPO_{70}PEO_{20}$), P85 ($PEO_{39}PPO_{52}PEO_{39}$) and F127 ($PEO_{107}PPO_{70}$ PEO_{107})] evidenced that the micellar rupture occurs with extremely fast kinetics, thus excluding the possibility of poly*pseudo*rotaxane formation *via* inclusion complexation (Valero et al. 2012). Interestingly, in the case of Pluronic P123, the authors reported a possible restructuration of the micelles toward swollen lamella, with an interlayer spacing that was much higher than the typical values reported in literature with conventional swelling agents (Holmqvist et al. 1998). However, the phenomenon of micellar rupture was shown to be highly sensitive to the substitution degree, nature and position of the modified groups within the cyclodextrin (Joseph et al. 2007; Valero et al. 2012). Thus, in contrast to the micellar rupture observed with the heptakis (2,6-di-o-methyl)-β-CD, micelles remained intact in the presence of other substituted β-cyclodextrin derivatives, such as the 2,3,6-trimethyl-β-CD, 2-hydroxyethyl-β-CD and 2-hydroxypropyl-β-CD (Valero et al. 2012).

The supramolecular assemblies formed between the Pluronic P123 ($PEO_{20}PPO_{70}PEO_{20}$) and the randomly methylated β-cyclodextrin (RaMe β-CD) were investigated with the attempt to be used as templates for the synthesis of porous inorganic solids (Bleta et al. 2013; 2014b). The RaMeβ-CD where methylation occurs at the C2, C3 or C6 positions with statistically 1.8 OH groups modified per glucopyranose unit, can be seen as a versatile structure directing agent. Indeed, this cyclodextrin is highly soluble in water, cheap, non-toxic, commercially available (Uekama and Irie 1987) and surface active (Leclercq et al. 2007). To investigate whether RaMeβ-CD has an impact on the copolymer micellar growth, dynamic light scattering (DLS) measurements were performed at 25 °C. Figure 3.13 a,b displays the DLS plots of the aqueous solutions prepared with 7.8 wt. % P123 and increasing amounts of RaMeβ-CD. In the range of 5–15 mg mL^{-1} RaMeβ-CD, the hydrodynamic radius remained constant ($R_H = 9.1$ nm), but the scattering intensity increased. This result was explained by the preferential location of these oligosaccharides in the water-rich corona of the P123 micelles, where they preferentially interact with the hydrophilic PEO blocks of the copolymer by hydrogen bonding. In contrast, for higher RaMeβ-CD concentrations, the size distribution plots indicated a progressive increase in the micellar hydrodynamic radius from 9.1 to 15.5 nm when the RaMeβ-CD concentration increased from 30 to 80 mg mL^{-1} (Fig. 3.13 b). Such variation was attributed to the ability of this cyclodextrin to act as swelling agent for the P123 micelles. Indeed, RaMeβ-CD with an average degree of substitution (DS) of about 12.6 tends to be more lipophilic than the native β-CD, showing a higher affinity for the hydrophobic core of the block copolymer micelles. Therefore,

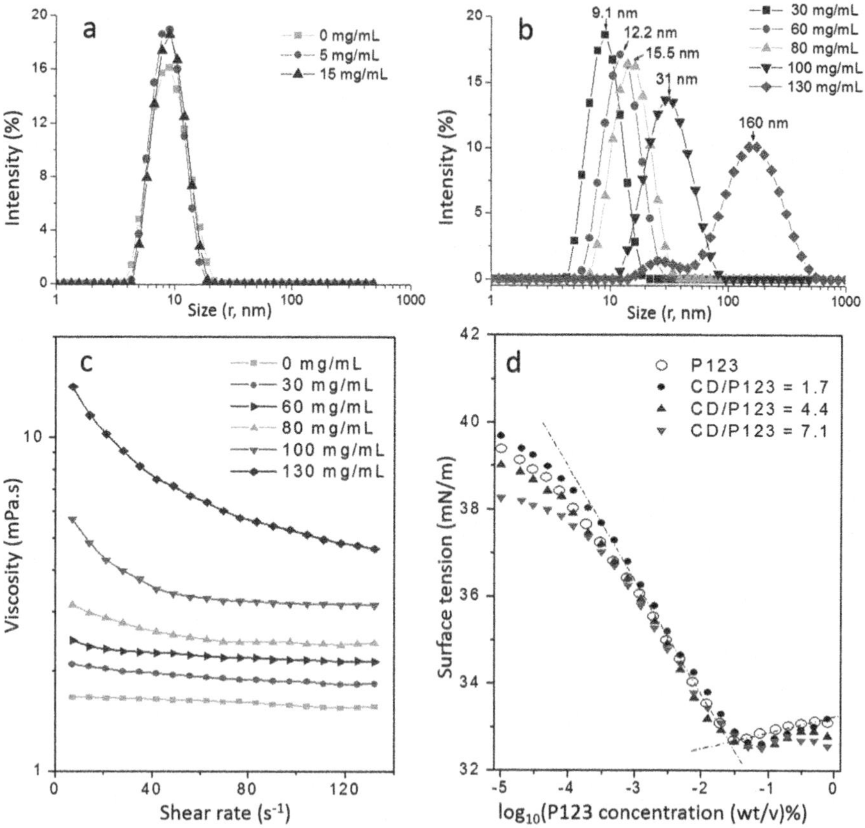

Fig. 3.13 Dynamic light scattering data (**a, b**), apparent viscosity *vs.* shear rate curves (**c**) and surface tension plots (**d**) of Pluronic P123 solutions prepared with various amounts of RaMeβ-CD. All measurements were performed at 25 °C. Adapted with permission from (Bleta et al. 2013). Copyright 2013 American Chemical Society

it was assumed that, in this concentration range, additional hydrophobic interactions occur between the OCH_3 groups of the cyclodextrin and the PPO blocks of the copolymer, resulting in the preferential location of RaMeβ-CD at the PPO-PEO interface layer. The scattering intensity of these assemblies gradually decreased with the CD concentration due to the formation of less well-defined objects with more flexible interfaces. A remarkable increase in the growth of these assemblies was noticed upon addition of larger amounts of RaMeβ-CD. Thus, the hydrodynamic radius abruptly shifted to 31 nm and 160 nm for 100 mg mL^{-1} and 130 mg mL^{-1} RaMeβ-CD respectively. These values far exceed the dimension of the P123 micelles even when the PEO and PPO chains are in a fully extended conformation, thus suggesting modification of the interfacial curvature of the RaMeβ-CD-loaded micelles and a shape transformation from spherical to ellipsoidal. Such shape transformation was found to occur for RaMeβ-CD/P123 molar ratios higher than 7.5 (Bleta et al. 2014b).

The viscosity of the RaMeβ-CD-loaded micelles was also measured. Figure 3.13c shows the apparent viscosity *vs.* shear rate plots recorded at 25 °C. The addition of 30 and 60 mg mL^{-1} RaMeβ-CD induced no significant modification of the rheological behavior of the micellar solutions since the apparent viscosity remained nearly constant (~1.6–2.2 mPa·s) over the entire range of shear rates (0–130 s^{-1}). Conversely, the addition of 80 and 100 mg mL^{-1} RaMeβ-CD induced a shear-thinning behavior (*i.e.*, the viscosity decreased with increasing the shear rate), as well as a more pronounced increase in viscosity to ~2.4 and ~3.2 mPa·s, respectively, as measured at 130 s^{-1}. Such a particular rheological behavior is usually observed in thixotropic systems comprised of particles with strong shape anisotropy such as rods, discs or platelet shapes (Bleta et al. 2011). This behavior was therefore consistent with the formation of ellipsoidal structures, in agreement with the DLS data.

Finally information about the impact of RaMeβ-CD on the micellization of Pluronic P123 was provided by surface tension measurements. If the hydrophobic interactions between the PPO chains, which are the driving force for the micellization, are screened by the presence of the cyclodextrin, this may lead to a disruption of the copolymer micelles and an increase of the Critical Micelle Concentration (CMC) (Bernat et al. 2008; Mahata et al. 2010). Figure 3.13 d shows the surface tension plots recorded at 25 °C for Pluronic P123 solutions prepared without and with RaMeβ-CD (CD/P123 molar ratios in the range of 1.7–7.1). The surface tension of the copolymer was not significantly affected by the addition of RaMeβ-CD whatever the molar ratio of RaMeβ-CD to P123 used, giving further evidence that the interactions between Pluronic and RaMeβ-CD are relatively weak and do not affect significantly the micellization process.

3.4.2.2 Mesoporous γ-Al_2O_3 Materials with Tunable Porosity

When these supramolecular coassemblies were utilized as templates, mesoporous γ-alumina (γ-Al_2O_3) with a crystalline framework, large pore size and high surface area were obtained (Bleta et al. 2013). The synthetic procedure is schematized in Fig. 3.14. Briefly, in a first step, boehmite nanoparticles were synthesized in aqueous solution ($H_2O/Al \approx 100$) using aluminum tri-sec-butoxide $Al(OBu)_3$ (denoted ASB) and nitric acid as the metal oxide precursor and peptizing agent ($HNO_3/Al = 0.07$), respectively. In a second step, boehmite nanoparticles were allowed to self-assemble around the RaMeβ-CD/P123 assemblies at room temperature. The P123 concentration was fixed at 7.8% (PEO/Al = 1) whereas the RaMeβ-CD concentration was varied from 30 to 130 mg mL^{-1} (*i.e.* RaMeβ-CD/P123 molar ratio in the range of 1.7 to 7.1). After drying at 60 °C for 48 h, xerogels were calcined at 500 °C to remove the organic template and allow the transition from boehmite (AlO(OH)) to γ-Al_2O_3, which is assumed to occur at ~380 °C, as evidenced by thermal analysis and confirmed by XRD measurements.

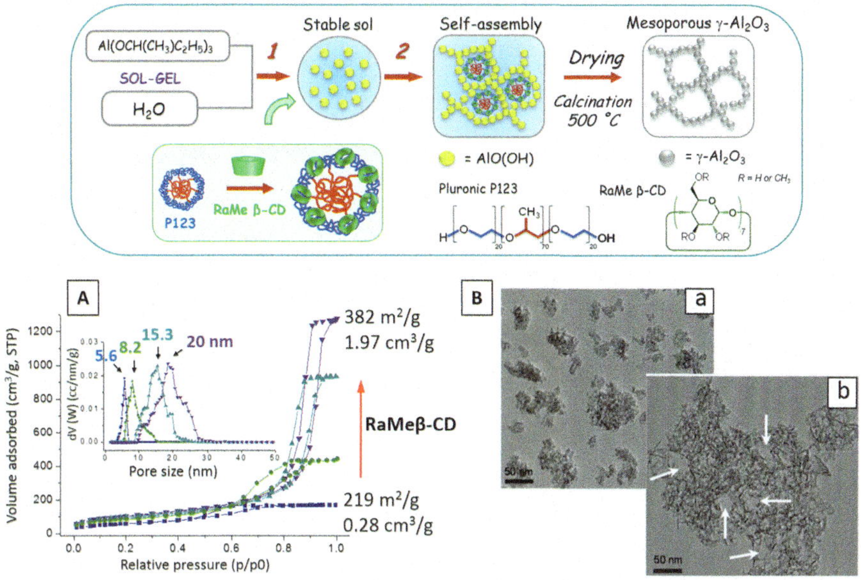

Fig. 3.14 Top: Schematic illustration of the template-directed synthesis of mesoporous γ-Al_2O_3 with controlled porosity where boehmite (AlO(OH)) colloids act as building blocks for the construction of the inorganic network. (A) N_2-adsorption isotherms and corresponding pore size distributions (inset) for the mesoporous γ-Al_2O_3 prepared without template, with Pluronic P123 (PEO/Al = 1) and with increasing amounts of RaMeβ-CD (30–130 mg mL^{-1}). Note that the pore size can be adjusted from 5.6 to 20 nm by addition of RaMeβ-CD to the micellar solution of Pluronic. (B) Transmission electron microscopy (TEM) images for mesoporous γ-Al_2O_3 prepared without template (**a**) and with RaMeβ-CD-loaded micelles (80 mg mL^{-1}) (**b**). Adapted with permission from (Bleta et al. 2013). Copyright 2013 American Chemical Society

To obtain information about the textural characteristics of the calcined γ-Al_2O_3, N_2 adsorption-desorption analyses were carried out. From Fig. 3.14a, it can be seen that all isotherms presented a distinct H1 hysteresis loop characteristic of mesoporous materials. The control sample, prepared without template, presented a capillary condensation step that started at a relative pressure (P/P_0) of about 0.4 indicating the presence of small mesopores. The corresponding pore size distribution (PSD) plot was relatively narrow and centered at 5.6 nm. The formation of such small pores was attributed to the assembly of several crystallites in compact rearrangements with "card-pack" microstructures. Upon addition of the block copolymer (P1-Al-T500 sample), a steep rise in the nitrogen uptake was observed at relative pressures (P/P_0) higher than 0.8 indicating the formation of large mesopores. Accordingly, the corresponding pore size distribution plot indicated a dramatic increase in the pore size (from 5.6 to 14.8 nm) and pore volume (from 0.28 to 1.37 $cm^3\ g^{-1}$) due to the formation of big micelles acting as soft templates. Notably, the textural characteristics were further improved upon addition of increasing amounts of RaMeβ-CD. Thus, the pore size increased from 14.8 to 19.7 nm and the pore volume from 1.45 to 1.97 $cm^3\ g^{-1}$ (Table 3.2). The most striking textural

Table 3.2 Textural characteristics of mesoporous γ-Al_2O_3 calcined at 500 °C

Sample	S_{BET}(m^2 g^{-1})	PV(cm^3 g^{-1})	S_{cum}(m^2 g^{-1})	V_{cum}(cm^3 g^{-1})	PS(nm)
Al-T500	219	0.28	227	0.26	5.6
P1-Al-T500	357	1.37	387	1.35	14.8
P1-CD30-Al-T500	354	1.45	387	1.43	14.9
P1-CD60-Al-T500	356	1.62	372	1.60	16.3
P1-CD80-Al-T500	382	1.97	427	1.94	19.3
P1-CD130-Al-T500	360	1.66	373	1.63	10.5 and 19.7

P = [PEO]/[Al], CD = RaMeβ-CD concentration in the sol (mg mL^{-1}), T = calcination temperature, S_{BET} = BET specific surface area determined in the relative pressure range 0.1–0.25, PV = pore volume calculated from adsorbed volume at $P/P_o = 0.995$, S_{cum}, V_{cum}, PS = cumulative surface area, cumulative volume and pore size resulting from non-local density functional theory (NLDFT) calculations. Reprinted with permission from (Bleta et al. 2013). Copyright 2013 American Chemical Society

characteristics were obtained for the material prepared with 80 mg mL^{-1} RaMeβ-CD (average pore size = 19.3 nm, pore volume = 1.97 cm^3 g^{-1} and specific surface area = 382 m^2 g^{-1}). Such an enhancement in the porosity can be directly attributed to the swelling effect of RaMeβ-CD, in line with the trend observed from DLS data.

The representative transmission electron microscopy (TEM) images (Fig. 3.14b) indicated that contrary to the non-templated alumina which was globally dense and made of aggregated particles, the templated material showed a well-defined fiberlike morphology, indicating the important role of the template in restructuring the particle network. Interestingly, in addition to the fiberlike morphology, several voids with an average diameter of ~20 nm, similar to the pore size determined from N_2-adsorption measurements, were also present at a very high yield throughout the nanoparticle network (see arrows). These results indicated that, after the thermal treatment at 500 °C, the material successfully adopted some characteristics of the supramolecular template. Hence, the void space formed between the nanoparticles may be seen as a solid replica of the original swollen micelles formed by the coassembly of the copolymer and the cyclodextrin.

3.4.2.3 Mesoporous TiO_2 Materials with Tunable Porosity and Crystallinity

The template-directed colloidal self-assembly approach was successfully extended to the synthesis of mesoporous titanium dioxide (TiO_2, titania) (Lannoy et al. 2014; Bleta et al. 2014a, b, c). TiO_2 has been applied as one of the most promising photocatalysts for the removal of industrial organic pollutants from water and air (Zhao et al. 2004; Gomathi Devi and Kavitha 2013), as well as for the photocatalytic water splitting for hydrogen production (Fujishima and Honda 1972). TiO_2 commonly crystallizes in three polymorphic forms, *i.e.* anatase (tetragonal, I41/amd), brookite (orthorhombic, Pbca) and rutile (tetragonal, P42/mnm). Bulk rutile is the only thermodynamically stable phase, while bulk anatase and bulk brookite are metastable (Zhang and Banfield 1998, 2000). Nevertheless, under controlled conditions, anatase and brookite can be thermodynamically stabilized when the particle size is below 11 nm for the former and between 11 and 35 nm for the latter (Zhang and Banfield 2000). Among the three polymorphs, anatase and rutile have received the greatest attention due to the facility of their synthesis. Anatase has a band gap of 3.2 eV with the absorption edge at 386 nm which lies in the near UV range, whereas rutile has a lower band gap of 3.02 eV with the adsorption edge in the visible range at 416 nm. The anatase polymorph is usually reported to be more active than rutile (Pillai et al. 2007), mainly because of the fast electron-hole recombination in the latter which results from its lower band gap (Periyat et al. 2008).

It is today well-accepted that the photocatalytic reactions mainly take place on the surface of the irradiated semiconductor (Linsebigler et al. 1995). Consequently, in addition to the effect of the crystal phase composition mentioned above, other factors such as the crystallite size, the surface area, the pore volume, the orientation of the active faces and the adsorption properties of the pollutant are also likely to affect the photocatalytic activity (Ovenstone 2001; Yang et al. 2008; Chen and Caruso 2013; Nursam et al. 2015; Wang et al. 2013). Therefore, the impact of the RaMeβ-CD-P123 micelles on the structural, textural and morphological characteristics of titania was investigated. In a first step, a translucent hydrosol made-up of crystalline TiO_2 nanoparticles was synthesized in water/isopropanol solution ($H_2O/Ti = 88$) by a sol-gel method using titanium isopropoxide ($Ti(OiPr)_4$) as inorganic precursor and nitric acid as peptizing agent ($HNO_3/Ti = 0.2$). In a second step, the RaMeβ-CD-P123 supramolecular assemblies were used as templates to direct the self-assembly of the pre-synthesized TiO_2 nanoparticles (Fig. 3.15). This second step of synthesis was performed at 25 °C because at this temperature, the RaMeβ-CD-P123 solutions present the lowest viscosity, which facilitates the structuration of the nanoparticles around the supramolecular template. After drying, the recovered xerogels were calcined at 500 °C to remove the template and allow further crystallization of the material.

To gain information about the impact of the template on the textural characteristics of titania, N_2-adsorption analyses were performed. All calcined samples display type IV isotherms which are typical features of mesoporous materials (Fig. 3.15A). The control sol-gel titania presented a capillary condensation step that started at a relative pressure (P/P_0) of about 0.4, indicating the presence of small mesopores with a diameter of 5.3 nm ascribed to the holes formed between the close packed crystallites. The addition of copolymer (P27 sample) strongly, but predictably, improved the textural characteristics of the material, showing an abrupt increase of the pore size to 9.2 nm (Table 3.3). Interestingly, when the cyclodextrin was added to the block copolymer solution, the textural characteristics were further improved. Therefore, for RaMeβ-CD/Ti molar ratios comprised between 0.046 and 0.198, the pore size increased progressively from 9.4 to 12.1 nm and the pore volume from 0.25 to 0.35 $cm^3\ g^{-1}$. No notable evolution occurred for RaMeβ-CD/Ti molar ratios higher than 0.198 (Table 3.3) for which the isotherms became quite different in the

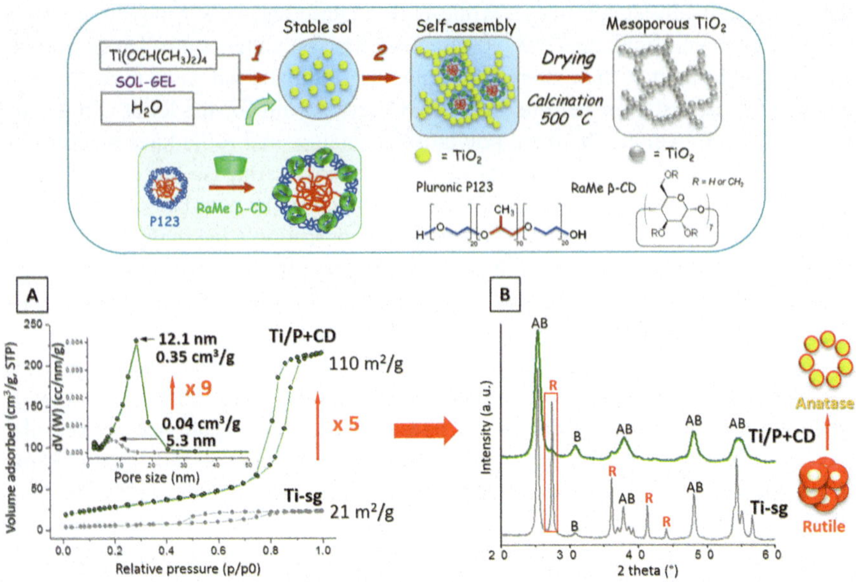

Fig. 3.15 Top: Schematic illustration of the template-directed synthesis of mesoporous crystalline titania where TiO_2 colloids act as building blocks for the construction of a nanostructured framework around the organic template. (**A**) N_2 adsorption isotherms and pore size distributions (inset) for the sol-gel TiO_2 prepared without template, and nanostructured TiO_2 prepared with RaMeβ-CD-P123 assemblies. (**B**) X-ray diffraction (XRD) patterns of the corresponding materials. "A", "B" and "R" denote the anatase, brookite and rutile phases respectively. Adapted with permission of the Royal Society of Chemistry from (Lannoy et al. 2014)

region of relative pressures higher than 0.9 where the N_2 adsorption continued to increase, indicating also the presence of some macropores.

Before any thermal treatment, the sol-gel TiO_2 contained 68% anatase (A) (JCPDS card no. 00–021-1272) and 32% brookite (B) (JCPDS card no. 01–076-1934). The crystallite sizes determined from the Scherrer formula were 6.6 ± 0.8 nm (A) and 5.3 ± 0.6 nm (B). Such a small particle size was explained by the high hydrolysis ratio employed which favors fast nucleation rates producing small and well-crystallized nanoparticles. However, upon calcination at 500 °C, the small particles agglomerated and the increased degree of nanoparticle packing facilitated the phase transformation. Therefore, from the diffraction diagram of the control sol-gel TiO_2 (Fig. 3.15B), one can note the appearance of an intense sharp peak at $2\theta = 27.4°$ corresponding to the (110) plane of the rutile (R) (JCPDS card no. 00–034-0180), arising from the transformation of both anatase and brookite during calcination. The contents of anatase and brookite, determined from the Rietveld refinement, dropped to ~35% and ~27% respectively, while ~38% rutile formed as a result of the sintering (Table 3.3). Meanwhile, the size of these three polymorphs grew to 36 nm (A), 19 nm (B) and 60 nm (R) as the phase transformation progressed. When the Pluronic alone was utilized as template (P27 sample), the rutile reflections

Table 3.3 Structural and textural characteristics of the different mesoporous TiO_2 prepared through template-directed colloidal self-assembly

Sample			Anatase		Brookite		Rutile		N_2 adsorption		
			cs[c] (nm)	ct[d] (%)	cs[c] (nm)	ct[d] (%)	cs[c] (nm)	ct[d] (%)	S_{BET}[e] ($m^2\ g^{-1}$)	pv[f] ($cm^3\ g^{-1}$)	ps[g] (nm)
	P + RB/Ti[a]	RB/P[b]	Effect of addition of P123-RaMeβ-CD assemblies								
Sol-gel TiO_2	–		36.4±2.9	35.3±2.8	18.7±1.5	26.5±2.1	60.4±4.8	38.2±3.1	21±1	0.03±0.00	5.3±1.1
P27	0.027	0	17.2±1.4	54.5±4.4	13.8±1.1	42.1±3.4	20.8±1.7	3.4±0.3	89±4	0.24±0.01	9.2±1.8
P27RB46	0.073	1.7	15.5±1.2	51.5±4.1	11.6±0.9	46.0±3.7	16.6±1.3	2.5±0.2	109±5	0.25±0.01	9.4±1.9
P27RB122	0.149	4.5	13.6±1.1	50.1±4.0	8.9±0.7	48.6±3.9	13.4±1.1	1.3±0.1	112±6	0.29±0.01	10.3±2.1
P27RB198	0.225	7.4	10.8±0.9	47.7±3.8	7.6±0.6	51.5±4.1	10.7±0.9	0.8±0.1	110±6	0.35±0.02	12.1±2.4
P27RB305	0.332	11.4	9.9±0.8	45.4±3.6	6.7±0.5	54.3±4.3	9.0±0.7	0.3±0.1	96±5	0.36±0.02	12.4±2.5
			Effect of the calcination temperature (P27RB198 sample)								
T400	0.225	7.4	9.0±0.7	90.4±7.2	4.4±0.4	9.6±0.8	–	–	158±8	0.45±0.02	9.3±1.9
T500	0.225	7.4	10.8±0.9	47.7±3.8	7.6±0.6	51.5±4.1	10.7±0.9	0.8±0.1	110±6	0.35±0.02	12.1±2.4
T600	0.225	7.4	17.7±1.4	48.1±3.8	8.5±0.7	35.2±2.8	23.6±1.9	16.7±1.3	58±3	0.35±0.02	21.0±4.2
T700	0.225	7.4	35.2±2.8	37.6±3.0	–	–	62.5±5.0	62.4±5.0	16±1	0.14±0.01	–
T800	0.225	7.4	–	–	–	–	82.8±6.6	100	–	–	–

[a](Pluronic P123 + RaMeβ-CD)/Ti molar ratio in the sol, [b]RaMeβ-CD/P123 molar ratio in the sol, [c]crystallite size calculated from the Scherrer formula, [d]polymorph content determined from Rietveld refinements, [e]specific surface area determined by BET method in the relative pressure range of 0.1–0.25, [f]pore volume computed by non-local density functional theory (NLDFT) calculations, [g]pore size determined by NLDFT. Reprinted with permission of the Royal Society of Chemistry from (Lannoy et al. 2014)

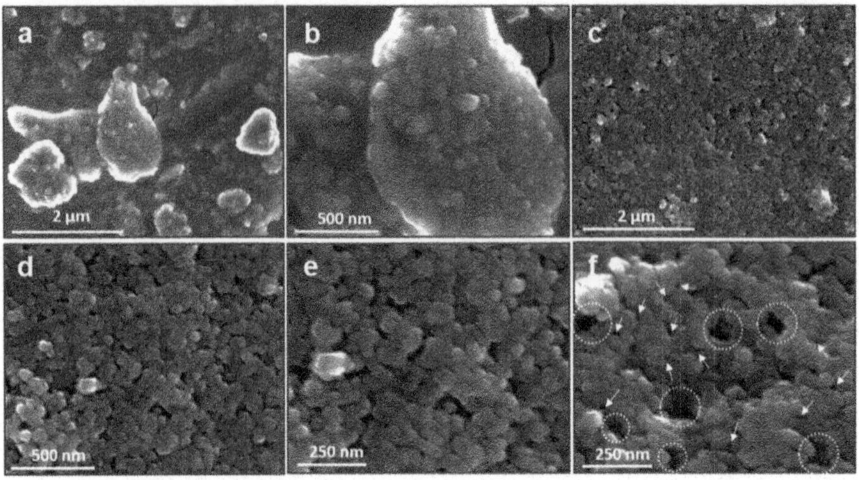

Fig. 3.16 Representative field emission scanning electron microscopy (FE-SEM) images taken at different magnifications for sol-gel titania prepared without template (a and b) and with P123-RaMeβ-CD assemblies (P27RB198 sample) (c-f). Note the presence of mesopores with an average diameter of 10–20 nm (see arrows) and macropores with an average diameter of 60–100 nm (see dotted circles). Reprinted with permission of the Royal Society of Chemistry from (Lannoy et al. 2014)

remarkably decreased in intensity indicating a delay in the phase transformation. Interestingly, this phenomenon became even more pronounced upon addition of increasing amounts of RaMeβ-CD. Indeed, the diffraction peaks of anatase and brookite became broader indicating smaller crystallites, while more brookite formed and rutile polymorph almost disappeared. Thus, for the sample prepared with a CD/Ti molar ratio of 0.198, the size of the crystallites decreased to 8–11 nm and the rutile content became negligible with respect to the contents of anatase (48%) and brookite (52%).

From the representative FE-SEM images of the materials prepared without and with template (Fig. 3.16), the effect of the supramolecular assemblies on the morphology of the network was clearly visualized. Therefore, the sol-gel titania prepared without template was comprised of rounded particles densely packed into large aggregates with no regular shape and very low interparticle porosity (Fig. 3.16a and b). In contrast, the material prepared using the supramolecular assemblies (P27RB198 sample) showed uniform particles with spherical shape indicating the important role of the template in restructuring the particle network (Fig. 3.16c-f). Moreover, several mesopores were also noticed at a high yield with an average diameter of 10–20 nm (see arrows) as well as some macropores with an average diameter of 60–100 nm (see dotted circles), in accordance with the shape of the N_2-adsorption isotherm at high relative pressures. This indicated that the material had adopted some characteristics of the supramolecular template maintaining a nano-structured network even after calcination at 500 °C.

3.4.3 *Mesoporous Nanocomposites from Cyclodextrins or RaMeβ-CD/Pluronic P123 Assemblies*

3.4.3.1 Using Native and Modified Cyclodextrins as Structure Directing Agents

The binding affinity of cyclodextrin on titania nanoparticles is highly dependent on their nature, natives or modified. Thus, the adsorption isotherms have been shown to follow the Langmuir model, and adsorption capacities as high as 33 μmol g^{-1} were obtained with the native β-CD compared to 15.2 μmol g^{-1} with the permethylated 2-O-methyl-β-CD (DS = 4) and 0 μmol g^{-1} with the 2,6-di-O-methyl-β-CD (DS = 14) (Zhang et al. 2013). Such adsorption was proposed to occur predominantly through the -OH groups located at the secondary ring face of β-CD, which also caused the selective photodegradation of a series of bisphenols by preferential inclusion complexation with the primary ring side.

In the case of the native α-CD, β-CD and γ-CD, the numerous hydroxyl groups, located on both the narrow and the wider ring faces, may favor the interaction of the macrocycle with the surface -OH groups of titania. Conversely, the 2-hydroxypropyl β-CD (HPβ-CD) and the randomly methylated β-CD (RaMeβ-CD), possessing less surface hydroxyl groups on the ring, are likely to form less hydrogen bonds. Additionally, the adsorption capacity of these modified oligosaccharides may be hindered by the steric constraint created by the 2-hydroxyproyl and methoxy groups, thus leaving less place for the interaction with the non-substituted hydroxyl groups.

The changes occurring in the structure of cyclodextrin assemblies after interaction with titania colloids and after solvent evaporation were followed by XRD measurements. From Fig. 3.17, it can be seen that the native α-CD and β-CD, before being introduced to the titania hydrosol, present several sharp diffraction lines characteristic of their cage-type crystalline microstructure (Harada et al. 2009). After interaction with titania colloids, the disappearance of the most intense reflections observed at 12.2°, 14.3° and 21.6° with the neat α-CD and at 9.0° and 12.5° with the neat β-CD suggested the disruption of the cage-type microstructure due to the adsorption of these oligosaccharides on the titania surface. On the other hand, the neat β-CD derivatives (HPβ-CD and RaMeβ-CD) presented only two broad peaks due to their amorphous character. Interestingly, from the patterns of the hybrid HPβ-CD/TiO_2 and RaMeβ-CD/TiO_2 materials, it was noticed that these reflections were still intense, indicating weaker interactions with the titania surface. Similar results were obtained with carbon materials which presented a significantly lower adsorption capacity towards HPβ-CD and RaMeβ-CD compared to the native cyclodextrins, but a higher capacity to improve the dispersion of carbon particles in water (Okumura et al. 2001).

The substitution of hydroxyl groups by a relatively large number of methoxy (-OCH_3) or 2-hydroxypropyl (-$OCH_2CH(CH_3)OH$) groups is one of the main factors that affects the physicochemical properties of the cyclodextrins, implying changes in both the solubility profile (due to the disruption of intermolecular hydrogen

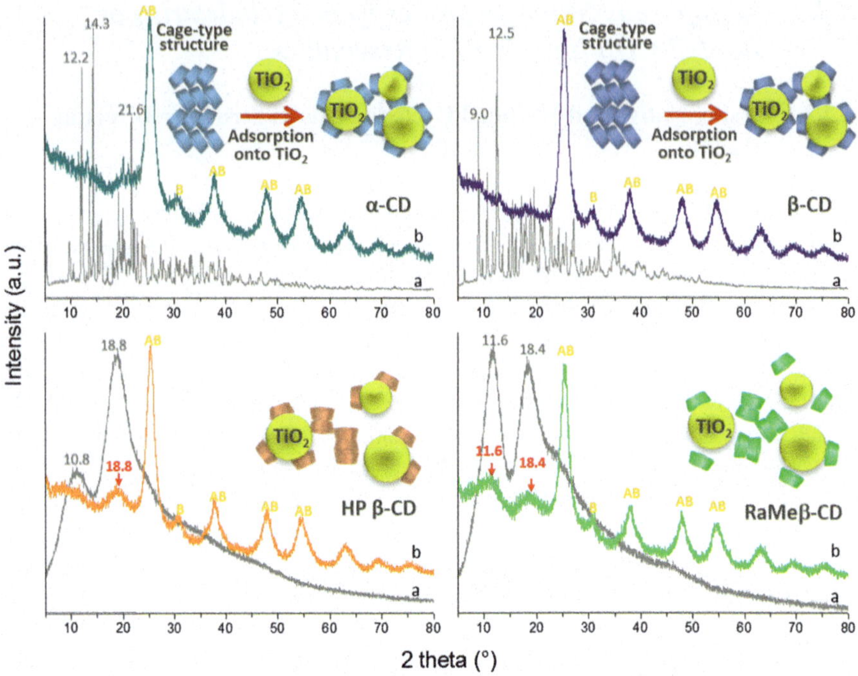

Fig. 3.17 X ray diffraction patterns of the neat cyclodextrins (a) and corresponding cyclodextrin/TiO_2 hybrid xerogels (b) prepared with aCD/Ti molar ratio of 0.076 for α-CD, HPβ-CD, and RaMeβ-CD and a molar ratio of 0.032 for β-CD. The xerogels were dried at 60 °C. Note that after interaction with titania colloids, the most intense reflections observed with the neat α-CD and β-CD disappear suggesting the disruption of the cage-type microstructure due to the adsorption of these cyclodextrins on the TiO_2 surface. Reprinted with permission from (Bleta et al. 2014a). Copyright 2014 American Chemical Society

bonds) and the interfacial behavior (due to the presence of more marked hydrophobic and hydrophilic microenvironments). Thus, the surface tension data shown in Fig. 3.18A indicate that, in contrast to the native cyclodextrins, which are almost not surface-actives, HPβ-CD and RaMeβ-CD present surface tension values of 59.9 and 56.8 mN m^{-1} respectively at 38 mM (concentration utilized for the preparation of the mesoporous titania materials). The slightly lower surface activity of HPβ-CD compared to RaMeβ-CD may be explained by the lower lipophilic character of the 2-hydroxypropyl groups compared to the methoxy ones. In this sense, the relatively higher surface activity of RaMeβ-CD may offer a means to reduce the surface energy of titania nanocrystals, thus facilitating their movement which is critical for their self-assembly. Significant modifications were also observed on the textural characteristics of the photocatalysts (Fig. 3.18B). Thus, using α-CD, the specific surface area increased from 21 to 68 m^2 g^{-1}, the pore volume from 0.03 to 0.13 cm^3 g^{-1} and the pore size from 5.3 to 7.0 nm. Similar results were also obtained with β-CD and γ-CD. Interestingly, the porosity was further enhanced when the modified cyclodextrins were used as structure directing agents and the most relevant

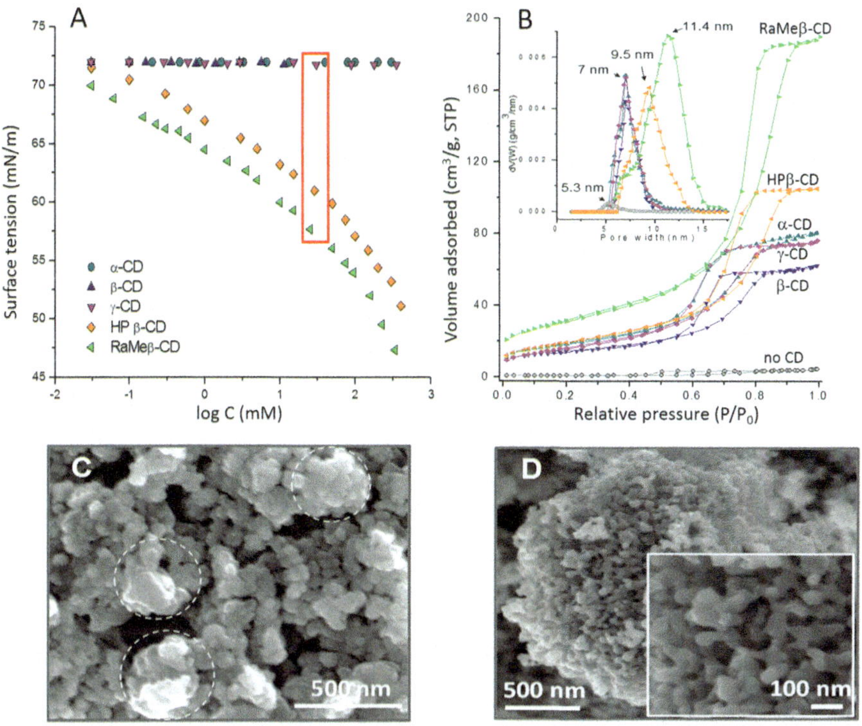

Fig. 3.18 (A) Surface tension plots of various cyclodextrins in water at 25 °C. The solid rectangle shows the domain of cyclodextrin concentrations under which TiO_2 materials have been prepared. (**B**) N_2-adsorption isotherms and corresponding pore size distributions (inset) of TiO_2 prepared without template and with various CDs. Representative field emission scanning electron microscopy (FE-SEM) images for sol-gel TiO_2 prepared with α-CD (C) and RaMeβ-CD (D). Note the formation of highly interconnected pore network with the methylated CD. Adapted with permission from (Bleta et al. 2014a). Copyright 2014 American Chemical Society

textural characteristics were obtained for titania prepared with RaMeβ-CD presenting a specific surface are of 115 $m^2\ g^{-1}$, a pore volume of 0.3 $cm^3\ g^{-1}$ and a pore size of 11.4 nm.

Moreover, from the representative field emission scanning electron microscopy (FE-SEM) images, it was noticed that the cyclodextrin had also an impact on the morphology of titania catalysts (Fig. 3.18C, D). Therefore, compared to native β-CD which produced some local agglomeration of the nanoparticles (see white circles in Fig. 3.18C), RaMeβ-CD gave rise to a highly interconnected pore network (Fig. 3.18D).

The overall picture emerging from these experimental data is that among the five cyclodextrins investigated, the RaMeβ-CD presents the best combination of surface active properties and weak CD-CD intermolecular interactions to efficiently direct the self-assembly of titania nanoparticles in a uniform network. The mechanism suggested for this self-assembly is shown in Fig. 3.19. In the presence of the native

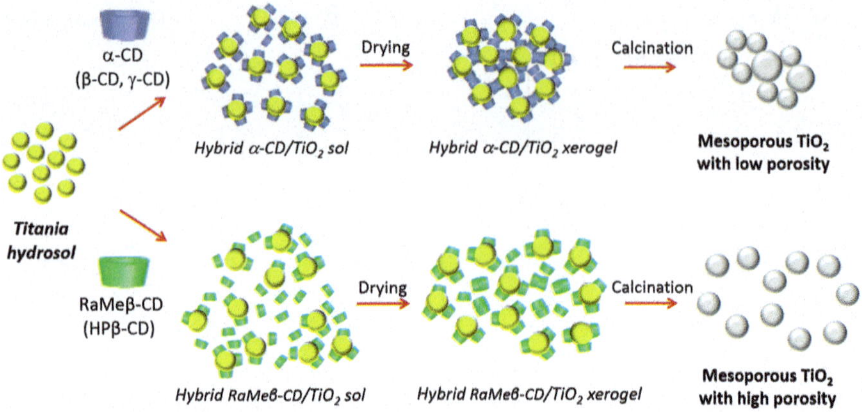

Fig. 3.19 Schematic illustration of the mechanism proposed for the self-assembly of titania colloids in the presence of native cyclodextrins (α-CD, β-CD, and γ-CD) and modified cyclodextrins (RaMeβ-CD and HPβ-CD). Reprinted with permission from (Bleta et al. 2014a). Copyright 2014 American Chemical Society

cyclodextrins (α-CD, β-CD and γ-CD), the colloid interface is rather rough (due to its high surface energy) and the intermolecular interactions are stronger (due to the numerous -OH groups). By consequence, the interactions between adsorbed cyclodextrins should favor the local agglomeration of titania nanoparticles during solvent evaporation, thus resulting in a less porous network. By contrast, in the presence of HPβ-CD and RaMeβ-CD, smoother interfaces are created by the lipophilic groups present in the macrocycle, leading to a decrease in the surface energy of titania nanocrystals and a reorganization of the colloids in a more homogeneous and porous network with small sizes and a fine morphology.

3.4.3.2 Mesoporous UV-Light Responsive TiO_2 Photocatalysts

To evaluate the photocatalytic activity of the titania materials prepared using the different cyclodextrins as structure directing agents, phenoxyacetic acid (PAA), a toxic herbicide, was chosen as probe molecule for degradation under UV light (360 nm). PAA is a parent molecule of the well-known 2,4-dichlorophenoxyacetic acid (2,4-D) and 2,4,5-trichlorophenoxyacetic acid (2,4,5-T) herbicides (Singh et al. 2007). The PAA photodegradation rates obtained with the different photocatalysts after 7 hours of exposure under UV-light illumination (360 nm) (Fig. 3.20) show that the nature of the cyclodextrin has an impact on the photocatalytic activity of titania. Thus, TiO_2 materials prepared from α-CD, β-CD and γ-CD were all photoactive under UV irradiation and gave a PAA degradation rate in the range of 58–64%, which was almost 45% higher than that of the sol-gel TiO_2 (43%). Further enhancement in the photocatalytic activity was noticed with the material prepared from HPβ-CD showing an intermediate PAA degradation rate of 72%

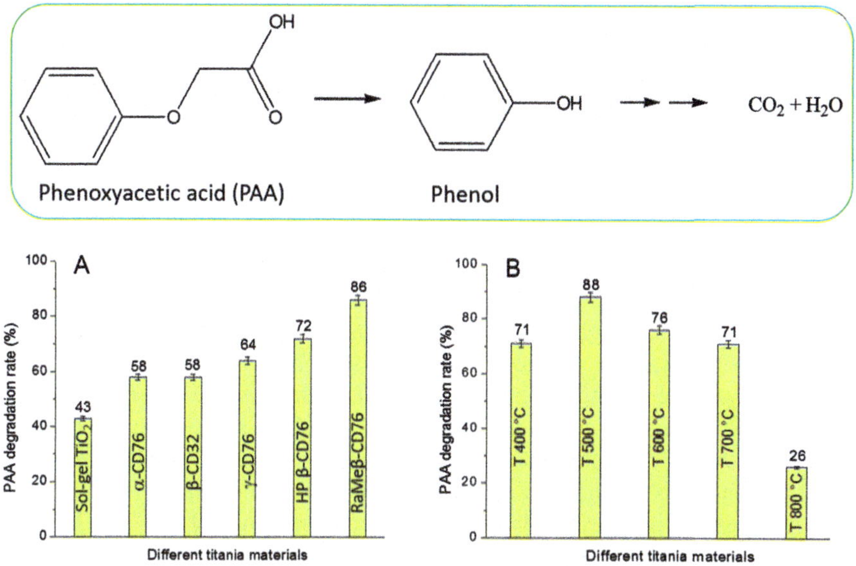

Fig. 3.20 Photocatalytic degradation rate of the phenoxyacetic acid (PAA) under UV-light irradiation after 7 h on titania materials prepared (**A**) with various cyclodextrins at a fixed CD/Ti molar ratio: 0.076 for α-CD, γ-CD, HPβ-CD, and RaMeβ-CD, and 0.032 for β-CD. Adapted with permission from (Bleta et al. 2014a). Copyright 2014 American Chemical Society. (**B**) Effect of calcination temperature on the photocatalytic activity of TiO_2 materials prepared from RaMeβ-CD/P123 assemblies. Adapted with permission of the Royal Society of Chemistry from (Lannoy et al. 2014)

between γ-CD (66%) and RaMeβ-CD (86%) (Fig. 3.20A). Moreover, the photocatalytic activity increased progressively upon addition of increasing amounts of RaMeβ-CD and a maximum degradation rate of 86% was reached for a RaMeβ-CD/Ti molar ratio of 0.076, which was twice that of the sol-gel TiO_2. Beyond this optimum, a gradual decrease in the photocatalytic activity to 77% and 68% was noticed.

It is worth emphasizing that the combination of good structural and textural properties requires the choice of a correct temperature of calcination. For instance, the TiO_2 material prepared from the RaMeβ-CD/P123 assemblies and calcined at 400 °C presented better textural characteristics ($S_{BET} = 158\ m^2\ g^{-1}$; $pv = 0.45\ cm^3\ g^{-1}$; ps = 9.3 nm) compared with the material calcined at 500 °C ($S_{BET} = 110\ m^2\ g^{-1}$; $pv = 0.35\ cm^3\ g^{-1}$; ps = 12.1 nm) (Table 3.3), but it gave lower PAA degradation rates due to the high density of defects at 400 °C (Fig. 3.20B) which should favor the electron-hole recombination. Conversely, above 500 °C, the catalyst was well-crystallized but, the size of the crystallites increased very rapidly (62.5 nm for rutile at 700 °C) resulting in a degradation of the textural characteristics ($S_{BET} = 16\ m^2\ g^{-1}$; $pv = 0.14\ cm^3\ g^{-1}$ at 700 °C) (Table 3.3). So, the temperature of 500 °C represented the conditions under which an optimal balance of high pore volumes, large surface areas and high crystallinity may be obtained to achieve the optimum photocatalyst for the photodegradation of PAA in water.

The origin of the enhancement of the photocatalytic activity was related to a combined effect of improved textural characteristics and controlled crystalline properties. The high surface areas, which may be correlated with the small size of the crystallites (8–16 nm), should provide a large number of adsorption sites and active centers surrounding the electron-hole pairs, thus facilitating the first step of the photocatalytic reaction. On the other hand, the high pore volumes may allow for more PAA to be adsorbed on the internal surface of the pores, thus improving the diffusion of the substrate to the adsorption sites during the photocatalytic process. Finally, the low density of crystalline defects obtained for the material calcined at 500 °C may produce less grain boundaries and thus, a larger amount of charge carriers should reach the surface of the crystal to initiate the redox reactions.

Taken together, these results showed that the photocatalytic activity of TiO_2 materials may be correlated with their structural and textural characteristics, both of which depend on the concentration and chemical nature of the cyclodextrin employed. Indeed, the strongest effects in the pore volume, surface area, phase composition and photocatalytic activity were observed when RaMeβ-CD was used as structure directing agent and for a calcination temperature of 500 °C. Overall, it was shown that all the above parameters are interlinked and a harmonization between them is necessary to obtain an efficient photocatalyst.

3.4.3.3 Mesoporous Visible-Light Responsive au/TiO_2 Photocatalysts

In recent years, Au/TiO_2 composites have attracted much interest as efficient plasmonic photocatalysts owing to the ability of Au nanoparticles to absorb light in the visible region and TiO_2 to efficiently separate the photogenerated electrons and holes at the metal-semiconductor interface (Kowalska et al. 2009; Wang and Caruso 2011; Wang et al. 2012; Naya et al. 2014). The redox ability of gold nanoparticles actually originates from their localized surface plasmon resonance (LSPR) arising from the collective oscillations of electrons on the nanoparticle surface under light irradiation (Lin et al. 2015) and depends both on particle size and shape (Park et al. 2007). Highly active visible-light Au/TiO_2 photocatalysts were prepared by taking advantage of the ability of cyclodextrins to direct the self-assembly of TiO_2 colloids in a porous network over which Au nanoparticles can be uniformly dispersed (Lannoy et al. 2017). The overall procedure employed for the preparation of Au-modified TiO_2 is schematically illustrated in Fig. 3.21. In a first step, a stable sol made-up of TiO_2 nanoparticles crystallized in anatase (70%) and brookite (30%) is synthesized in water/isopropanol solution (H_2O/Ti = 88) by sol-gel process using titanium isopropoxide ($Ti(OiPr)_4$) as inorganic precursor and nitric acid as peptizing agent (HNO_3/Ti = 0.2) (Bleta et al. 2010). In a second step, the gold salt precursor (chloroauric acid, $HAuCl_4$) was introduced in the titania sol together with various cyclodextrins (α-CD, β-CD, γ-CD, RaMeβ-CD or HPβ-CD). After drying at 60 °C and calcination at 500 °C, a composite material made-up of metallic Au nanoparticles (2.5 wt%,) dispersed on the TiO_2 surface was recovered. In this approach, the cyclodextrin had a dual role, *i.e.* it acted as a structure directing agent to guide the

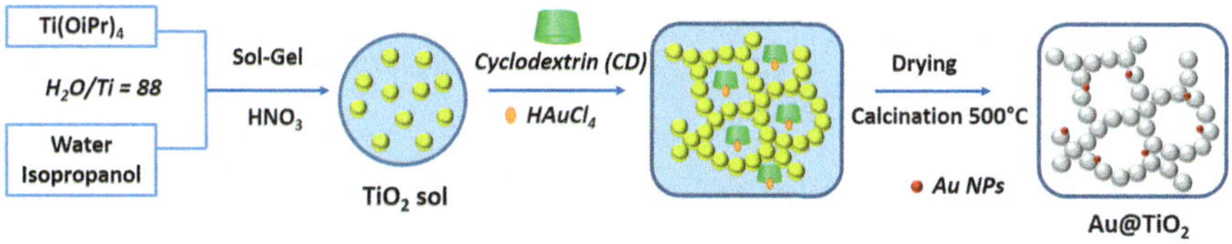

Fig. 3.21 Schematic illustration of the synthesis of mesoporous Au/TiO_2 catalysts where the cyclodextrin acts as structure directing agent to direct the self-assembly of gold and TiO_2 colloids in a nanostructured framework. Reprinted with permission from (Lannoy et al. 2017). Copyright 2017 American Chemical Society

self-assembly of TiO_2 colloids in a nanostructured network and, at the same time, it ensured uniform dispersion of gold nanoparticles over the mesoporous support.

Evidence for the key role of the cyclodextrins in the morphology of Au/TiO_2 composites was provided by electron microscopy. Fig. 3.22 depicts typical field emission scanning electron microscopy (FE-SEM) images of two selected composites prepared with the native β-CD and RaMeβ-CD. Gold nanoparticles were clearly distinguished in all micrographs revealing intimate contact with the mesoporous TiO_2 support. Moreover, it was noticed that, in contrast to β-CD which gave rise to dense and compact structures with sharp angular domains, over which small Au particles were uniformly dispersed (Fig. 3.22a–d), RaMeβ-CD exerted an opposing action, producing a highly porous framework over which Au nanoparticles with larger sizes which sometimes reached or exceeded 200 nm diameter were formed (Fig. 3.22f–i). The average diameter of Au nanoparticles prepared from β-CD was about 15–20 nm Fig. 3.22e), in agreement with XRD data, while two types of particles with average diameters of approximately 15–30 nm and 100–150 nm formed in the presence of RaMeβ-CD (Fig. 3.22j).

The corresponding transmission electron microscopy (TEM) images confirmed the existence of monomodal particles with β-CD (Fig. 3.23a, b) and bimodal particles with RaMeβ-CD (Fig. 3.23c, d). On the other hand, the natives α-CD and γ-CD presented a similar behavior to β-CD, giving rise to small monodisperse particles, while bigger and more polydisperse particles formed with HPβ-CD.

From Fig. 3.24a, b it can be noticed that all CD-derived photocatalysts present enhanced photocatalytic performances compared with the corresponding sol-gel materials. RaMeβ-CD gave rise to the most efficient photocatalysts for the PAA degradation under both UV and visible-light irradiation. The reaction mechanism in these photocatalysts is different depending on whether excitation occurs on the semi-conductor bandgap (bare TiO_2) or on the surface plasmon of gold nanoparticles (Au-modified TiO_2). Thus, in the case of bare TiO_2 which is composed of anatase (Eg = 3.2 eV) and brookite (Eg = 3.3 eV), irradiation with UV-light is necessary to produce positive holes in the valence band (h^+_{VB}) and electrons in the conduction band (e^-_{CB}) (Fig. 3.24c). The photogenerated holes can then oxidize adsorbed water and surface -OH groups (and eventually some PAA molecules), while the photogenerated electrons can reduce molecular oxygen adsorbed on the TiO_2 surface. These reactions can finally lead to the production of hydroxyl radicals

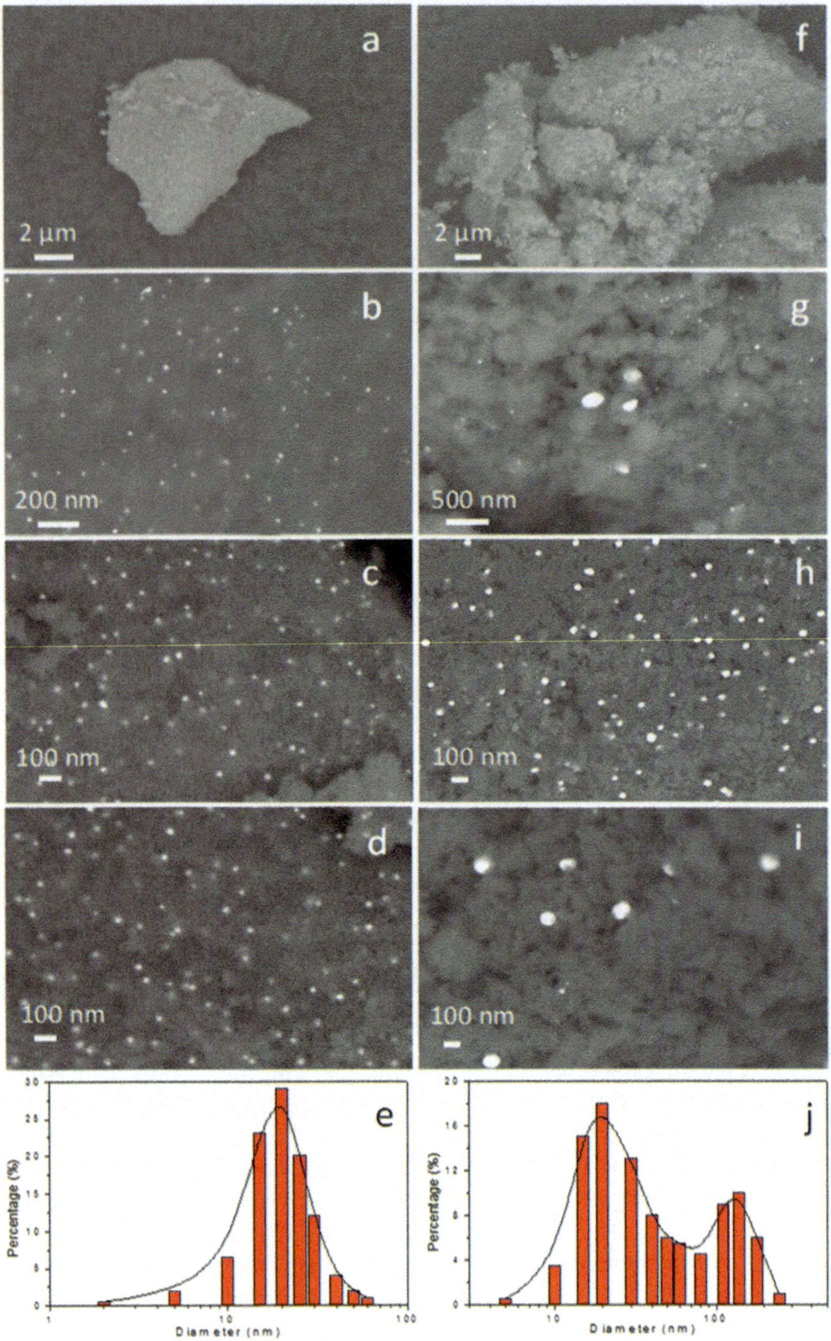

Fig. 3.22 Field emission-scanning electron microscopy (FE-SEM) images recorded with back-scattered electrons on β-CD derived Au/TiO_2 (**a–d**) and RaMeβ-CD derived Au/TiO_2 composites (**f–i**). Corresponding Au particle size distributions for the photocatalysts prepared with β-CD (e) and RaMeβ-CD (j). Reprinted with permission from (Lannoy et al. 2017). Copyright 2017 American Chemical Society

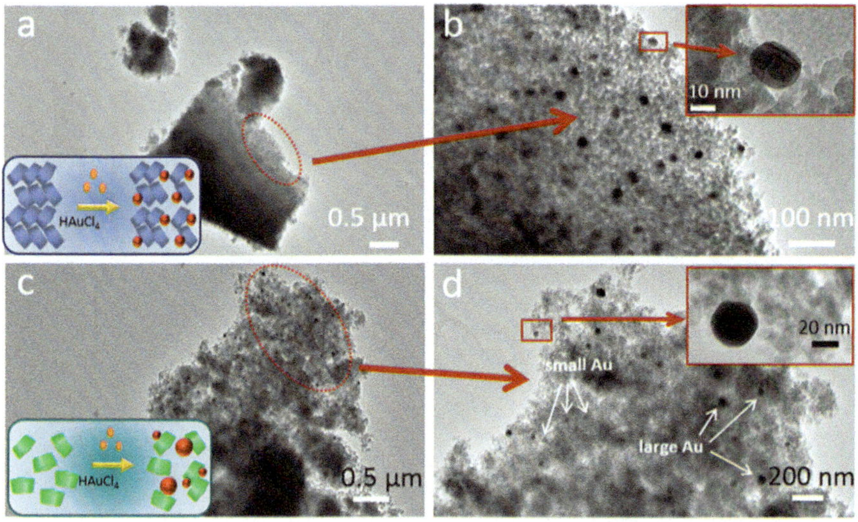

Fig. 3.23 Transmission electron microscopy (TEM) bright-field images of the Au/TiO_2 composites prepared with β-CD (**a, b**) and RaMeβ-CD (**c, d**). Note the formation of monomodal particles with β-CD and bimodal particles with RaMe β-CD. Reprinted with permission from (Lannoy et al. 2017). Copyright 2017 American Chemical Society

($^{\bullet}OH$) and superoxide radical anions ($O_2^{\bullet-}$) respectively, which are powerful oxidants for the degradation of organic pollutants in water. On the other hand, in the case of Au/TiO_2 composites, the electronic charge carriers responsible for the PAA degradation are likely to come mainly from the LSPR excitation of gold nanoparticles under visible light irradiation (Fig. 3.24d). Moreover, as the lifetime of electrons generated in the localized surface plasmon resonance (LSPR) process is very short (less than 10^{-3} ns for Au nanoparticles), their intimate contact with the titania surface should favor the transfer of the hot electrons from the plasmonic Au nanoparticles to the TiO_2 conduction band, thus hindering the e^-/h^+ recombination and enhancing the photocatalytic efficiency (Clavero 2014). On the basis of these results, it was concluded that the effect of cyclodextrins in increasing the porosity of both TiO_2 and Au/TiO_2 photocatalysts follows the order RaMeβ-CD > HPβ-CD > γ-CD ≈ α-CD > β-CD, while their effect in decreasing the size of gold nanoparticles follows an opposite trend, i.e. β-CD > γ-CD > α-CD > RaMeβ-CD > HPβ-CD (Fig. 3.24e, f).

The high photocatalytic activity of Au/TiO_2 prepared from RaMeβ-CD was ascribed to the presence of marked hydrophobic and hydrophilic microenvironments located on both the narrow and wider faces of the ring, conferring to this cyclodextrin an amphiphilic behavior and self-organizing properties at solid-liquid interfaces. Thus, in contrast to native cyclodextrins which are not surface-active and tend to form close-packed crystallites by intermolecular hydrogen bonding (Bleta et al. 2014a; b; c), RaMeβ-CD presents a good combination of surface active properties and weak intermolecular interactions to efficiently direct the self-assembly of titania

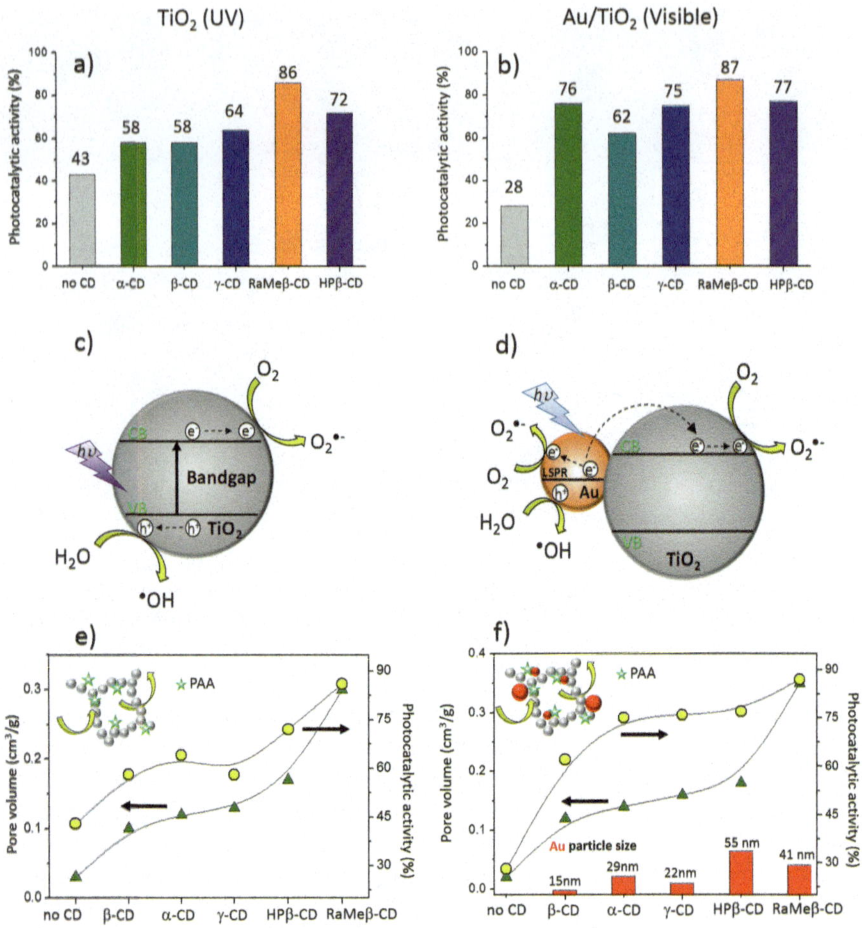

Fig. 3.24 Comparison of the phenoxyacetic acid (PAA) degradation rate after 7 h under UV-light irradiation (360 nm) on bare TiO_2 (**a**), and under visible-light irradiation (420 nm) on Au/TiO_2 (**b**). UV-induced photocatalytic activity by bandgap excitation of TiO_2 (**c**), and visible light-induced photocatalytic activity driven by localized surface plasmon resonance (LSPR) excitation of Au nanoparticles followed by interfacial electron transfer to the conduction band of TiO_2 (**d**). Effect of the pore volume on the photocatalytic activity of bare TiO_2 materials under UV-light irradiation (360 nm) (**e**). Effect of the pore volume and Au particle size on the photocatalytic activity of Au/TiO_2 composites under visible-light irradiation (420 nm) (**f**). Reprinted with permission from (Lannoy et al. 2017). Copyright 2017 American Chemical Society

colloids in a highly porous network over which gold nanoparticles can be uniformly dispersed. Moreover, as a result of the higher lipophilic character of the methoxy groups (-OCH_3) compared to the 2-hydroxypropyl ones (-$OCH_2CH(CH_3)OH$), smoother interfaces are likely to be created with RaMeβ-CD compared to HPβ-CD, thus facilitating the colloidal self-assembly. Consequently, the TiO_2 and Au/TiO_2

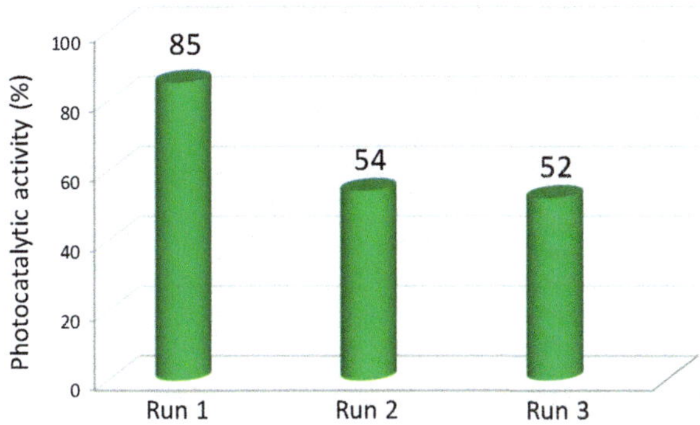

Fig. 3.25 Reusability tests of the RaMeβ-CD derived Au/TiO_2 photocatalyst. The phenoxyacetic acid (PAA) degradation rate was measured after 7 h of exposure under visible-light irradiation (420 nm). Reprinted with permission from (Lannoy et al. 2017, ESI). Copyright 2017 American Chemical Society

photocatalysts prepared from RaMeβ-CD yielded an enhancement of the photocatalytic activity due to a combined effect of good textural characteristics and high crystallinity. Indeed, the large surface area (110 $m^2\ g^{-1}$) should provide a large number of adsorption sites and active centers surrounding the e^-/h^+ pairs, while the high pore volume (0.35 $cm^3\ g^{-1}$) should facilitate the diffusion of PAA molecules towards the adsorption sites during the photocatalytic process. Moreover, the larger Au particles formed within the RaMeβ-CD-derived Au/TiO_2 photocatalysts are likely to improve the photoabsorption in a wider wavelength range, thus increasing the number of absorbed photons (Kowalska et al. 2009). In addition, the existence of gold particles with various dimensions within the same photocatalysts should also facilitate the electron transport from small to large particles through the conduction band of TiO_2 (Naya et al. 2014). Finally, the good crystallinity of both TiO_2 and Au particles may produce less defects and thus enhance the interfacial electron transfer rate between Au nanoparticles and the TiO_2 semi-conductor.

Based on these results, the recyclability and reuse of the most efficient photocatalyst (*i.e.* RaMeβ-CD-derived Au/TiO_2) was evaluated in the degradation of the phenoxyacetic acid (PAA), under visible light irradiation. The photocatalytic activity markedly decreased from 85% to 54% during the second run, then remained almost stable to 52% during the third run. This result was explained by the competitive adsorption and agglomeration of the PAA molecules (as well as the by-products) on the photocatalyst surface through formation of hydrogen bonds between the surface hydroxyls and the carboxylic acid groups of the PAA. Thus, because of the surface coverage with organics after each run, new parts of the photocatalyst surface become unavailable, thus reducing the number of the photocatalytic sites involved in the photoreaction and hindering the adsorption of water molecules, which are key factors in the production of radicals. These results confirmed the fact that the water/semiconductor interface is very important in photocatalysis and the masking of the photocatalytic sites may have a strong impact on the photodegradation efficiency (Fig. 3.25).

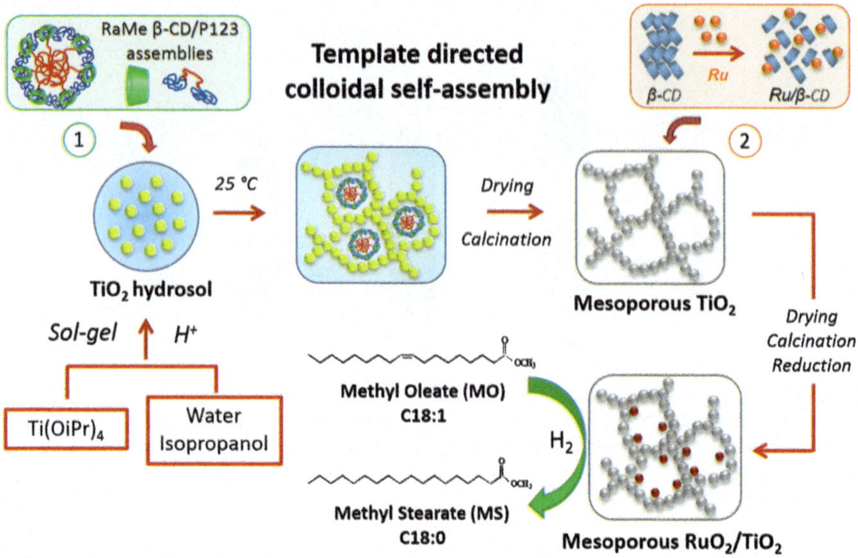

Fig. 3.26 Schematic illustration of the synthesis of RuO_2/TiO_2 composites by RaMeβ-CD-P123-directed colloidal self-assembly. Catalysts were calcined at 400 °C, then reduced under H_2 before being evaluated in the hydrogenation of methyl oleate. Reprinted with permission of the Royal Society of Chemistry from (Bleta et al. 2016)

3.4.3.4 Mesoporous RuO_2/TiO_2 Composites with High Catalytic Performance in the Hydrogenation of Oils

In heterogeneous catalysis, the design of nanostructured support materials with high specific surface area and tailored porosity is of crucial importance for enhancing the catalyst effectiveness (Bell 2003) (Patent No. WO 2015/055672 A1, 2015). Randomly methylated β-CD (RaMeβ-CD) derived mesoporous titania materials were used as efficient matrices for the preparation of RuO_2/TiO_2 composites with high catalytic activity in the hydrogenation of methyl oleate (MO, C18:1) to methyl stearate (MS, C18:0) under mild conditions (Bleta et al. 2016). The mesoporous catalyst were prepared using a two steps synthesis procedure (Fig. 3.26). In a first step the supramolecular assemblies formed between the block copolymer Pluronic P123 and RaMeβ-CD were used as template to prepare a highly porous TiO_2 support. Then, in a second step, Ru nanoparticles were dispersed over the support by impregnation with the assistance of the native β-CD followed by drying and calcination at 400 °C to produce a RuO_2/TiO_2 composite. Before catalytic tests, RuO_2 was reduced in metallic Ru under H_2 flow at 400 °C for 4 h.

To understand the effect of the CD-based assemblies, four RuO_2/TiO_2 composites (2.5 wt% Ru) were prepared using both conventional sol-gel TiO_2 (TiO_2-sg) and nanostructured RaMeβ-CD-P123-templated TiO_2 (TiO_2-ns) as supports, without or with assistance of β-CD as dispersing agent. XRD patterns of the different catalysts (Fig. 3.27A) indicated that similarly to the undoped TiO_2-ns support, Ru2.5TiO_2-ns

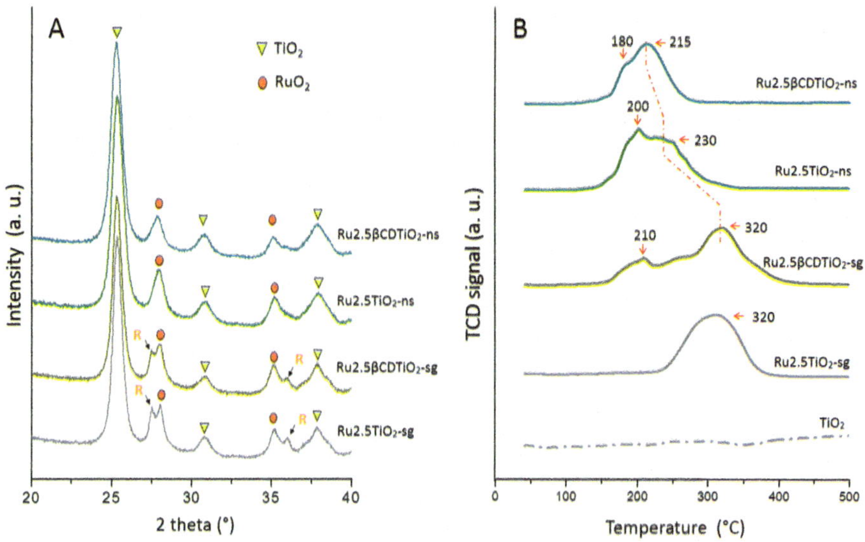

Fig. 3.27 (**A**) X-ray diffraction (XRD) patterns and (**B**) H_2-temperature programed reduction (H_2-TPR) profiles of RuO_2/TiO_2 composites (2.5 wt% Ru) supported over conventional sol-gel TiO_2 (TiO_2-sg) and nanostructured RaMeβ-CD-P123-templated TiO_2 (TiO_2-ns), with or without the assistance of native β-CD as dispersing agent. R denotes the rutile phase. The H_2-TPR profile of the non-doped TiO_2 is added for comparison. Reprinted with permission of the Royal Society of Chemistry from (Bleta et al. 2016)

and Ru2.5β-CDTiO_2-ns catalysts contain only anatase and brookite implying that no anatase-to-rutile or brookite-to-rutile phase transformation occurs during the second calcination at 400 °C. On the other hand, three distinct crystalline phases (anatase, brookite and rutile) were clearly observed for the Ru2.5TiO_2-sg and Ru2.5β-CDTiO_2-sg catalysts. Furthermore, for all materials, additional reflexion lines were clearly observed at $2\theta = 27.8°$ and 34.9° which were indexed to the (110) and (101) planes respectively of the tetragonal RuO_2 (JCPDS 00–043-1027). For the TiO_2-sg supported catalysts, the (110) and (101) planes of RuO_2 and rutile-TiO_2 overlapped due to the lattice matching between these two phases both of which adopt a tetragonal structure. Moreover, it was shown that both the support and the dispersing agent could affect the size of RuO_2 crystallites. Thus, for the Ru2.5 and Ru2.5β-CD catalysts deposited over TiO_2-ns, the crystallite sizes determined from the line broadening of the (110) diffraction peak were 9.8 nm and 9.2 nm respectively, while the average sizes obtained for the catalysts deposited over TiO_2-sg were beyond 10 nm. The H_2-temperature programed reduction (H_2-TPR) profiles indicated that the catalysts supported over nanostructured TiO_2 globally display lower reduction temperatures compared to those supported over conventional sol-gel TiO_2 materials (Fig. 3.27B). This means that the templated-support promotes the formation of small and readily reducible Ru nanoparticles which do not interact strongly with the porous support.

Evidence for the key role of RaMeβ-CD-based assemblies on the dispersion of RuO_2 nanoparticles was provided by high resolution transmission electron microscopy (HR-TEM) combined with high angle annular dark-field scanning transmission electron microscopy (HAADF-STEM). Observations were performed on two selected catalysts, *i.e.* Ru2.5β-CDTiO$_2$-ns and Ru2.5β-CDTiO$_2$-sg. From the TEM image shown in Fig. 3.28a, it can be noticed that the surface of the RaMeβ-CD-P123-templated TiO_2 support is decorated with RuO_2 nanoparticles which can be directly visualised by the bright spots resulting from the difference in Z-contrast between the two oxides (Fig. 3.28b). Moreover, energy dispersive spectroscopy (EDS) elemental mapping on different regions of the TEM micrograph (Fig. 3.28c-f) allows to further differentiate the two phases, *i.e.* TiO_2 (white regions) and RuO_2 nanoparticles (dark regions). TiO_2 nanoparticles in this sample contains tetragonal anatase (~15 nm) and orthorhombic brookite (~10 nm) (Fig. 3.28g–h). Interestingly, mesopores of 15–20 nm diameter, resulting from the self-assembly of these nanoparticles around the supramolecular template, can also be clearly visualised in this micrograph (pink circles), in agreement with N_2-adsorption analyses. On the other hand, RuO_2 nanoparticles are more or less spherical and approximately 10–12 nm in size (Fig. 3.28i), in agreement with XRD results. Conversely, the Ru2.5β-CDTiO$_2$-sg catalyst shows mainly a dense and less porous network (Fig. 3.28j) made-up of anatase (~35%), brookite (~26%) and rutile (~39%) and over which RuO_2 particles tend to form elongated heterogeneous structures (Fig. 3.28k–n). Such unusual epitaxial growth of RuO_2 nanoparticles over rutile-TiO_2 was also observed by Xiang *et al.* (Xiang et al. 2012) and was suggested to result from the structural matching between these two phases which actually share the same type of lattice symmetry (tetragonal) and have very similar lattice parameters (a = 4.59 Å, c = 2.96 Å for rutile-TiO_2 and a = 4.49 Å, c = 3.10 Å for RuO_2).

The performance of the different supported Ru/TiO_2 catalysts was evaluated in the liquid phase hydrogenation of methyl oleate (MO, C18:1) to methyl stearate (MS, C18:0) under mild conditions (40 bar H_2, 50 °C) (Fig. 3.29). Methyl esters of vegetable oils, derived from polyunsaturated fatty acids, can be transformed through partial or complete hydrogenation into saturated fatty acids with improved physical properties (*e.g.* high melting point and oxidative stability) (Philippaerts et al. 2013). Methyl stearate is an important saturated oil obtained from catalytic hydrogenation of unsaturated methyl oleate and widely used as a feedstock for catalytic hydroge-

Fig. 3.28 (continued) catalyst showing uniform dispersion of RuO_2 nanoparticles over the mesoporous RaMeβ-CD-P123-templated TiO_2 material; energy dispersive spectroscopy (EDS) spectra with the electron beam focused first on a white region (TiO_2) (**e**), and then in a dark region (RuO_2) (**f**) of the catalyst images. HR-TEM micrographs of the Ru2.5β-CDTiO$_2$-ns catalyst indicating uniform dispersion of TiO_2 nanoparticles crystallised in anatase and brookite (**g** and **h**), as well as RuO_2 nanoparticles (5–10 nm) (i) TEM and high resolution (HR) TEM images of the Ru2.5β-CDTiO$_2$-sg catalyst composed mostly of a nonporous support material showing TiO_2 and RuO_2 nanoparticle aggregation (j-n). Reprinted with permission of the Royal Society of Chemistry from (Bleta et al. 2016)

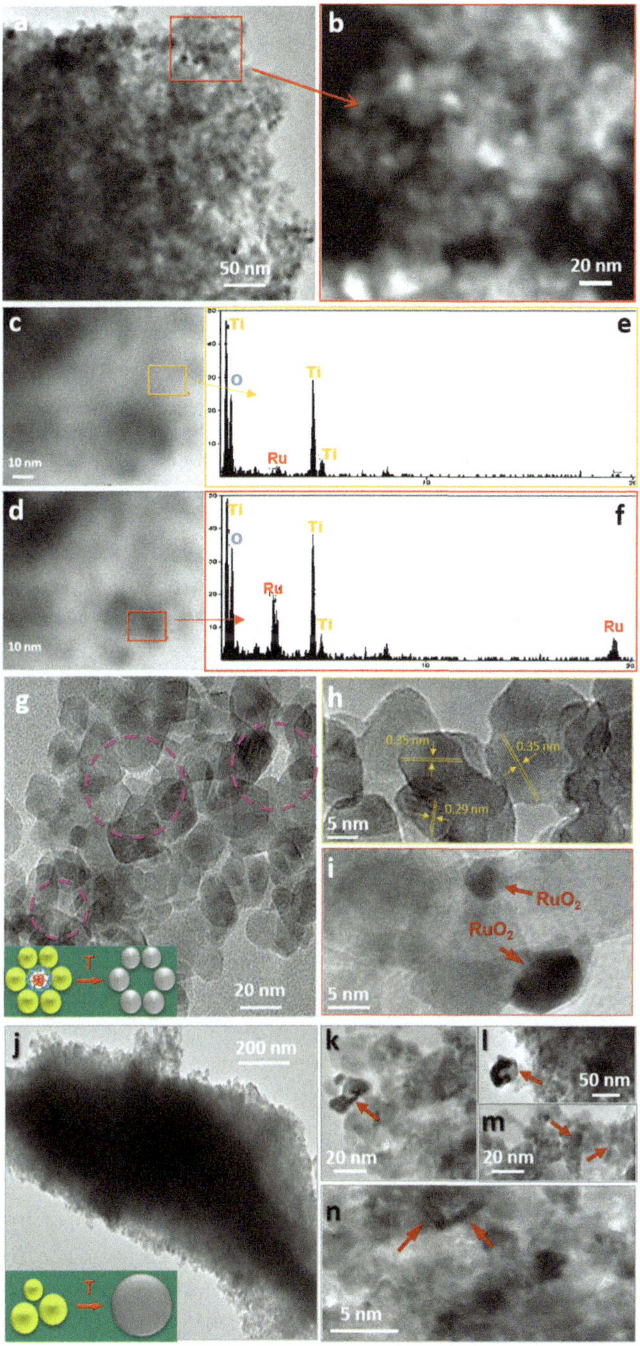

Fig. 3.28 Transmission electron microscopy (TEM) (**a**) and high angle annular dark-field scanning transmission electron microscopy (HAADF-STEM) images (**b–d**) of the Ru2.5β-CDTiO_2-ns

nolysis into its corresponding saturated fatty alcohol (stearyl alcohol, C18:0) (Pritchard et al. 2015). Ru-based catalysts have been shown to be effective for low temperature hydrogenation of carboxylic acid esters, especially for selective hydrogenation of the olefinic C=C bond without affecting the carbonyl C=O bond.

Prior to catalytic tests, RuO_2/TiO_2 composites were pre-treated under hydrogen flow at 400 °C for 4 h to reduce the RuO_2 in metallic Ru. The results obtained after 2 h of reaction (40 bar H_2, 50 °C) with the four catalysts, prepared using either TiO_2-sg or TiO_2-ns support, with the assistance or not of the native β-CD (Ru/β-CD = 11) as dispersing agent, are shown in Fig. 3.29. The results obtained with the non-doped supports (TiO_2-sg and TiO_2-ns) as well as with the Ru2.5 catalyst supported over commercial TiO_2 (Ru2.5TiO_2-A and Ru2.5TiO_2-P25) are added for comparison.

Depending on the preparation method, clear differences can be observed in both the catalytic activity (Fig. 3.29a) and selectivity (Fig. 3.29b). Unlike the control TiO_2-sg and TiO_2-ns supports, which showed almost no activity in the hydrogenation of methyl oleate, all supported Ru/TiO_2 catalysts were selective in the hydrogenation of the olefinic C=C bond producing the fully saturated ester, *i.e.* methyl stearate (C18:0) as the main product and the methyl elaïdate (C18:1, trans-9) as the only byproduct. Moreover, the catalytic performance was strongly affected by the support (TiO_2-sg or TiO_2-ns) and the impregnation method (with or without assistance of native β-CD). Thus, when Ru was deposited over nanostructured RaMeβ-CD-P123-templated TiO_2, the catalytic activity increased sharply from 36% (Ru2.5/TiO_2-sg) to 82% (Ru2.5/TiO_2-ns). Interestingly, the conversion efficiency was further improved when the native β-CD was used as dispersing agent. Thus, the conversion rate increased by ~14% (from 36% to 50%) for the Ru2.5 catalyst deposited over TiO_2-sg and by ~12% (from 82% to 94%) for the same catalyst deposited over TiO_2-ns support. Among the four supported catalysts prepared using the colloidal approach, Ru2.5β-CDTiO_2-ns was the fastest one, achieving almost full MO conversion (~94%) in 2 h. Interestingly, the activity of this catalyst was almost 3-fold higher than that of ruthenium deposited over commercial anatase-TiO_2 (Ru2.5TiO_2-A) and P25-TiO_2 (Ru2.5TiO_2-P25).

Similarly to the catalytic activity, clear differences were observed also in the selectivity toward methyl stearate which increased from 47% to 80% when moving from Ru2.5TiO_2-sg to Ru2.5TiO_2-ns. However, although a beneficial effect of β-CD was noticed for the Ru2.5TiO_2-sg catalyst (MS selectivity increased from 47% to 67%), no clear difference was observed between the Ru2.5TiO_2-ns catalysts prepared without or with assistance of β-CD (MS selectivity was the same ~80%). Over the investigated mesoporous Ru/TiO_2 catalysts, the catalytic performance agreed well with the textural characteristics of the support and the degree of dispersion of the ruthenium species. Indeed, the highest performance of the Ru2.5β-CDTiO_2-ns catalyst resulted from a combined effect of improved textural characteristics of the support and uniform dispersion of ruthenium nanoparticles. Thus, the high surface area of the RaMeβ-CD-P123-templated TiO_2 material (80 m^2 g^{-1}) should provide a high level of dispersion of ruthenium nanoparticles,

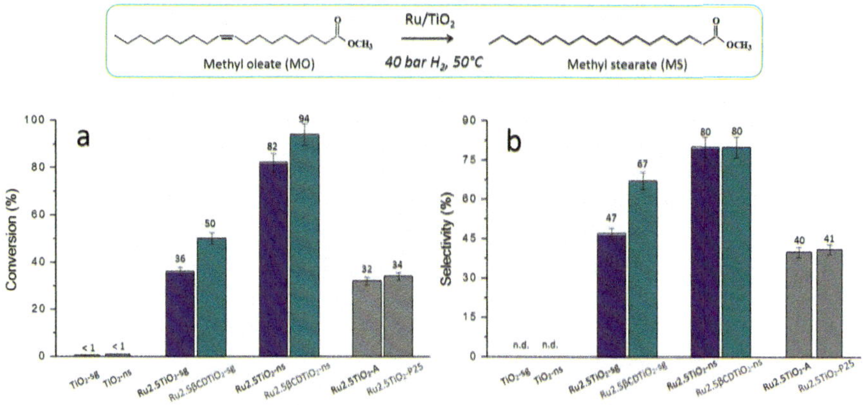

Fig. 3.29 Comparison of the catalytic activity (**a**) and selectivity (**b**) of different supported catalysts in the hydrogenation of methyl oleate (MO, C18:1) to methyl stearate (MS, C18:0). Reaction conditions: Ru (0.02 mmol), methyl oleate (MO/Ru molar ratio = 100), H_2 (40 bar), solvent (heptane, 10 mL), stirring rate (750 rpm), temperature (50 °C), reaction time (2 h). Reprinted with permission of the Royal Society of Chemistry from (Bleta et al. 2016)

facilitating the contact of methyl oleate with the catalyst surface during the hydrogenation reaction. On the other hand, the high pore volume (0.36 $cm^3\ g^{-1}$) should allow for more reactant molecules to be adsorbed to the internal surface of the pores, thus improving the diffusion of reactants and products toward the active phase during the catalytic process. Finally, the use of native β-CD as dispersing agent may allow for reducing the interparticle aggregation thanks to the ability of this cyclodextrin to interact with both $RuNO(NO_3)_3$ and TiO_2 nanoparticles. Overall, these results revealed a dual role played by the CD-based assemblies, *i.e.* an enhancement of the porosity and surface area of the RuO_2/TiO_2 composites and a better dispersion of Ru nanoparticles over the support, parameters that appear to be essential for enhancing the efficacy of the catalyst in the liquid phase hydrogenation of methyl oleate.

Moreover, as leaching and deactivation are common phenomena encountered in any liquid phase hydrogenation reaction, the ease of recovery and reusability of the catalyst are important features that need to be considered. For this purpose, the recyclability of the most efficient catalyst, *i.e.* Ru2.5β-CDTiO_2-ns, was investigated in three successive runs. Remarkably, no loss of activity and selectivity was observed during the consecutive cycles, indicating high stability of the supported catalyst under the employed reaction conditions (Fig. 3.30). The high surface area and large pore volume generated by the template-directed self-assembly of titania colloids should allow for a good dispersion of nanosized ruthenium particles and, therefore, effectively prevent aggregation while improving the chemical stability of the catalyst. Moreover, the thermal treatment at 400 °C should improve fixation of ruthenium species over the TiO_2 material, preventing leaching from the matrix to the reaction media during the catalytic process. The high efficacy of this catalyst

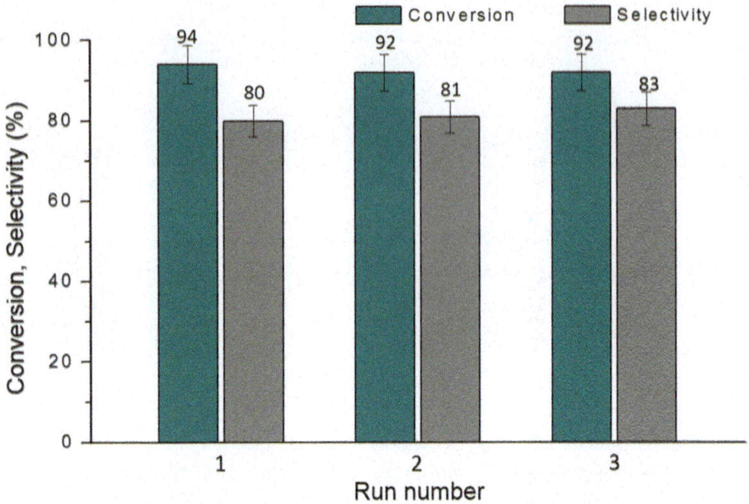

Fig. 3.30 Reusability tests of the Ru2.5β-CDTiO_2-ns catalyst. Reaction conditions: methyl oleate/ruthenium molar ratio (MO/Ru = 100), H_2 (40 bar), solvent (heptane, 10 mL), stirring rate (750 rpm), temperature (50 °C), reaction time (2 h). Reprinted with permission of the Royal Society of Chemistry from (Bleta et al. 2016)

combined with ease of recovery and reuse, without loss of conversion and selectivity, makes the colloidal self-assembly a simple and versatile approach for the synthesis of supported catalysts with high performance in liquid phase hydrogenation reactions.

3.5 Conclusion

In this chapter, we presented an overview of the different types of porous inorganic materials that can be prepared from cyclodextrins or cyclodextrin-based assemblies, ranging from microporous to mesoporous materials and further to hierarchically structured porous materials with a three-dimensional interconnected network. From the perspective of the synthesis, the approach that involves hydrolysis and catalytic polycondensation of a silicon alkoxide precursor around a supramolecular template is most commonly used for the fabrication of silica materials. On the other hand, the template-directed colloidal self-assembly approachen ables rational design of a broad range of nanostructured transition metal oxides with very high specific surface areas (380 $m^2\ g^{-1}$ for γ-Al_2O_3 and 110 $m^2\ g^{-1}$ for TiO_2), tailored porosities in the mesoscale (2–20 nm), controlled crystal phase compositions and various morphologies ranging from spheres, to rod-like particles and fibers. Moreover, metal nanoparticles can also be uniformly dispersed within these materials providing nanostructured composites with tunable properties. With respect to catalytic applications, the cyclodextrin-derived nanostructured porous materials and composites hold great promise owing to their attractive and unique characteristics which are key factors in obtaining highly active, selective, stable and recyclable heterogeneous catalysts.

References

Alphonse P, Bleta R, Lavergne D, Barrault J, Maciuca A, Tatibouët JM, Rossignol C (2015) Patent No. WO 2015/055672 A1

Antonietti M (2006) Silica nanocasting of lyotropic surfactant phases and organized organic matter: material science or an analytical tool? Philos Trans R Soc Lond Ser A 364:2817–2840

Armbrust EV (2009) The life of diatoms in the world's oceans. Nature 459:185–192

Armbrust E, Berges J, Bowler C, Green B, Martinez D, Putnam N, Zhou S, Allen AE, Apt KE, Bechner M, Brzezinski MA, Chaal BK, Chiovitti A, Davis AK, Demarest MS, Detter JC, Glavina T, Goodstein D, Hadi MZ, Hellsten U, Hildebrand M, Jenkins BD, Jurk J, Kapitonov VV, Kröger N, Lau WW, Lane TW, Larimer FW, Lippmeier JC, Lucas S, Medina M, Montsant A, Obornik M, Parker MS, Palenik B, Pazour GJ, Richardson PM, Rynearson TA, Saito MA, Schwartz DC, Thamatrakoln K, Valentin K, Vardi A, Wilkerson FP, Rokhsar D (2004) The genome of the diatom Thalassiosira Pseudonana: ecology, evolution, and metabolism. Science 306:79–86

Beck J, Vartuli J, Roth W, Leonowicz M, Kresge C, Schmitt K, Chu CTW, Olson DH, Sheppard EW, McCullen SB, Higgins JB, Schlenker JL (1992) A new family of mesoporous molecular sieves prepared with liquid crystal templates. J Am Chem Soc 114:10834–10843

Bell A (2003) The impact of nanoscience on heterogeneous catalysis. Science 299:1688–1691

Bernat V, Ringard-Lefebvre C, Bas G, Perly B, Djedaïni-Pilard F, Lesieur S (2008) Inclusion complex of n-octyl β-d-glucopyranoside and α-cyclodextrin in aqueous solutions: thermodynamic and structural characterization. Langmuir 24:3140–3149

Bleta R, Blin J, Stebe M (2006) Solubilization of various fluorocarbons in a fluorinated surfactant/water system: relation with the design of porous materials. J Phys Chem B 110:23547–23556

Bleta R, Alphonse P, Lorenzato L (2010) Nanoparticle route for the preparation in aqueous medium of mesoporous TiO2 with controlled porosity and crystalline framework. J Phys Chem C 114:2039–2048

Bleta R, Jaubert O, Gressier M, Menu M (2011) Rheological behaviour and spectroscopic investigations of cerium-modified AlO(OH) colloidal suspensions. J Colloid Int Sci 363:557–565

Bleta R, Alphonse P, Pin L, Gressier M, Menu M (2012) An efficient route to aqueous phase synthesis of nanocrystalline -Al2O3 with high porosity: from stable boehmite colloids to large pore mesoporous alumina. J Colloid Int Sci 367:120–128

Bleta R, Machut C, Leger B, Monflier E, Ponchel A (2013) Coassembly of block copolymer and randomly methylated β-cyclodextrin: from swollen micelles to mesoporous alumina with tunable pore size. Macromolecules 46:5672–5683

Bleta R, Lannoy A, Machut C, Monflier E, Ponchel A (2014a) Understanding the role of cyclodextrins in the self-assembly, crystallinity, and porosity of titania nanostructures. Langmuir 30:11812–11822

Bleta R, Machut C, Leger B, Monflier E, Ponchel A (2014b) Investigating the effect of randomly methylated β-cyclodextrin/block copolymer molar ratio on the template-directed preparation of mesoporous alumina with tailored porosity. J Incl Phenom Macrocycl Chem 80:323–335

Bleta R, Menuel S, Leger B, Da Costa A, Monflier E, Ponchel A (2014c) Evidence for the existence of crosslinked crystalline domains within cyclodextrin-based supramolecular hydrogels through sol-gel replication. RSC Adv 4:8200–8208

Bleta R, Noel S, Addad A, Ponchel A, Monflier E (2016) Mesoporous RuO_2/TiO_2 composites prepared by cyclodextrin-assisted colloidal self-assembly: towards efficient catalysts for the hydrogenation of methyl oleate. RSC Adv 6:14570–14579

Blin J, Bleta R, Ghanbaja J, Stebe M (2006) Fluorinated emulsions: templates for the direct preparation of macroporous-mesoporous silica with a highly ordered array of large mesopores. Micropor Mesopor Mater 94:74–80

Boal A, Ilhan F, DeRouchey J, Thurn-Albrecht T, Russell T, Rotello V (2000) Self-assembly of nanoparticles into structured spherical and network aggregates. Nature 404:746–748

Born M, Ritter H (1995) Side-chain polyrotaxanes with a tandem structure based on cyclodextrins and a polymethacrylate main chain. Angew Chem Int Ed Engl 34:309–311

Breslow R, Dong S (1998) Biomimetic reactions catalyzed by cyclodextrins and their derivatives. Chem Rev 98:1997–2011

Chen D, Caruso R (2013) Recent progress in the synthesis of spherical titania nanostructures and their applications. Adv Funct Mater 23:1356–1374

Clavero C (2014) Plasmon-induced hot-electron generation at nanoparticle/metal-oxide interfaces for photovoltaic and photocatalytic devices. Nat Photonics 8:95–103

Corma A (1997) From microporous to mesoporous molecular sieve materials and their use in catalysis. Chem Rev 97:2373–2420

Corma A, Moliner M, Diaz-Cabanas MJ, Serna P, Femenia B, Primo J, Garcia H (2008) Biomimetic synthesis of microporous and mesoporous materials at room temperature and neutral pH, with application in electronics, controlled release of chemicals, and catalysis. New J Chem 32:1338–1345

Davis M (2002) Ordered porous materials for emerging applications. Nature 417:813–821

Dreiss C, Nwabunwann E, Liu R, Brooks N (2009) Assembling and de-assembling micelles: competitive interactions of cyclodextrins and drugs with Pluronics. Soft Matter 5:1888–1896

Fujishima A, Honda K (1972) Electrochemical photolysis of water at a semiconductor electrode. Nature 238:37–38

Gaitano G, Brown W, Tardajos G (1997) Inclusion complexes between cyclodextrins and triblock copolymers in aqueous solution: a dynamic and static light-scattering study. J Phys Chem B 101:710–719

Gomathi Devi L, Kavitha R (2013) A review on non-metal ion doped titania for the photocatalytic degradation of organic pollutants under UV/solar light: role of photogenerated charge carrier dynamics in enhancing the activity. Appl Catal B 140-141:559–587

Hamm CE, Merkel R, Springer O, Jurkojc P, Maier C, Prechtel K, Smetacek V (2003) Architecture and material properties of diatom shells provide effective mechanical protection. Nature 421:841–843

Han BH, Antonietti M (2002) Cyclodextrin-based pseudopolyrotaxanes as templates for the generation of porous silica materials. Chem Mater 14:3477–3485

Han BH, Smarsly B, Gruber C, Wenz G (2003) Towards porous silica materials via nanocasting of stable pseudopolyrotaxanes from α-cyclodextrin and polyamines. Microporous Mesoporous Mater 66:127–132

Hao E, Lian T (2000) Layer-by-layer assembly of CdSe nanoparticles based on hydrogen bonding. Langmuir 16:7879–7881

Harada A (1996) Preparation and structures of supramolecules between cyclodextrins and polymers. Coord Chem Rev 148:115–133

Harada A (2001) Cyclodextrin-based molecular machines. Acc Chem Res 34:456–464

Harada A, Kamachi M (1990) Complex formation between poly (ethylene glycol) and α-cyclodextrin. Macromolecules 23:2821–2823

Harada A, Hashidzume A, Yamaguchi H, Takashima Y (2009) Polymeric rotaxanes. Chem Rev 109:5974–6023

Herrmann W, Keller W, Wenz G (1997) Kinetics and thermodynamics of the inclusion of ionene-6, 10 in α-cyclodextrin in an aqueous solution. Macromolecules 30:4966–4972

Hildebrand M (2008) Diatoms, biomineralization processes, and genomics. Chem Rev 108:4855–4874

Holmqvist P, Alexandridis P, Lindman B (1998) Modification of the microstructure in block copolymer-water-"oil" systems by varying the copolymer composition and the "oil" type: small-angle X-ray scattering and deuterium-NMR investigation. J Phys Chem B 102:1149–1158

Imhof A, Pine D (1997) Ordered macroporous materials by emulsion templating. Nature 389:948–951

Jiang B, Guo D, Liu Y (2010) Self-assembly of amphiphilic perylene-cyclodextrin conjugate and vapor sensing for organic amines. J Org Chem 75:7258–7264

Joseph J, Dreiss C, Cosgrove T, Pedersen J (2007) Rupturing polymeric micelles with cyclodextrins. Langmuir 23:460–466

Kooistra W, Gersonde R, Medlin L, Mann D (2007) In: Falkowski PG, Knoll AH (eds) The origin and evolution of diatoms: their adaptation to a planktonic existence. Evolution of primary producers in the sea. Elsevier, Boston, pp 207–249

Kowalska E, Abe R, Ohtani B (2009) Visible light-induced photocatalytic reaction of gold-modified titanium(IV) oxide particles: action spectrum analysis. Chem Commun:241–243

Kresge CT, Leonowicz ME, Roth WJ, Vartuli JC, Beck JS (1992) Ordered mesoporous molecular sieves synthesized by a liquid-crystal template mechanism. Nature 359:710–712

Kröger N, Poulsen N (2008) Diatoms from cell wall biogenesis to nanotechnology. Annu Rev Genet 42:83–107

Kröger N, Deutzmann R, Sumper M (1999) Polycationic peptides from diatom biosilica that direct silica nanosphere formation. Science 286:1129–1132

Lannoy A, Bleta R, Machut C, Monflier E, Ponchel A (2014) Block copolymer-cyclodextrin supramolecular assemblies as soft templates for the synthesis of titania materials with controlled crystallinity, porosity and photocatalytic activity. RSC Adv 4:40061–40070

Lannoy A, Bleta R, Machut-Binkowski C, Addad A, Monflier E, Ponchel A (2017) Cyclodextrin-directed synthesis of gold-modified TiO_2 materials and evaluation of their photocatalytic activity in the removal of a pesticide from water: effect of porosity and particle size. ACS Sustain Chem Eng 5:3623–3630

Lazzara G, Milioto S (2008) Copolymer-cyclodextrin inclusion complexes in water and in the solid state. A physico-chemical study. J Phys Chem B 112:11887–11895

Leclercq L, Bricout H, Tilloy S, Monflier E (2007) Biphasic aqueous organometallic catalysis promoted by cyclodextrins: can surface tension measurements explain the efficiency of chemically modified cyclodextrins? J Colloid Int Sci 307:481–487

Li J, Harada A, Kamashi M (1994) Sol-gel transition during inclusion complex formation between α-cyclodextrin and high molecular weight poly (ethylene glycol)s in aqueous solution. Polym J 26:1019–1026

Lin Z, Wang X, Liu J, Tian Z, Dai L, He B, Han C, Wu Y, Zeng Z, Hu Z (2015) On the role of localized surface plasmon resonance in UV-vis light irradiated au/TiO2 photocatalysis systems: pros and cons. Nanoscale 7:4114–4123

Linsebigler A, Lu G, Yates J (1995) Photocatalysis on TiO_2 surfaces: principles, mechanisms, and selected results. Chem Rev 95:735–758

Liu Y, Goebl J, Yin Y (2013) Templated synthesis of nanostructured materials. Chem Soc Rev 42:2610–2653

Lu AH, Schüth F (2006) Nanocasting: a versatile strategy for creating nanostructured porous materials. Adv Mater 18:1793–1805

Lu C, Wu N, Jiao X, Luo C, Cao W (2003) Micropatterns constructed from au nanoparticles. Chem Commun:1056–1057

Mahata A, Bose D, Ghosh D, Jana B, Bhattacharya B, Sarkar D, Chattopadhyay N (2010) Studies of triton X-165-β-cyclodextrin interactions using both extrinsic and intrinsic fluorescence. J Colloid Int Sci 347:252–259

Naya S, Niwa T, Kume T, Tada H (2014) Visible-light-induced electron transport from small to large nanoparticles in bimodal gold nanoparticle-loaded titanium(IV) oxide. Angew Chem 126:7433–7437

Nogueiras-Nieto L, Alvarez-Lorenzo C, Sandez-Macho I, Concheiro A, Otero-Espinar F (2009) Hydrosoluble cyclodextrin/poloxamer polypseudorotaxanes at the air/water interface, in bulk solution, and in the gel state. J Phys Chem B 113:2773–2782

Nursam N, Wang X, Caruso R (2015) Macro–/mesoporous titania thin films: analysing the effect of pore architecture on photocatalytic activity using high-throughput screening. J Mater Chem A 3:24557–24567

Okumura H, Kawaguchi Y, Harada A (2001) Preparation and characterization of inclusion complexes of poly(dimethylsiloxane)s with cyclodextrins. Macromolecules 34:6338–6343

Ovenstone J (2001) Preparation of novel titania photocatalysts with high activity. J Mater Sci 36:1325–1329

Park J, Momma T, Osaka T (2007) Spectroelectrochemical phenomena on surface plasmon resonance of au nanoparticles immobilized on transparent electrode. Electrochim Acta 52:5914–5923

Patolsky F, Weizmann Y, Lioubashevski O, Willner I (2002) Au-nanoparticle nanowires based on DNA and polylysine templates. Angew Chem Int Ed 41:2323–2327

Periyat P, Pillai S, McCormack D, Colreavy J, Hinder S (2008) Improved high-temperature stability and sun-light-driven photocatalytic activity of sulfur-doped anatase TiO_2. J Phys Chem C 112:7644–7652

Philippaerts A, Jacobs P, Sels B (2013) Is there still a future for hydrogenated vegetable oils ? Angew Chem Int Ed 52:5220–5226

Pillai S, Periyat P, George R, Colreavy J, George R, Hayden H, Seery M, McCormack DE, Corr D, Hinder S (2007) Synthesis of high-temperature stable anatase TiO_2 photocatalyst. J Phys Chem C 111:1605–1611

Polarz S, Antonietti M (2002) Porous materials via nanocasting procedures: innovative materials and learning about soft-matter organization. Chem Commun:2593–2604

Polarz S, Smarsly B, Bronstein L, Antonietti M (2001) From cyclodextrin assemblies to porous materials by silica templating. Angew Chem Int Ed 40:4417–4421

Poulsen N, Kröger N (2004) Silica morphogenesis by alternative processing of silaffins in the diatom Thalassiosira Pseudonana. J Biol Chem 279:42993–42999

Pritchard J, Filonenko G, van Putten R, Hensen E, Pidko E (2015) Heterogeneous and homogeneous catalysis for the hydrogenation of carboxylic acid derivatives: history, advances and future directions. Chem Soc Rev 44:3808–3833

Rajh T, Micic O, Nozik A (1993) Synthesis and characterization of surface-modified colloidal CdTe qunatum dots. J Phys Chem 97:11999–12003

Saenger W (1980) Cyclodextrin inclusion compounds in research and industry. Angew Chem Int Ed Engl 19:344–362

Singh H, Saquib M, Haque M, Muneer M, Bahnemann D (2007) Titanium dioxide mediated photocatalysed degradation of phenoxyacetic acid and 2,4,5-trichlorophenoxyacetic acid in aqueous suspensions. J Mol Catal A Chem 264:66–72

Su B, Sanchez C, Yang X (2012) Hierarchically structured porous materials, From Nanoscience to catalysis, separation, optics, energy, and life science. Wiley-VCH, Weinheim

Sumper M, Hett R, Lehmann G, Wenzl S (2007) A code for lysine modifications of a silica biomineralizing silaffin protein. Angew Chem Int Ed 46:8405–8408

Sun M, Huang S, Chen L, Li Y, Yang X, Yuan Z, Su B (2016) Applications of hierarchically structured porous materials from energy storage and conversion, catalysis, photocatalysis, adsorption, separation, and sensing to biomedicine. Chem Soc Rev 45:3479–3563

Szejtli J (1998) Introduction and general overview of cyclodextrin chemistry. Chem Rev 98:1743–1753

Travelet C, Schlatter G, Hebraud P, Brochon C, Lapp A, Anokhin D, Ivanov DA, Gaillard C, Hadziioannou G (2008) Multiblock copolymer behaviour of α-CD/PEO-based polyrotaxanes: towards nano-cylinder self-organization of α-CDs. Soft Matter 4:1855–1860

Travelet C, Schlatter G, Hebraud P, Brochon C, Lapp A, Hadziioannou G (2009) Formation and self-organization kinetics of α-CD/PEO-based pseudo-polyrotaxanes in water. A specific behavior at 30 °C. Langmuir 25:8723–8734

Travelet C, Hebraud P, Perry C, Brochon C, Hadziioannou G, Lapp A, Schlatter G (2010) Temperature-dependent structure of α-CD/PEO-based polyrotaxanes in concentrated solution in DMSO: kinetics and multiblock copolymer behavior. Macromolecules 43:1915–1921

Tsai C, Zhang W, Wang C, Van Horn R, Graham M, Huang J, Chen Y, Guo M, Cheng S (2010) Evidence of formation of site-selective inclusion complexation between β-cyclodextrin and poly(ethylene oxide)-block-poly(propylene oxide)-block poly(ethylene oxide) copolymers. J Chem Phys 132:1–9

Uekama K, Irie T (1987) Cyclodextrins and their industrial uses. In: Duchêne D (ed) Pharmaceutical applications of methylated cyclodextrin derivatives. Editions de la Santé, Paris, pp 395–439

Valero M, Grillo I, Dreiss C (2012) Rupture of Pluronic micelles by di-methylated β-cyclodextrin is not due to polypseudorotaxane formation. J Phys Chem B 116:1273–1281

Van Der Voort P, Vercaemst C, Schaubroeck D, Verpoort F (2008) Ordered mesoporous materials at the beginning of the third millennium: new strategies to create hybrid and non-siliceous variants. Phys Chem Chem Phys 10:347–360

Velev O, Kaler E (2000) Structured porous materials via colloidal crystal templating: from inorganic oxides to metals. Adv Mater 7:531–534

Walcarius A (2013) Mesoporous materials and electrochemistry. Chem Soc Rev 42:4098–4140

Wan Y, Zhao D (2007) On the controllable soft-templating approach to mesoporous silicates. Chem Rev 107:2821–2859

Wang X, Caruso R (2011) Enhancing photocatalytic activity of titania materials by using porous structures and the addition of gold nanoparticles. J Mater Chem 21:20–28

Wang X, Blackford M, Prince K, Caruso RA (2012) Preparation of boron-doped porous titania networks containing gold nanoparticles with enhanced visible-light photocatalytic activity. ACS Appl Mater Interfaces 4:476–482

Wang X, Cao L, Chen D, Caruso R (2013) Engineering of monodisperse mesoporous titania beads for photocatalytic applications. ACS Appl Mater Interfaces 5:9421–9428

Weickenmeier M, Wenz G (1997) Threading of cyclodextrins onto a polyester of octanedicarboxylic acid and polyethylene glycol. Macromol Rapid Commun 18:1109–1115

Wenz G (1994) Cyclodextrins as building blocks for supramolecular structures and functional units. Angew Chem Int Ed Engl 33:803–822

Xiang G, Shi X, Wu Y, Zhuang J, Wang X (2012) Size effects in atomic-level epitaxial redistribution process of RuO_2 over TiO_2. Sci Rep 2(801):1–6

Yang H, Sun C, Qiao S, Zou J, Liu G, Smith S, Cheng HM, Lu G (2008) Anatase TiO2 single crystals with a large percentage of reactive facets. Nature 453:638–641

Yang X, Chen L, Li Y, Rooke J, Sanchez C, Su B (2017) Hierarchically porous materials: synthesis strategies and structure design. Chem Soc Rev 46:481–558

Yu W, Peng X (2002) Formation of high-quality CdS and other II–VI semiconductor nanocrystals in noncoordinating solvents: tunable reactivity of monomers. Angew Chem Int Ed 41:2368–2371

Zhang H, Banfield J (1998) Thermodynamic analysis of phase stability of nanocrystalline titania. J Mater Chem 8:2073–2076

Zhang H, Banfield J (2000) Understanding polymorphic phase transformation behavior during growth of nanocrystalline aggregates: insights from TiO2. J Phys Chem B 104:3481–3487

Zhang X, Yang Z, Li X, Deng N, Qian S (2013) β-cyclodextrin's orientation onto TiO_2 and its paradoxical role in guest's photodegradation. Chem Commun 49:825–827

Zhao W, Ma W, Chen C, Zhao J, Shuai Z (2004) Efficient degradation of toxic organic pollutants with Ni2O3/TiO2-x Bx under visible irradiation. J Am Chem Soc 126:4782–4783

Zhu C, Du D, Eychmüller A, Lin Y (2015) Engineering ordered and nonordered porous noble metal nanostructures: synthesis, assembly, and their applications in electrochemistry. Chem Rev 115:8896–8943

Chapter 4
Thermal Analyses of Cyclodextrin Complexes

Nicoleta G. Hădărugă, Geza N. Bandur, and Daniel I. Hădărugă

Contents

Abstract Human health is the most important issue in the society. There are many compounds such as food additives and ingredients or medicinal compounds that can alter human health. Cyclodextrins can protect these compounds against light and oxidative degradation by molecular encapsulation. Moreover, cyclodextrins can also enhance properties of biologically active compounds such as better water solubility and bioavailability, and controlled release. Cyclodextrins are cyclic oligosaccharides comprising of six to eight α-(1→4)-linked D-glucopyranose units corresponding to the natural α-, β- and γ-cyclodextrin, respectively. Cyclodextrins have a truncated cone-like structure, with a hydrophobic inner cavity and highly hydrophilic exterior. This property allows molecular encapsulating of geometrically

N. G. Hădărugă
Department of Food Science, Banat's University of Agricultural Sciences and Veterinary Medicine "King Michael I of Romania" from Timişoara, Timişoara, Romania

G. N. Bandur · D. I. Hădărugă (✉)
Department of Applied Chemistry, Organic and Natural Compounds Engineering, Polytechnic University of Timişoara, Timişoara, Romania
e-mail: geza.bandur@upt.ro; daniel.hadaruga@upt.ro

S. Fourmentin et al. (eds.), *Cyclodextrin Fundamentals, Reactivity and Analysis*, Environmental Chemistry for a Sustainable World,
https://doi.org/10.1007/978-3-319-76159-6_4

compatible hydrophobic compounds for obtaining host-guest supramolecular systems. Among many characterization methods that are applicable in both solution and solid state, thermal techniques were widely used for analysis and stability evaluation of cyclodextrin complexes.

We review the use of thermal methods for the analysis of cyclodextrin complexes and non-complexed cyclodextrins. We discuss the applications of thermogravimetry-differential thermogravimetry, differential thermal analysis, differential scanning calorimetry, hot stage microscopy, thermogravimetry-mass spectrometry, gas chromatography-time-of-flight-mass spectrometry, and isothermal titration calorimetry. Cyclodextrin complexes are classified according to the types of biological activity of the guest compound, e.g. drugs, odorants, essential oils and vegetable extracts, antioxidants, fatty acids, oils and fatty acid based derivatives, and other organic, organometallic and inorganic compounds. The formation of cyclodextrin inclusion complexes is evidenced by disappearance of the thermal characteristics of the guest compound after nanoencapsulation. It is the case of melting or boiling points. Information on the nanoencapsulation process is obtained from the behavior of hydration water molecules of complexes, and from the stability of the guest compound during heating up to the decomposition.

4.1 Introduction

Thermal analyses are valuable techniques for studying liquid and solid materials. They comprise of many methods such as thermogravimetry-differential thermogravimetry, differential thermal analysis, differential scanning calorimetry, thermomechanical analysis, thermodilatometry, dynamic mechanical analysis, dielectric analysis, micro- and nanoscale local thermal analysis, such as atomic force and scanning thermal microscopies (Biliaderis 1983; Lelievre and Liu 1994; Ozawa 2000; Wunderlich 2007). Furthermore, there are thermal analyses methods that are coupled with other techniques such as thermogravimetry-Fourier transform-infrared spectroscopy, thermogravimetry-mass spectrometry, simultaneous differential scanning calorimetry-thermogravimetry or differential scanning calorimetry-X-ray diffractometry (Paulik and Paulik 1978; Meisel 1982; Giron 2002).

There are different properties that are measured by thermal analyses (Meisel 1982). Thermogravimetry can provide information about phase transitions, such as vaporization, sublimation, absorption/desorption. This technique also provide information on chemical phenomena, which involves mass change. Such phenomena are dehydration, decomposition or oxidation. On the other hand, differential thermal analysis and differential scanning calorimetry can provide information on physical and/or chemical endothermic and exothermic processes that occur during heating of a material, such as glass transition, crystallization, or melting (Teramoto 1990; Mathot and Benoist 1994; Wunderlich 2007; Menczel and Bruce Prime 2009). Thermomechanical analysis, thermodilatometry and dynamic mechanical analysis are more related to the changes on macroscopic properties of solid materials such as dimensions or mechanical properties (Teramoto 1990). Coupled thermal analysis techniques are powerful tools for material analysis. Thermogravimetry-differential thermal analysis-mass spectrometry is helpful for elucidation of reaction mechanisms and for the determination of chemical composition by means of reaction gases that are analyzed by mass spectrometry (Mathot and Benoist 1994). Evolved gases can be better analyzed by gas chromatography-Fourier transform-infrared spectroscopy or gas chromatography-mass spectrometry using a coupled equipment (Giron 2002).

4.2 Cyclodextrins

Cyclodextrins are appropriate compounds for applying thermal analyses, especially thermogravimetry, differential thermal analysis and differential scanning calorimetry. They can be coupled or not with other techniques. These analyses are helpful because cyclodextrins have structural architectures resembling with truncated cones with wholes, which allows forming supramolecular structures (Kurkov and Loftsson 2013; Duchêne and Bochot 2016). Natural cyclodextrins are cyclic oligosaccharides having six, seven or eight α-(1→4)-linked glucopyranose units. They corresponds to α-, β- and γ- cyclodextrin (Fig. 4.1). Primary and secondary hydroxyl groups are oriented to the exterior of the structure, which correspond to primary and secondary faces, respectively. These orientations provide water solubility (Brewster and Loftsson 2007; Carrier et al. 2007). On the other hand, the inner cavity is built by pyrane moieties of glucose units having hydrophobic properties (Lima et al. 2016). Consequently, solid cyclodextrins are obtained as hydrates and allows obtaining host-guest supramolecular structures with hydrophobic molecules (Loftsson and Duchêne 2007). During the interaction of hydrophobic compounds with the inner cavity of cyclodextrins, water molecules are replaced (Fig. 4.2). Dissociation of both water/solvent molecules and hydrophobic guest compounds can be evaluated by thermal methods (Mura 2015; Yuan et al. 2015; Manakov et al. 2016). Three main regions are considered for thermal analysis of cyclodextrins and their complexes (Meier et al. 2001; Sreenivasan 2001; Mura 2015): the first interval corresponds to water and/or solvent dissociation and vaporization up to ~130 °C; the last interval appears after ~270 °C and is attributed to the decomposition of cyclodextrins and possible the guest molecules; dissociation of the guest molecules generally appears in the range between these two temperatures. Furthermore, other physical processes such as minimizing the phase transition or glass transition state effects can be evaluated in the above mentioned region.

4.2.1 *α-Cyclodextrin*

Cyclodextrins are obtained by enzymatic processes (using cyclodextrin-glucosyltransferase). They are further separated and purified by crystallization from aqueous solutions. Generally, complexation agents are used for purification and crystallization. Consequently, cyclodextrins are obtained as hydrates.

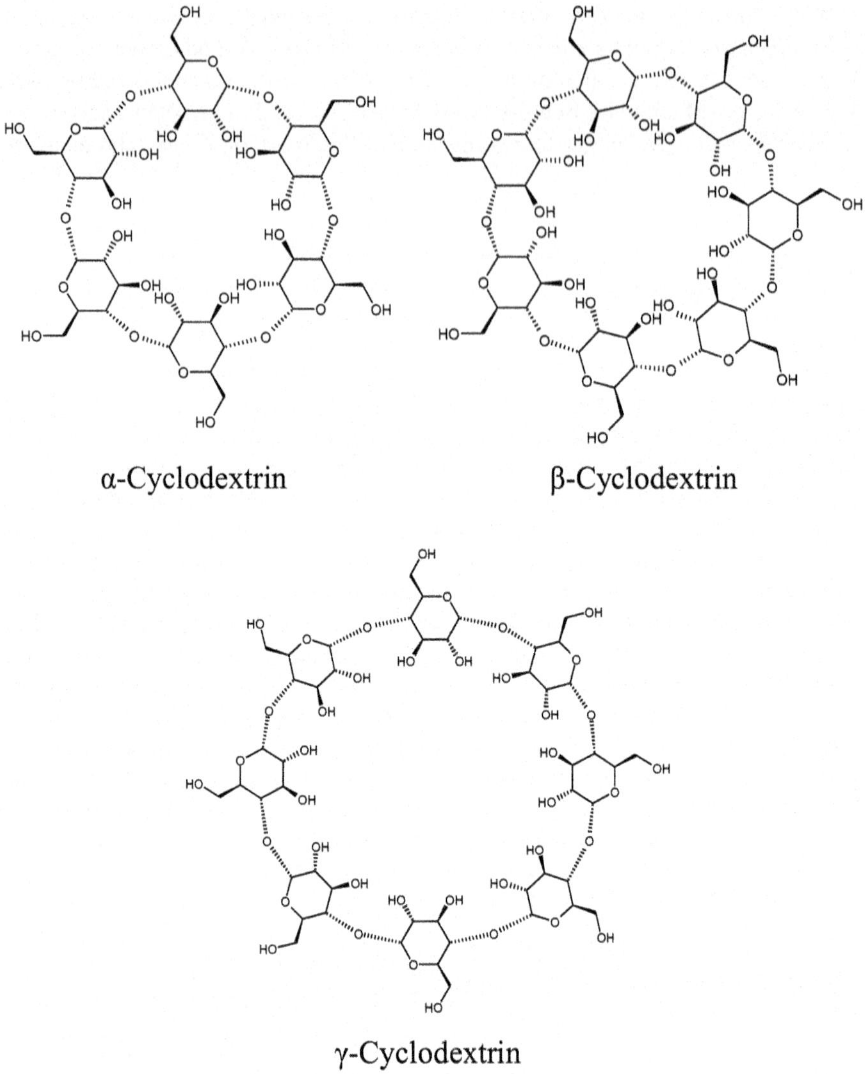

Fig. 4.1 Structures of natural cyclodextrins; α-, β- and γ-cyclodextrin comprise of six, seven and eight α-(1→4)-linked glucopyranose units, respectively

α-Cyclodextrin hydrate have three crystalline forms named form I, form II and form III. They depend on the method of preparation and purification. The first two forms contain six water molecules and the third 7.5 water molecules (Duchêne and Bochot 2016; Manakov et al. 2016; Szejtli 2004). Thermogravimetric analysis of commercial α-cyclodextrin indicates an important mass loss up to 105 °C of 9.3% (Fig. 4.3) that corresponds to water release. Only 0.7% mass loss was observed for the next interval of 105–270 °C. The decomposition of α-cyclodextrin occurs after

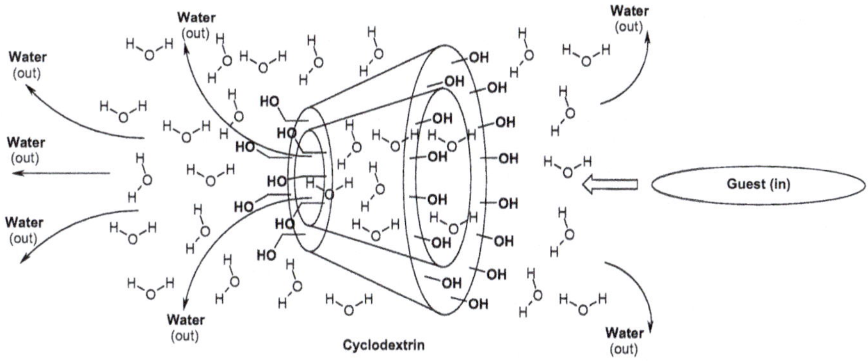

Fig. 4.2 The schematic mechanism of host-guest molecular encapsulation process. Cyclodextrin is the host molecule and have a truncated cone-like structure with primary and secondary hydroxyl groups (bolded, left and right, respectively). During the nanoencapsulation process hydration water molecules inside the cyclodextrin cavity are replaced by a hydrophobic guest molecule ("Guest-in" and "Water-out")

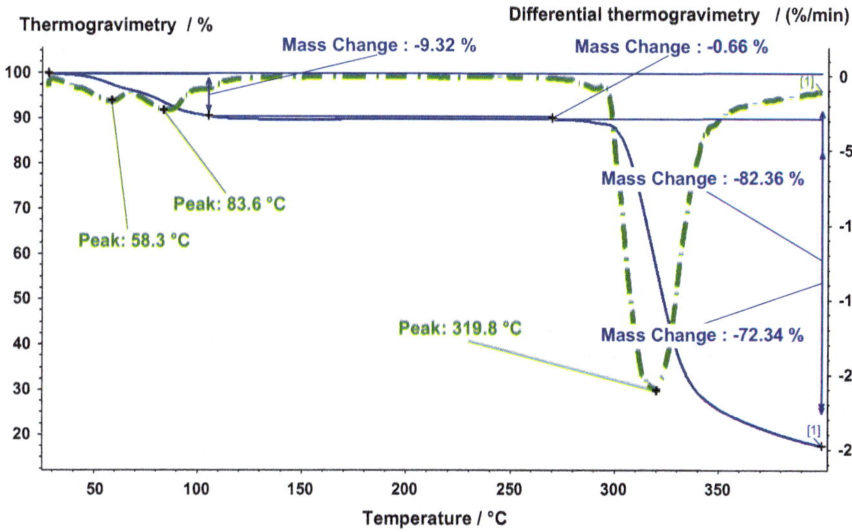

Fig. 4.3 Thermogravimetry-differential thermogravimetry analysis of α-cyclodextrin. Mass loss corresponding to water release of 9.3% have two significant temperature peaks of 58.3 °C and 83.6 °C. They correspond to different bonding of water molecules in the α-cyclodextrin hydrate. No other significant variation appears up to α-cyclodextrin decomposition starting at ~270 °C. The decomposition mass loss is 72.34% and have a peak temperature of 319.8 °C. The overall mass loss is 82.36% (Hădărugă et al. 2016, unpublished)

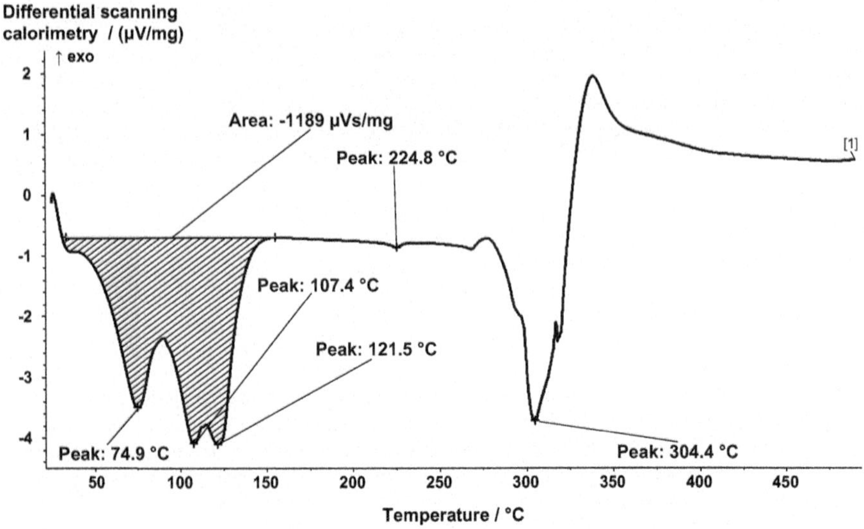

Fig. 4.4 Differential scanning calorimetry analysis of α-cyclodextrin. The main endothermic effect appears up to 150 °C and corresponds to water release. Three temperature peaks of 74.9 °C, 107.4 °C and 121.5 °C related to various physically bonded water molecules can be observed. A small endothermic state transition effect appears at 224.8 °C. α-Cyclodextrin decomposition has a calorimetric peak at 304.4 °C (Hădărugă et al. 2016, unpublished)

270 °C. The differential thermogravimetric peak appears at ~320 °C. Differential thermogravimetry reveals at least two peaks corresponding to water release at 58.3 °C and 83.6 °C. They suggest at least two types of hydration water molecules, named "surface water" and "strongly-retained water". However, the last type of water molecules are released even at 120 °C from the α-cyclodextrin hydrate.

Differential scanning calorimetry analysis of α-cyclodextrin provide information that is consistent with thermogravimetric data. Thus, "surface" water is released at lower temperatures with a maximum calorimetric effect at ~75 °C. On the other hand, two peaks appear for the release of "strongly-retained" water molecules at 107 °C and 121.5 °C (Fig. 4.4). The total calorimetric effect corresponding to water dissociation from α-cyclodextrin hydrate is about 1189 µV·s/mg or 338 J/g. There is a weak endothermic effect in the second differential scanning calorimetry interval, with a maximum at 224.8 °C. It is not clearly established the nature of this effect, but it can be related to transitions of α-cyclodextrin from a form to another. Rearrangement of the crystal structure after the loss of water molecules can be such physical process (Giordano et al. 2001).

Manakov and collaborators studied the α-cyclodextrin behavior during the dehydration of its hexahydrate form (Manakov et al. 2016). The temperature - composition phase diagram of α-cyclodextrin-water binary system was investigated by differential thermal analysis and differential scanning calorimetry. The structure of different α-cyclodextrin hydrates was determined by powder X-ray diffractometry. It

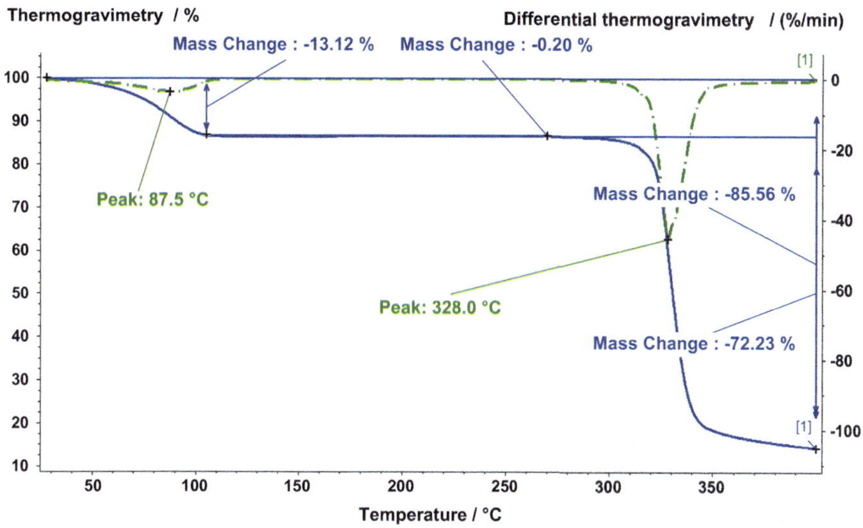

Fig. 4.5 Thermogravimetry-differential thermogravimetry analysis of β-cyclodextrin. A mass loss of 13.12% up to 110 °C corresponds to water release. Only one temperature peak of 87.5 °C appears in this region. The mass loss is not significant up to the decomposition of β-cyclodextrin that have a temperature peak of 328 °C. The overall mass loss is almost 85.6% (Hădărugă et al. 2016, unpublished)

was observed that α-cyclodextrin·6.1H_2O form I structure is transformed to α-cyclodextrin·5.3H_2O structure during the temperature range of 78–86 °C. On the other hand, Muñoz-Ruiz, Neoh and their collaborators observed another weak endothermic peak at 138–139 °C. It was attributed to an unknown structural modification of α-cyclodextrin than the water dissociation and evaporation (Muñoz-Ruiz and Paronen 1997; Neoh et al. 2008).

4.2.2 *β-Cyclodextrin*

β-Cyclodextrin is the most studied and used natural cyclodextrin. It has the lowest water solubility of ~1.8%. Despite of low water solubility, it is appropriate for pharmaceutical formulations. Moreover, β-cyclodextrin is very cheap. Consequently, many studies on thermal behavior of β-cyclodextrin and its complexes were performed. It is the case of thermogravimetry, differential thermal analysis and differential scanning calorimetry. β-Cyclodextrin has similar thermogravimetric behavior such as α-cyclodextrin. However, only one differential thermogravimetric large peak corresponding to water dissociation and evaporation appears at 87.5 °C (Fig. 4.5). The mass loss of the commercial β-cyclodextrin hydrate is about 13.1% up to 105 °C. Almost no variation in the second interval exists. The oxidation/

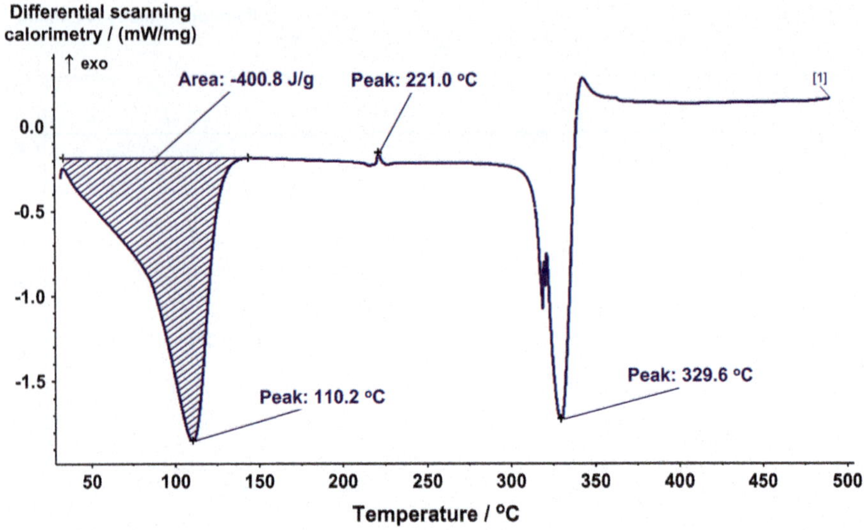

Fig. 4.6 Differential scanning calorimetry analysis of β-cyclodextrin. Only one endothermic peak of 110.2 °C was observed for the water release. The state transition process for the anhydrous β-cyclodextrin has an endothermic-exothermic calorimetric effect, with a maximum at 221 °C. The decomposition of β-cyclodextrin was observed at 329.6 °C in this measurement (Hădărugă et al. 2016, unpublished)

degradation and ignition of β-cyclodextrin starts from ~270 °C, with a differential thermogravimetric peak at 328 °C.

Differential scanning calorimetry analysis of commercial β-cyclodextrin hydrate also reveals an important endothermic peak corresponding to water release at ~110 °C (Fig. 4.6). No specific peaks for "surface" and "strongly-retained" water molecules such as in the case of α-cyclodextrin was observed. The 401 J/g calorimetric effect of β-cyclodextrin degradation appears at the temperature of 329.6 °C, which is consistent with thermogravimetric-differential thermogravimetric data. On the temperature range of 130–270 °C almost no calorimetric effect appear with one exception at ~221 °C. This exo-endothermic effect is attributed to structural modifications of anhydrous β-cyclodextrin crystal structure, because no mass loss appears in thermogravimetric analysis.

4.2.3 γ-Cyclodextrin

γ-Cyclodextrin hydrates generally have the lowest water content of all natural cyclodextrins, depending on the method of preparation. However, water molecules have the highest mobility in γ-cyclodextrin in comparison with α- and β-cyclodextrin. Thermogravimetric analysis of commercial γ-cyclodextrin indicates a mass loss of 7.2% up to 105 °C and almost no variation until decomposition (Fig. 4.7). The high

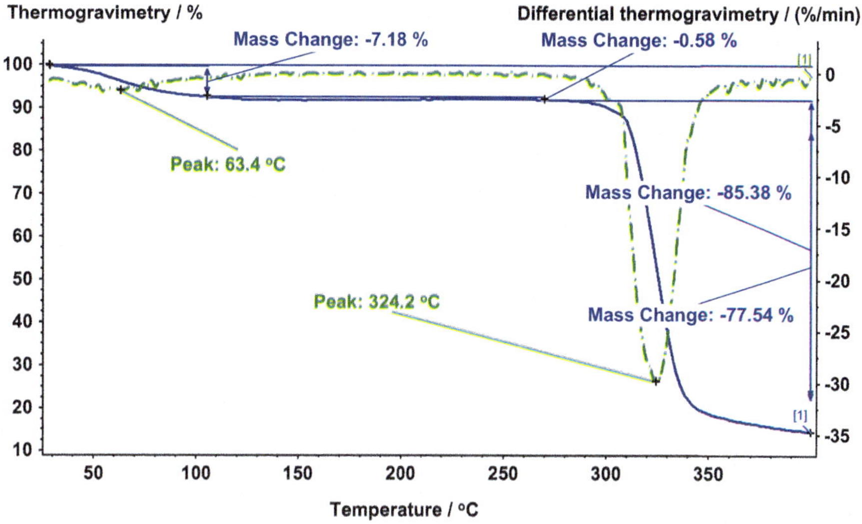

Fig. 4.7 Thermogravimetry-differential thermogravimetry analysis of γ-cyclodextrin. The mass loss of 7.18% corresponding to water release appears at 63.4 °C. The mass loss up to the decomposition of γ-cyclodextrin is only 0.58%. The decomposition of γ-cyclodextrin appears at 324.2 °C and have a mass loss of 77.54%. The overall mass loss is about 85.4% (Hădărugă et al. 2016, unpublished)

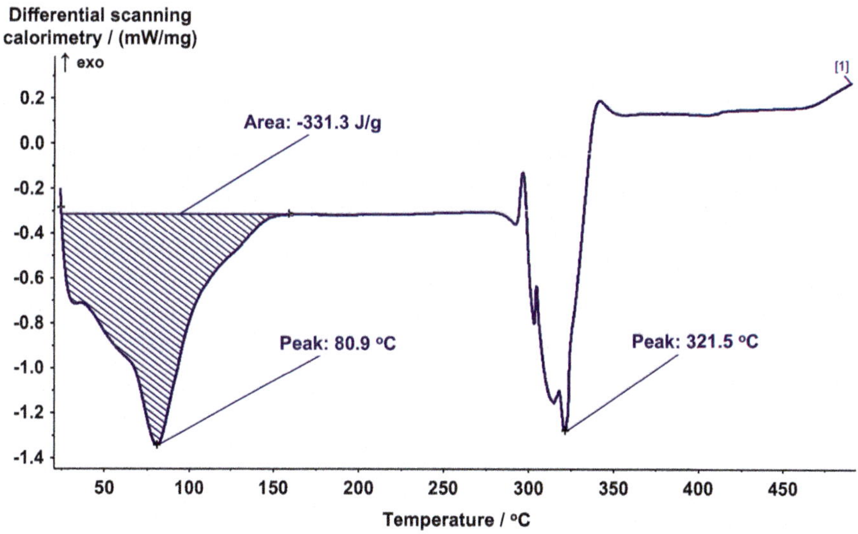

Fig. 4.8 Differential scanning calorimetry analysis of γ-cyclodextrin. The water release appears at 80.9 °C and the γ-cyclodextrin decomposition has a temperature peak of 321.5 °C. No other calorimetric effects were observed between these two processes (Hădărugă et al. 2016, unpublished)

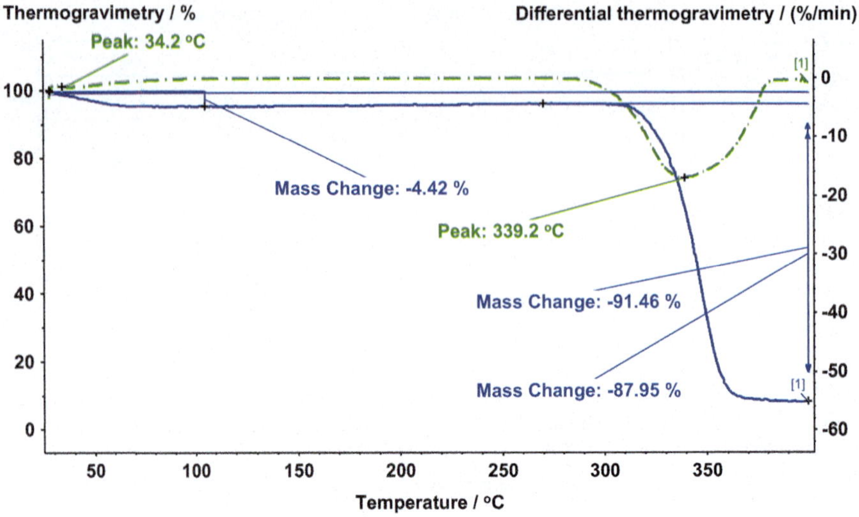

Fig. 4.9 Thermogravimetry-differential thermogravimetry analysis of 2-hydroxypropyl-β-cyclodextrin. A weak mass loss of 4.42% was determined for the water release from this hydrophilic cyclodextrin derivative. The decomposition of the 2-hydroxypropyl-β-cyclodextrin was determined at 339.2 °C, with a mass loss of 87.95%. The overall mass loss was 91.46% (Hădărugă et al. 2016, unpublished)

water mobility is supported by the differential thermogravimetric peak temperature corresponding to dissociation of water molecules (63.4 °C). This is at the lowest value for all natural cyclodextrins. On the other hand, the decomposition of γ-cyclodextrin starts at approximately the same temperature, with a maximum rate at ~324 °C. This is consistent with the differential scanning calorimetric peak temperature corresponding to the calorimetric effect of degradation of 321.5 °C (Fig. 4.8). The water release from the γ-cyclodextrin hydrate have an important calorimetric effect of ~331 J/g with a differential scanning calorimetric peak temperature of 80.9 °C (Fig. 4.8). The shape of the calorimetric peak is more related to α-cyclodextrin than β-cyclodextrin, revealing that it is a sum of peaks corresponding to different types of water molecules. Furthermore, no crystal modification of the anhydrous γ-cyclodextrin was observed.

4.2.4 Cyclodextrin Derivatives

2-Hydroxypropyl-β-cyclodextrin is one of the most studied cyclodextrin derivative. It has high water solubility, which can be >60% at higher degree of substitution. Consequently, the water molecules in 2-hydroxypropyl-β-cyclodextrin hydrate are weak physically bound in the crystal structure and released at lower temperature.

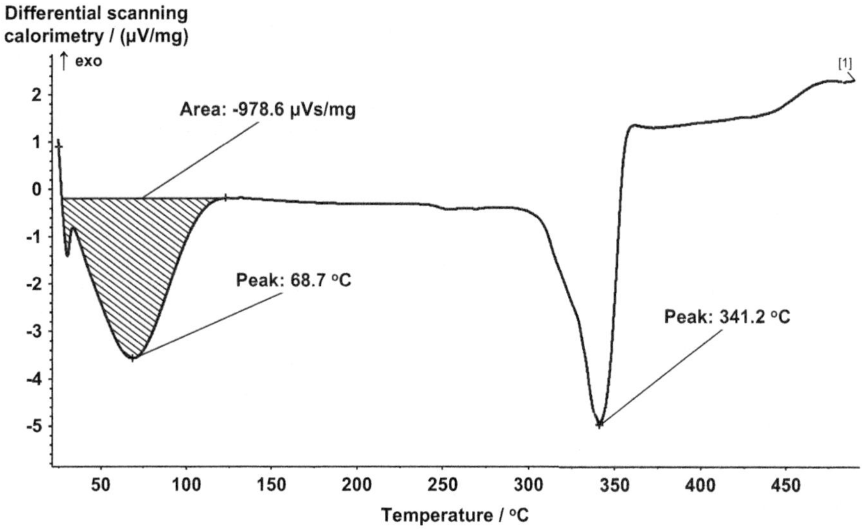

Fig. 4.10 Differential scanning calorimetry analysis of 2-hydroxypropyl-β-cyclodextrin. The water release process has an endothermic peak at 68.7 °C. No other significant calorimetric effect appear up to the decomposition of the cyclodextrin derivative at 341.2 °C (Hădărugă et al. 2016, unpublished)

The corresponding differential thermogravimetric peak temperature was ~34 °C (Fig. 4.9). Moreover, the water content is relatively low. The thermogravimetric mass loss was 4.4% up to 105 °C. No other physical chemical effects were observed up to the decomposition of 2-hydroxypropyl-β-cyclodextrin. Thus, differential thermogravimetric peak temperature was ~339 °C and the corresponding differential scanning calorimetric value was 341.2 °C (Figs. 4.9 and 4.10).

2-Hydroxypropyl-β-cyclodextrin was also evaluated by thermal methods in relation with the degree of substitution values (Yuan et al. 2015). Differential thermal analysis and differential scanning calorimetric data were compared with β-cyclodextrin results. It was observed that the decomposition temperature linearly increases with the degree of substitution of 2-hydroxypropyl-β-cyclodextrin.

2-Hydroxypropyl-β-cyclodextrin was analyzed by thermogravimetry-mass spectrometry combined technique, together with other cyclodextrin derivatives named 2-hydroxyethyl-β-cyclodextrin, heptakis(2,6-di-*O*-ethyl)-β-cyclodextrin, heptakis(2,6-di-*O*-methyl)-β-cyclodextrin, heptakis(2,3,6-tri-*O*-methyl)-β-cyclodextrin, partially methylated β-cyclodextrin having a degree of substitution of 4–6 and randomly methylated β-cyclodextrin (Éhen et al. 2005). The fragmentation behavior during heating have been evaluated. Results were compared with the all-important native cyclodextrins. No important fragmentation differences between native cyclodextrins and cyclodextrin derivatives were observed. However, characteristic mass spectrometric signals were identified for cyclodextrin derivatives, especially for ethylated and methylated cyclodextrins.

A comprehensive study on thermal analysis of cyclodextrin derivatives was performed by Veiga and collaborators (Veiga et al. 2002). Acetyl-α-, acetyl-β- and acetyl-γ-cyclodextrin, 6-*O*-maltosyl-β-cyclodextrin, methyl-β-cyclodextrin, hydroxypropyl-β- and hydroxypropyl-γ-cyclodextrin were characterized by hot stage microscopy, differential scanning calorimetry and thermogravimetry. No change of the particle appearance of acetylated cyclodextrins was observed by hot stage microscopy up to the fusion process that take place at 200–230 °C. This process depends on the type of cyclodextrin. On the other hand, methylated β-cyclodextrin showed solid-liquid transition at higher temperatures of 270–298 °C, while maltosyl-β-cyclodextrin had very low temperature of the transition phase of 171–185 °C. Relatively wide fusion temperature ranges for hydroxypropyl-β-cyclodextrin and hydroxypropyl-γ-cyclodextrin were observed. Values were in ranges of 220–257 °C and 230–260 °C, respectively. Furthermore, differential scanning calorimetry analysis of acetylated cyclodextrins reveals a slow oxidative decomposition. The thermogravimetric mass loss corresponding to dehydration of the studied cyclodextrin derivative hydrates was lower than 6.5%, excepting for maltosyl-β-cyclodextrin, where the mass loss was 8.4%. Thermogravimetric results are in good agreement with the differential scanning calorimetric effect of dehydration that was lower than 198 J/g.

4.3 Cyclodextrin Complexes

Cyclodextrin complexes are generally analyzed by thermal methods in order to evaluate the possibility to obtain the inclusion compound. This process can be suggested by the difference between the thermal behavior of cyclodextrin complexes and raw materials, named cyclodextrin, guest compound and physical mixture. There are particular aspects that qualitatively supports the formation of the molecular inclusion compound/complex: lowering the water content of the cyclodextrin/guest complex hydrate. They are generally obtained using water-hydrophilic solvent mixture by various methods such as crystallization from alcohol-water solution, kneading, spray-drying etc. The reducing of the "strongly-retained" water content after complexation or the "surface"/"strongly-retained" water ratio, the disappearance/reducing of the transition phase effect of the guest molecule, as well as the modification of the degradation behavior of cyclodextrin and guest molecules can be revealed by thermogravimetry, differential thermal analysis or differential scanning calorimetry. They are the most used thermal techniques for the evaluation of the host/guest molecular inclusion process.

Cyclodextrin/drug complexes were the most studied supramolecular systems, including analyses by thermal methods. Other guest compounds used for cyclodextrin nanoencapsulation belong to odorant, essential oil, fatty acid and oil classes as well as various organic, organometallic and inorganic compounds. Consequently, a synopsis on the thermal analysis of cyclodextrin complexes that are classified according to the incidence of the guest compounds in the literature is presented.

4.3.1 Cyclodextrin/Drug Complexes

Cyclodextrin/drug complexes were classified according to Anatomical Therapeutic Chemical classification of the guest molecule. They are relatively well distributed from the cyclodextrin complexation point of view (Table 4.1). Thermal analysis techniques used for the analysis of cyclodextrin/drug complexes are especially differential scanning calorimetry, thermogravimetry, differential thermal analysis, hot stage microscopy and even isothermal titration calorimetry.

Natamycin is used as anti-infective and antiseptic for local oral treatment, antibiotic against intestinal, gynecological and ophthalmological disorders and antifungal for dermatological preparations. Its β-cyclodextrin, γ-cyclodextrin and 2-hydroxypropyl-β-cyclodextrin complexes were studied by thermogravimetry for the water release behavior (Koontz and Marcy 2003). However, cyclodextrin/natamycin complexes have similar thermogravimetric behavior that are close to the starting cyclodextrins, suggesting the difficulty of complete molecular inclusion of this large guest, even in the γ-cyclodextrin cavity (Fig. 4.11).

Dexamethasone (Fig. 4.12) is a corticosteroid having various biological activities such as vasoprotective, anti-acne, decongestant activity, as well as in otological and ophthalmological preparations. It was investigated as γ-cyclodextrin and its alkylcarbonate derivative complexes by differential scanning calorimetric analysis demonstrating the disappearance of the melting peak of the drug (Cavalli et al. 2007). Ranitidine and nateglinide are anti-ulcerous and glucose lowering drugs, respectively (Fig. 4.12). They have been studied for their capacity to be included in β-cyclodextrin. The first complex was analyzed by differential scanning calorimetry and thermogravimetry that reveals the photostability of the complex (Jamrógiewicza et al. 2014). On the other hand, nateglinide was studied for the possibility to enhance the water solubility by β-cyclodextrin complexation (Ikeda et al. 2016). The isothermal titration calorimetric analysis reveals the inclusion of the isopropylcyclohexyl moiety in the β-cyclodextrin cavity rather than phenyl group.

α-Lipoic, nicotinic (vitamin B_3), folic (vitamin B_9) and orotic (vitamin B_{13}) acids belong to vitamin B related compounds class (Fig. 4.13). They have been investigated for their complexation ability by natural and semi-synthetically modified cyclodextrins. Characterization of complexes were performed using various methods such as classical methods X-ray diffractometry, Fourier transform-infrared spectroscopy, or ^{1}H–nuclear magnetic resonance, as well as thermal methods: thermogravimetry, differential thermal analysis, differential scanning calorimetry and isothermal titration calorimetry. It was demonstrated that the S-S linkage from α-lipoic acid is protected against degradation after inclusion in the α- or β-cyclodextrin cavity, according to differential thermal analysis data below 100 °C (Lin-Hui et al. 1995). Nicotinic acid was complexed by β-cyclodextrin and heptakis(2,6-*O*-dimethyl)-β-cyclodextrin at host:guest molar ratios of 1:1 and 1:2 using freeze-drying and co-precipitation methods (Funk et al. 1993). The formation of the crystalline β-cyclodextrin/nicotinic acid inclusion complex obtained by co-precipitation was demonstrated by X-ray diffractometry, thermogravimetry, differ-

Table 4.1 Cyclodextrin/drug complexes analyzed by thermal methods (Anatomical Therapeutic Chemical classification of the guest compound, the type of cyclodextrins used for obtaining complexes and the corresponding thermal methods used for characterization were included in the table)

N°	Guest compound (drug)	Anatomical Therapeutic Chemical class	Host compound (cyclodextrin)[a]	Thermal analyses[b]	References
1	Natamycin	A01AB; A07AA; D01AA; G01AA; S01AA	β-CD, HP-β-CD, γ-CD	TG	Koontz and Marcy 2003
2	Dexamethasone	A01AC; C05AA; D07AB; D07CB; D07XB; D10AA; H02AB; R01AD; S02BA; S03BA; S03CA	γ-CD	DSC	Cavalli et al. 2007
3	Ranitidine	A02BA	β-CD	DSC, TG	Jamrógiewicza et al. 2014
4	Nateglinide	A10BX	β-CD	ITC	Ikeda et al. 2016
5	α-Lipoic acid	A11EA	α-CD, β-CD	TG, DTA	Lin-Hui et al. 1995
6	Nicotinic acid (vitamin B_3)	A11EA	β-CD, DIMEB	DSC, TG	Funk et al. 1993
7	Orotic acid (vitamin B_{13})	A11EA	β-CD	DSC	Kacso et al. 2009
8	Folic acid (vitamin B_9)	A11EA	α-CD, β-CD, γ-CD	ITC	Ceborska et al. 2016
9	Phenylalanine	B05BA	α-CD, β-CD	TG, DTA	Song et al. 2006
10	Ibuprofen	C01EB; G02CC; M01AE; M02AA	β-CD	DSC, TG	Pereva et al. 2016
11	Indometacin	C01EB; M01AB; M02AA	α-CD, β-CD, γ-CD	DSC	Makhlof et al. 2008
12	Bosentan	C02KX	HP-β-CD	DSC	Jadhav and Pore 2016
13	Furosemide	C03CA; C03CB; C03EB	HP-β-CD, DIMEB, RAMEB	TG, DTG, DTA	Kreaz et al. 1999
14	Nitrendipine	C08CA	HP-β-CD	DSC	Wang et al. 2007
15	Fosinopril	C09AA; C09BA	HP-β-CD	TG	Sbârcea et al. 2015

(continued)

Table 4.1 (continued)

N°	Guest compound (drug)	Anatomical Therapeutic Chemical class	Host compound (cyclodextrin)[a]	Thermal analyses[b]	References
16	Lisinopril	C09AA; C09BA	β-CD	TG, DTA	Sbârcea et al. 2016
17	Zofenopril	C09AA; C09BA	β-CD	TG, DTA	Sbârcea et al. 2016
18	Valsartan	C09CA; C09DA	β-CD	DSC	de Matos Jensen et al. 2010
19	Amikacin	D06AX; J01GB; S01AA	β-CD polymer	DSC	Skiba et al. 2005
20	Acyclovir	D06BB; J05AB; S01AD	β-CD	DSC, TG	Rossel et al. 2000
21	Tamibarotene (retinobenzoic acid)	D10AD	HP-β-CD	DSC	Yang et al. 2017
22	Naproxen	G02CC	HP-β-CD, SBE-β-CD, Ac-β-CD, Ac-γ-CD	DSC, TG, cyclic DSC	Bettinetti et al. 2002
23	Naproxen and various aminoacids (L-Lys, L-Val, L-Leu, L-Arg)	G02CC	HP-β-CD	DSC, HSM	Mura et al. 2003
24	Naproxen	G02CC; M01AE; M02AA	β-CD, γ-CD	ITC	Todorova and Schwarz 2007
25	Progesterone	G03DA	HP-β-CD	DSC	Cerchiara et al. 2003
26	Progesterone	G03DA	γ-CD	DSC	Cavalli et al. 2007
27	Tadalafil	G04BE	β-CD, HP-β-CD, DIMEB	DSC	Badr-Eldin et al. 2008
28	Cefuroxime axetil	J01DC	β-CD	DSC	Sapte and Pore 2016
29	Meropenem	J01DH	β-CD	DSC	Paczkowska et al. 2016
30	Norfloxacin	J01MA; S01AX	β-CD, HP-β-CD	DSC	Loh et al. 2016
31	Enrofloxacin	J01MA; S01AX; S02AA	α-CD, β-CD, γ-CD, HP-β-CD	DSC, TG	Calsavara et al. 2012
32	Itraconazole	J02AC	β-CD, HP-β-CD, RAMEB, CRYSMEB, SBE-β-CD	DSC	Thiry et al. 2017
33	2-Phenyl-5-{[(2-phenyl-1,3-dioxolan-2-yl) methyl] sulfanyl}-1,3,4-oxadiazole	J04A	β-CD	DSC	Boldescu et al. 2012

(continued)

Table 4.1 (continued)

N°	Guest compound (drug)	Anatomical Therapeutic Chemical class	Host compound (cyclodextrin)[a]	Thermal analyses[b]	References
34	Rifampicin	J04AB	HP-β-CD	DSC	He et al. 2013
35	Efavirenz	J05AG	β-CD, HP-β-CD, RAMEB	DSC	Sathigari et al. 2009
36	Paclitaxel	L01CD	HP-β-CD-graphene oxide conjugate	TG	Tan et al. 2016
37	Celecoxib	L01XX	β-CD	DSC	Reddy et al. 2004
38	Cyclosporine A	L04AD	α-CD, HP-β-CD (mixture)	DSC	Malaekeh-Nikouei et al. 2007
39	Oxyphenbutazone	M01AA	β-CD, γ-CD	DSC, TG, HSM	Veiga and Merino 2002
40	Lornoxicam	M01AC	β-CD, HP-β-CD	DSC	Ammar et al. 2012
41	Oxaprozin	M01AE	α-CD, β-CD, γ-CD, DIMEB, RAMEB, CRYSMEB	DSC	Maestrelli et al. 2009
42	Flurbiprofen	M01AE; M02AA	β-CD, γ-CD	ITC	Todorova and Schwarz 2007
43	Flurbiprofen	M01AE; M02AA	γ-CD	DSC	Cavalli et al. 2007
44	Flurbiprofen	M01AE; M02AA	RAMEB	DSC	Rudrangi et al. 2016
45	Ketoprofen	M01AE; M02AA	HP-β-CD	DSC	Wang et al. 2007
46	Nabumetone	M01AX	β-CD, γ-CD	ITC	Todorova and Schwarz 2007
47	Acetylsalicylic acid	N02BA; A01AD; B01AC; M01BA	β-CD	TG, DTA	Szafranek and Szafranek 1993
48	Diazepam	N05BA	γ-CD	DSC	Cavalli et al. 2007
49	ITH12674 (3-(2-Isothiocyanato-ethyl)-5-methoxy-1*H*-indole)	N05CH	HP-β-CD	DSC	Michalska et al. 2017
50	Tryptophan	N06AX	α-CD, β-CD	TG, DTA	Song et al. 2006
51	Nicotine	N07BA	β-CD	TG	Hădărugă et al. 2010
52	Riluzole	N07XX	HP-β-CD	DSC, TG	Wang et al. 2016

(continued)

Table 4.1 (continued)

Nº	Guest compound (drug)	Anatomical Therapeutic Chemical class	Host compound (cyclodextrin)[a]	Thermal analyses[b]	References
53	Pyrimethamine	P01BD	HP-β-CD	DSC	Onyeji et al. 2009
54	Praziquantel	P02BA	Me-β-CD	DSC	Gomes et al. 2011
55	Ephedrine	R01AA; R01AB; R01BA; R03CA	β-CD	ITC	Wszelaka-Rylik 2017
56	Salbutamol	R03AC; R03AK	α-CD, β-CD, γ-CD, DIMEB	DSC	Cabral Marques et al. 1990
57	Brompheniramine	R06AB	α-CD, β-CD	ITC	Wszelaka-Rylik and Gierycz 2013
58	Bis(*t*-butyl-*S*-acyl-2-thioethyl)-AraCMP (UA911)		HP-β-CD	DSC	Jordheim et al. 2009

[a]*α-CD* α-cyclodextrin; *β-CD* β-cyclodextrin; *γ-CD* γ-cyclodextrin; *HP-β-CD* 2-hydroxypropyl-β-cyclodextrin; *DIMEB* heptakis(2,6-*O*-dimethyl)-β-cyclodextrin; *RAMEB* randomly methylated β-cyclodextrin; *SBE-β-CD* β-cyclodextrin heptakis(sulfobutylether), as sodium salt; *Ac-β-CD* acetyl-β-cyclodextrin; *Ac-γ-CD* acetyl-γ-cyclodextrin; *CRYSMEB* low methylated β-cyclodextrin.
[b]*TG* thermogravimetry; *DSC* differential scanning calorimetry; *ITC* isothermal titration calorimetry; *DTA* differential thermal analysis; *HSM* hot stage microscopy

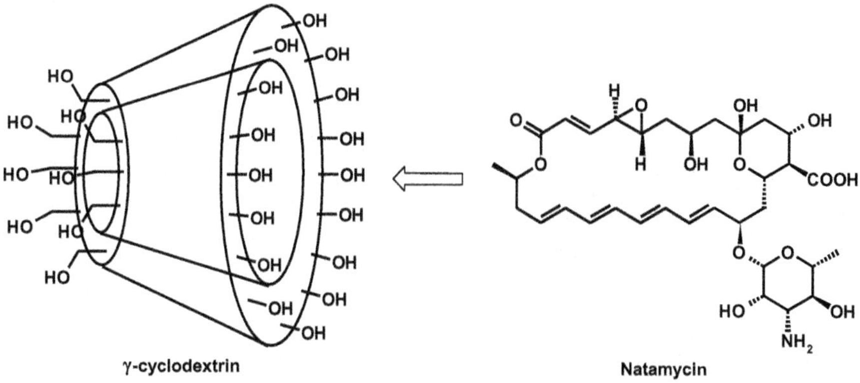

Fig. 4.11 Possible interaction mechanism of natamycin and γ-cyclodextrin. Molecular encapsulation of the drug compound take place by the more hydrophobic moiety oriented to the secondary face of the γ-cyclodextrin

ential scanning calorimetry and thermofractography. On the other hand, freeze-drying provide an amorphous β-cyclodextrin/nicotinic acid complex, but heptakis(2,6-*O*-dimethyl)-β-cyclodextrin poorly encapsulates this vitamin. Folic acid was used as guest molecule in natural cyclodextrin complexation (Ceborska et al. 2016).

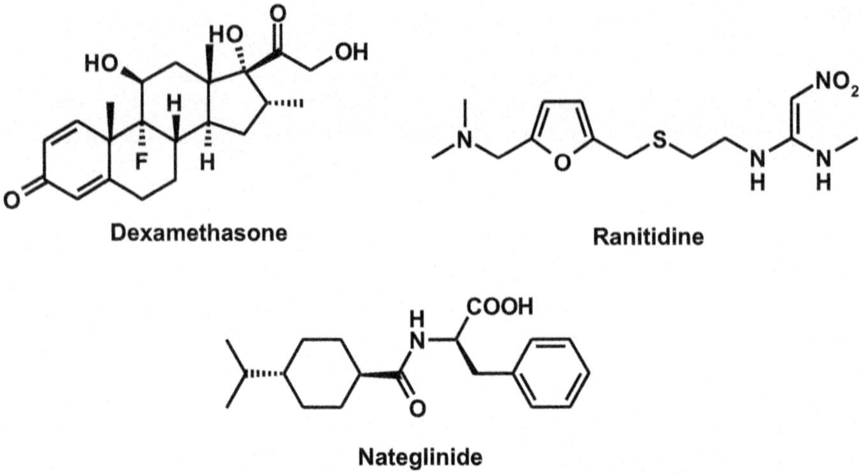

Fig. 4.12 Structures of dexamethasone, ranitidine and nateglinide used for cyclodextrin complexation

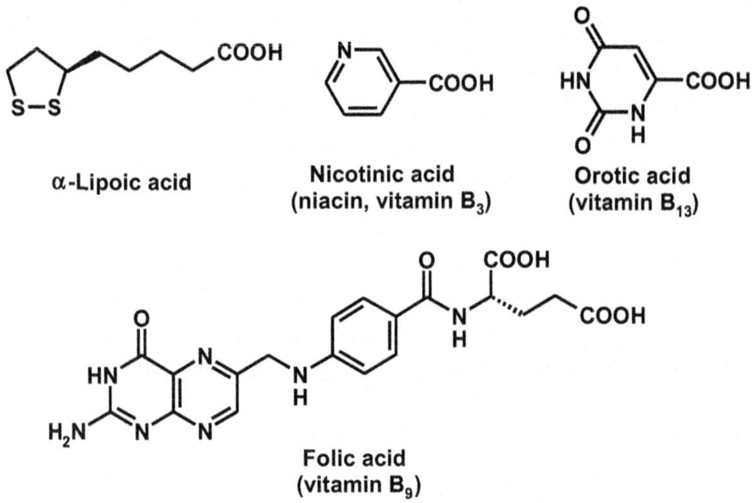

Fig. 4.13 Structures of the main vitamin B compounds used for cyclodextrin complexation

Complexes were analyzed by ^{1}H–nuclear magnetic resonance, electrospray ionization mass spectrometry and isothermal titration calorimetry. It was concluded that the stability of complexes increases in the following order: α-cyclodextrin/folic acid < γ-cyclodextrin/folic acid < β-cyclodextrin/folic acid complexes. Vitamin B_{13} is structurally related to vitamin B_3, but it is more hydrophilic (Fig. 4.13). Consequently, the guest compound was not completely included in the β-cyclodextrin cavity, as was revealed by differential scanning calorimetric analysis (Kacso et al. 2009). This

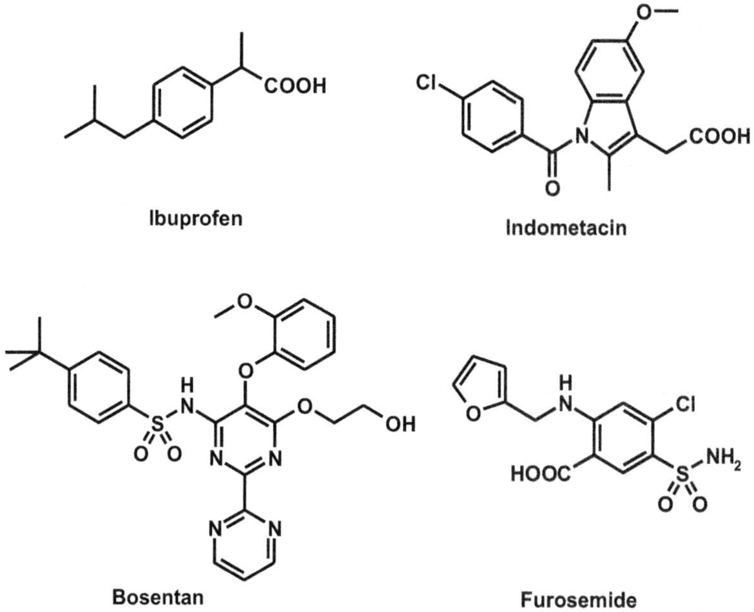

Fig. 4.14 Structures of anti-inflammatory drugs ibuprofen and indomethacin (top), as well as bosentan and furosemide (bottom) used for cyclodextrin complexation

aspect was evident in the kneading method, where the differential scanning calorimetric peak corresponding to the melting of orotic acid is still present at 146 °C.

Phenylalanine and tryptophan are two of aromatic amino acids investigated for their ability to be included in cyclodextrin cavities. All enantiomers as well as racemic mixtures were complexed with α- and β-cyclodextrin by co-precipitation method and analyzed by X-ray diffractometry, Fourier transform-infrared spectroscopy, electrospray ionization mass spectrometry, ^{1}H–nuclear magnetic resonance, thermogravimetry and differential thermal analysis (Song et al. 2006). The decomposition temperatures of complexes in comparison with the pure cyclodextrins were investigated by thermal analyses.

Ibuprofen and indomethacin are used as anti-inflammatory and anti-rheumatic agents, in muscular pain and even for some cardiac preparations (Fig. 4.14). They were complexed with natural cyclodextrins using various methods and the efficiency of encapsulation was evaluated by differential scanning calorimetry and thermogravimetry. An optimized ball milling method for obtaining β-cyclodextrins/ibuprofen complex provide a high degree of complexation with a low water content of the complex hydrate, as was determined by differential scanning calorimetry and thermogravimetry (Pereva et al. 2016). Cyclodextrin based nanocrystals containing indomethacin were obtained using emulsion solvent diffusion coupled with freeze-drying (Makhlof et al. 2008). Slight modification of the differential scanning calorimetric profile of nanocrystals in comparison with the starting cyclodextrins was observed, especially at the temperature interval corresponding to the melting of indomethacin. The lowering of the corresponding peak temperature from 159.4 °C for pure indomethacin to ~152 °C was determined.

Bosentan and furosemide (Fig. 4.14) were complexed with 2-hydroxypropyl-β-cyclodextrin and analyzed by differential scanning calorimetry, thermogravimetry and differential thermal analysis. The 2-hydroxypropyl-β-cyclodextrin/bosentan mixture was lyophilized and the formation of the inclusion complex was proved by differential scanning calorimetry, Fourier transform-infrared spectroscopy and X-ray diffraction. Furthermore, the solubility and dissolution properties of the complex were enhanced using L-arginine in the complexation process (Jadhav and Pore 2016). Cyclodextrin derivatives such as 2-hydroxypropyl-β-cyclodextrin, heptakis(2,6-di-*O*-methyl)-β-cyclodextrin or randomly methylated β-cyclodextrin were complexed with furosemide by freeze-drying at 1:1 and 2:1 molar ratios. The host-guest interaction was confirmed by X-ray diffractometry and the water content and dissociation from the complex were determined by thermogravimetry and differential thermal analysis (Kreaz et al. 1999).

Other studied cyclodextrin complexes with drugs having various vascular effects were nitrendipine, fosinopril, lisinopril, zofenopril and valsartan (Fig. 4.15). Nitrendipine was complexed by 2-hydroxypropyl-β-cyclodextrin using lyophilization of the corresponding *tert*-butanolic solution (Wang et al. 2007). The amorphous characteristic of the complex was proved by Fourier transform-infrared spectroscopy, X-ray diffraction and differential scanning calorimetric analyses. The same cyclodextrin derivative was used for the complexation of fosinopril, an angiotensin-converting-enzyme inhibitor. The 2-hydroxypropyl-β-cyclodextrin/fosinopril complexes were characterized by classical thermogravimetry-differential thermogravimetry, differential thermal analysis, and Fourier transform-infrared spectroscopy techniques and the possible interaction with common excipients was evaluated (Sbârcea et al. 2015). Lisinopril and zofenopril belong to the same Anatomical Therapeutic Chemical class as agents acting on the renin-angiotensin system, angiotensin-converting-enzyme inhibitors, and plain or in combinations. They were studied for the ability to form complexes with β-cyclodextrin by kneading method (Sbârcea et al. 2016). The stabilization of the drug compound in the presence of lactose was also investigated by thermogravimetry-differential thermogravimetry, differential thermal analysis and Fourier transform-infrared spectroscopy. Valsartan is an angiotensin II antagonist with low water solubility. β-Cyclodextrin/valsartan complexes were obtained using various techniques such as kneading, solid dispersion and freeze-drying methods and solid complexes were analyzed by differential scanning calorimetry, Fourier transform-infrared spectroscopy, X-ray diffractometry and nuclear magnetic resonance, as well as by isothermal titration calorimetry in solution. Analyses suggest the formation of a 1:1 inclusion compound (de Matos Jensen et al. 2010).

Amikacin is an antibiotic for systemic, ophthalmological and dermatological uses (Fig. 4.16). Obtaining of cyclodextrin-based microspheres containing amikacin for pulmonary drug delivery is challenging. Cross-linked β-cyclodextrin polymers were synthesized using terephthaloyl chloride and microcapsules containing amikacin

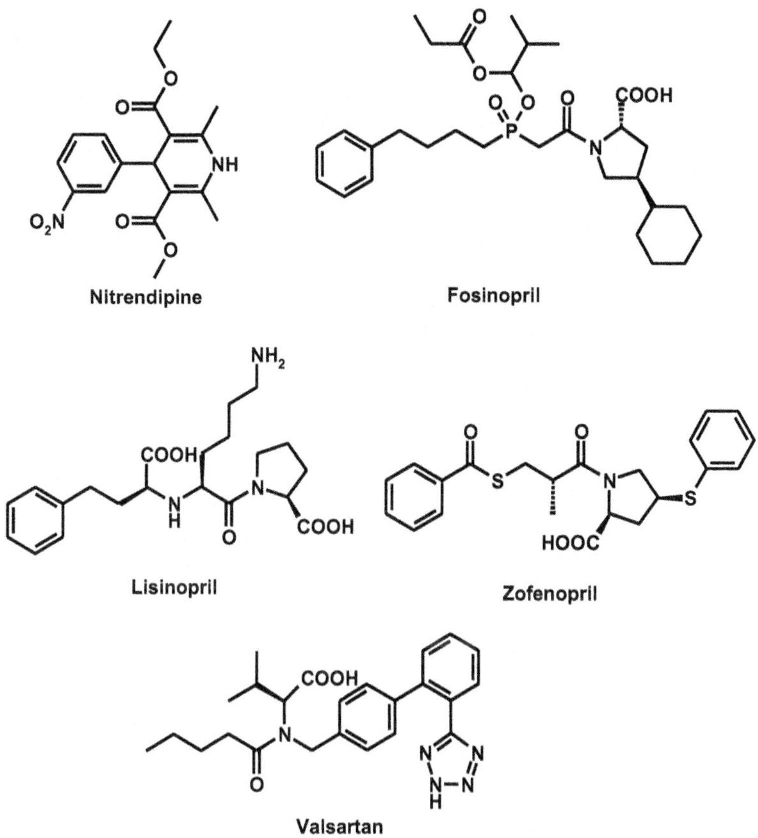

Fig. 4.15 The main structures of drugs with vascular effects that were used for obtaining cyclodextrin-based complexes with enhanced properties

were obtained by freeze-drying. These drug delivery systems were characterized by Fourier transform-infrared spectroscopy, differential scanning calorimetry and laser diffraction technique for the size distribution. The optimum microparticle diameter was about 5 μm for a proper encapsulation of amikacin (Skiba et al. 2005).

Antiviral drugs were also investigated for enhancing the bioavailability and water solubility, as well as stability, using cyclodextrin complexation. Acyclovir is a classical antiviral compound used for systemic, dermatological or ophthalmological applications (Fig. 4.16). It was nanoencapsulated in β-cyclodextrin by freeze-drying. The complex was classically analyzed by X-ray diffractometry, ^{1}H–nuclear magnetic resonance and differential scanning calorimetry revealing the disappearance of the peak corresponding to melting of acyclovir after complexation (Rossel et al. 2000).

Tamibarotene or retinobenzoic acid is a retinoid used in anti-acne preparations (Fig. 4.16). It was recently studied for enhancing its bioavailability and water solubility by 2-hydroxypropyl-β-cyclodextrin complexation. Inclusion complexes were

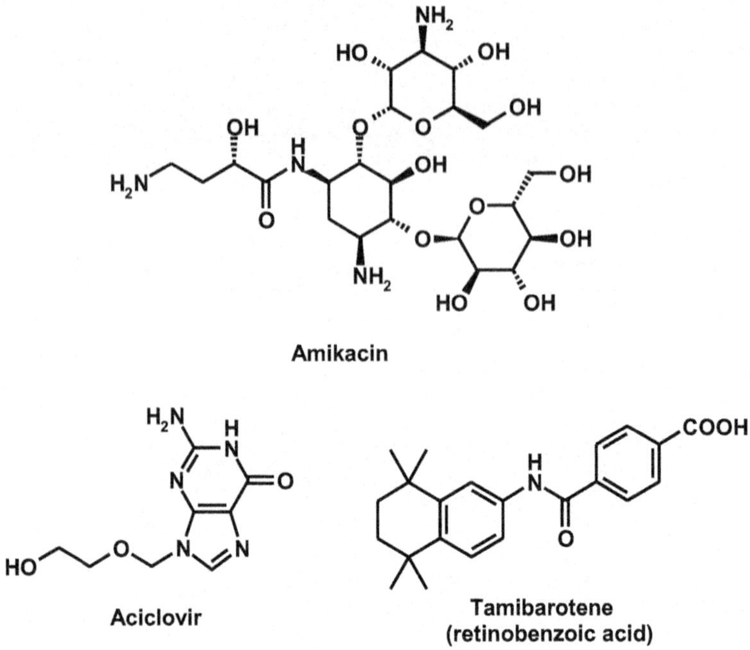

Fig. 4.16 The structures of amikacin, aciclovir and tamibarotene used for cyclodextrin complexations

obtained by freeze-drying and analyzed by Fourier transform-infrared spectroscopy and differential scanning calorimetry (Yang et al. 2017).

Naproxen, progesterone and tadalafil are some drugs designed for genital system. All these drugs have very low water solubility and bioavailability. Consequently, they were investigated for enhancing these properties by cyclodextrin complexation. Naproxen (Fig. 4.17) was studied for the possibility of nanoencapsulation by various cyclodextrin derivatives such as 2-hydroxypropyl-β-cyclodextrin, β-cyclodextrin heptakis(sulfobutylether) and acetylated β- and γ-cyclodextrin. Significant differences between charged cyclodextrin such as β-cyclodextrin heptakis(sulfobutylether) and the other cyclodextrin derivative complexes with naproxen by means of differential scanning calorimetric behavior were observed (Bettinetti et al. 2002). Cyclic heating-cooling differential scanning calorimetry for cyclodextrin derivative/naproxen complexes have been performed. Isothermal titration calorimetry was also applied to evaluates the binding of naproxen to β- and γ-cyclodextrin and the retention of bulk water in complexes (Todorova and Schwarz 2007). 2-Hydroxypropyl-β-cyclodextrin was used to obtain ternary supramolecular systems containing naproxen and various amino acids such as L-Lys, L-Val, L-Leu and L-Arg by co-grinding and co-evaporation methods (Mura et al. 2003). The synergistic effect of amino acids for enhancing the apparent water solubility of naproxen was evaluated. Solid ternary complexes were analyzed by differential scanning

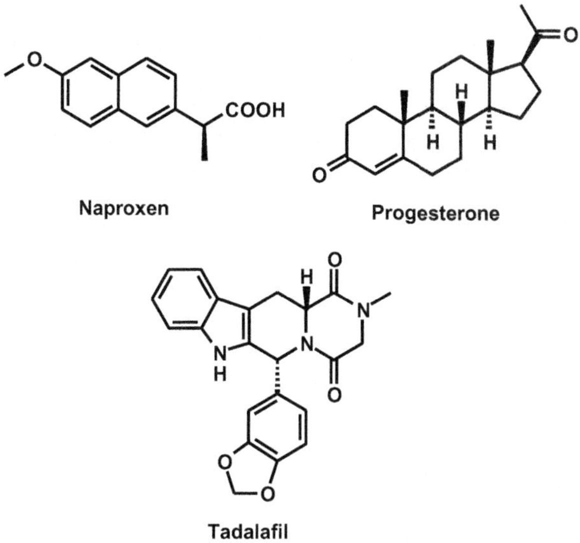

Fig. 4.17 Structures of genital system drugs used for cyclodextrin complexation

calorimetry, hot stage microscopy and scanning electron microscopy. It was demonstrated that 2-hydroxypropyl-β-cyclodextrin/naproxen/L-Arg ternary complex obtained by co-evaporation had enhanced properties such as bioavailability, reducing burning sensation and the bitter taste. Progesterone is a hydrophobic drug (Fig. 4.17). It belongs to progestogen anatomical therapeutic chemical class and have a rigid structure that allows molecular encapsulation in cyclodextrin cavities. Progesterone was nanoencapsulated in 2-hydroxypropyl-β-cyclodextrin and further incorporated into chitosan by spray-drying and freeze-drying methods (Cerchiara et al. 2003). The formation of the inclusion complex was evaluated by differential scanning calorimetry, through the disappearance of the melting calorimetric effect of progesterone after encapsulation at 125.3 °C. Other study related to progesterone encapsulation was focused on the effect of derivatization of γ-cyclodextrin with alkylcarbonate moieties on the progesterone complexation (Cavalli et al. 2007). Complexes were analyzed by X-ray diffractometry and differential scanning calorimetry and the effect of alkyl chain length on the complexation and release behavior was evaluated. It was evaluated the crystallinity of progesterone complexes obtained with these γ-cyclodextrin derivatives. A drug used for erectile dysfunction having very low water solubility is tadalafil (Fig. 4.17). The possibility to enhance the water solubility and dissolution rate was investigated by complexation with natural β-cyclodextrin and chemically modified cyclodextrins, named 2-hydroxypropyl-β-cyclodextrin and heptakis(2,6-*O*-dimethyl)-β-cyclodextrin (Badr-Eldin et al. 2008). The complexes were analyzed by differential scanning calorimetry, X-ray diffractometry and Fourier transform-infrared spectroscopy and the complexation efficiency was heptakis(2,6-*O*-dimethyl)-β-cyclodextrin >2-hydroxypropyl-β-cyclodextrin > β-cyclodextrin.

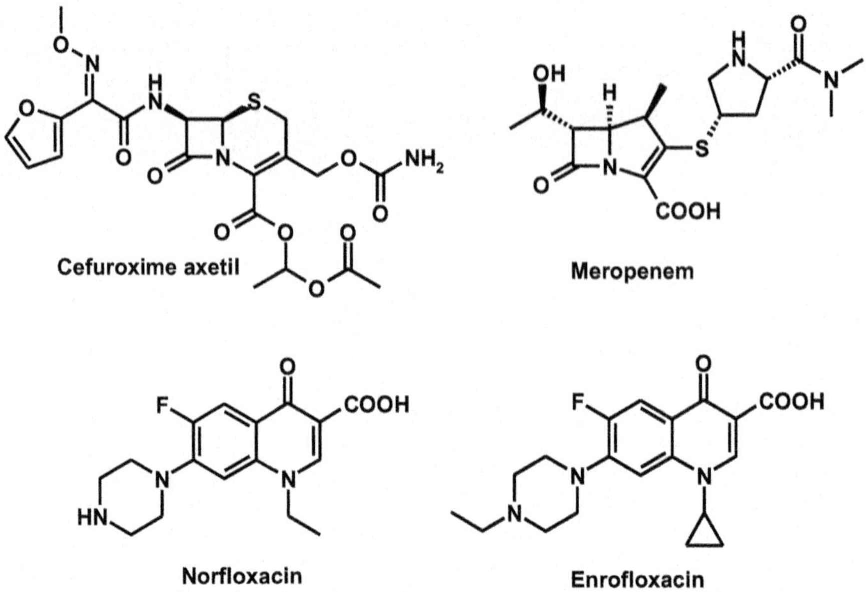

Fig. 4.18 Antibacterial structures used for cyclodextrin complexation

Antibacterial agents for systemic use were widely studied for their capability to be nanoencapsulated by cyclodextrins. It is the case of cefuroxime axetil and meropenem from β-lactam class (Fig. 4.18). Norfloxacin and enrofloxacin (Fig. 4.18) belong to fluoro-quinolone class, non-steroidal anti-inflammatory drugs. Cefuroxime axetil was complexed with β-cyclodextrin by spray-drying technique for enhancing the water solubility and bioavailability. Furthermore, the influence of L-Arg addition to the complexation process was also investigated by differential scanning calorimetry, X-ray diffractometry and scanning electron microscopy (Sapte and Pore 2016). Differential scanning calorimetric analysis revealed the disappearance of the glass transition peak for both binary and ternary complexes in comparison with the guest compound at 86.5 °C. It was established that the ternary complex has better physical chemical properties in comparison with the corresponding β-cyclodextrin complex. Meropenem was encapsulated in β-cyclodextrin for enhancing the bioavailability and the chemical stability such as the degradation of the amide bond from the β-lactam ring by hydrolysis (Paczkowska et al. 2016). Analyses by Raman spectroscopy, Fourier transform-infrared spectroscopy and differential scanning calorimetry confirmed the formation of the inclusion complex. The disappearance of the differential scanning calorimetric exothermic peak from 110–135 °C was attributed to the transition into crystalline state or decomposition of meropenem. The poorly water solubility of norfloxacin of 0.45 mg/mL and that for enrofloxacin of 0.15–0.17 mg/mL as well as their reduced bioavailability directed the researches on the nanoencapsulation in cyclodextrins. Loh and collaborators used β-cyclodextrin and 2-hydroxypropyl-β-cyclodextrin in order to improve

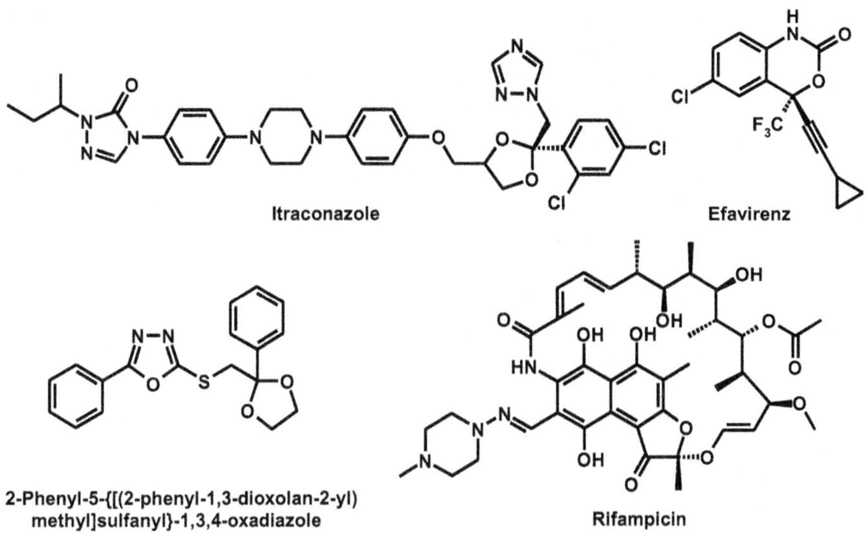

Fig. 4.19 Anti-mycotic, anti-mycobacterial and anti-viral structures used for cyclodextrin complexation

the water solubility and dissolution rate of norfloxacin (Loh et al. 2016). Complexes were obtained by three different methods at four molar ratios, namely 1:1, 2:1, 4:1 and 8:1. The conversion of the crystalline to amorphous form after complexation of norfloxacin was determined by differential scanning calorimetry. The water solubility of enrofloxacin was increased up to 1258% after complexation with 2-hydroxypropyl-β-cyclodextrin (Calsavara et al. 2012). However, all complexes obtained with the main natural cyclodextrins significantly enhanced the solubility. The formation of the molecular inclusion complexes was investigated by differential scanning calorimetry and thermogravimetry.

Anti-mycotic, anti-mycobacterial and anti-viral compounds are extensively studied for the cyclodextrin complexation possibility. Itraconazole (Fig. 4.19) is one of the most used anti-mycotic, but has poor solubility and dissolution rate. The enhancing of these properties by complexation with a wide range of cyclodextrins was investigated (Thiry et al. 2017). Binary complexes of itraconazole with β-cyclodextrin, 2-hydroxypropyl-β-cyclodextrin, randomly methylated β-cyclodextrin, low methylated β-cyclodextrin and β-cyclodextrin heptakis(sulfobutylether) have been obtained by a continuous hot-melt extrusion process. The influence of the Soluplus® amphiphilic copolymer addition during the complexation process was investigated by rheometry. The modification of the glass transition calorimetric effect from 60.6 °C as well as the disappearance of the effect corresponding to the melting point from 167 °C were evaluated by differential scanning calorimetry. New 5-aryl-2-thio-1,3,4-oxadiazole derivatives (Fig. 4.19) having anti-mycobacterial activity have been studied for β-cyclodextrin complexation capacity (Boldescu et al. 2012). β-Cyclodextrin/drug complex was obtained by lyophillization at molar ratios of 1:1 and 2:1. Complexes were analyzed by Fourier transform-infrared spectroscopy, X-ray diffractometry and differential scanning calorimetry. The last thermal method provided information on the disappearance of the peak corresponding to the melting of the anti-mycobacterial drug candidate at 121.8 °C, confirming the host-guest interaction. Rifampicin is an anti-mycobacterial drug that was nanoencapsulated in 2-hydroxypropyl-β-cyclodextrin in order to ameliorate its poor oral bioavailability. This is due to low water solubility and stability. The Fourier transform-infrared spectroscopy and differential scanning calorimetry confirm the formation of 2-hydroxypropyl-β-cyclodextrin/rifampicin inclusion complex, most probably by partial encapsulation of methyl-piperazinyl moiety through the secondary face of cyclodextrin. The anti-HIV drug efavirenz (Fig. 4.19) is almost insoluble in water and further, the gastrointestinal absorption is limited. In order to improve the apparent water solubility and dissolution of efavirenz, complexation by β-cyclodextrin, 2-hydroxypropyl-β-cyclodextrin and randomly methylated β-cyclodextrin have been performed (Sathigari et al. 2009). Freeze-drying method and β-cyclodextrin derivatives were more appropriate to enhance efavirenz properties, as it was proved by differential scanning calorimetry and dissolution studies.

Antineoplastic agents are more or less complex structures, but some of them have poor bioavailability. Consequently, complexation by cyclodextrins have been applied to enhance their properties. Paclitaxel (Fig. 4.20) is used to treat various cancer types, but its low water solubility suggest to obtain pharmaceutical formulations with ameliorated properties. Polyethoxylated castor oil or Cremophor EL have been widely used in this regards. Recently, novel 2-hydroxypropyl-β-cyclodextrin based drug carrier was studied. Paclitaxel-loaded 2-hydroxypropyl-β-cyclodextrin/carboxylated graphene oxide nanohybrid have been cross-linked using glutaryl-aldehyde in order to obtain nanospheres containing antineoplastic drug (Tan et al. 2016). Pharmaceutical formulation was characterized by X-ray diffractometry, Fourier transform-infrared spectroscopy, for particle size distribution, for the drug loading efficiency and entrapment efficiency, as well as by thermal methods, thermogravimetry-differential thermogravimetry, for evaluating the water loss and thermal decomposition of oxygen-containing groups. Celecoxib is another nonsteroidal anti-inflammatory drug with capacity to reduce the colon cancer progression (Fig. 4.20). It have very low water solubility of 3–7 μg/mL. Consequently, β-cyclodextrin/celecoxib complexation have been investigated using freeze-drying, co-evaporation and kneading methods. Complexes were analyzed by X-ray diffractometry, scanning electron microscopy and differential scanning calorimetry. For all complexes obtained by the above mentioned methods, the reduction of the calorimetric effect of the melting of celecoxib from 162.5 °C was observed (Reddy et al. 2004). Cyclosporine A is a macrocyclic undecapeptide having immunosuppressive activity (Fig. 4.20). Its structure cannot be completely encapsulated in cyclodextrin cavities, but it can form partial complexes due to the presence of a medium chain 1-hydroxy-2-methyl-4-hexenyl moiety with both hydrophobic and H-bonding properties. It seems that the complexation of cyclosporine A by α-cyclodextrin, 2-hydroxypropyl-β-cyclodextrin and their mixtures increased the water solubility of

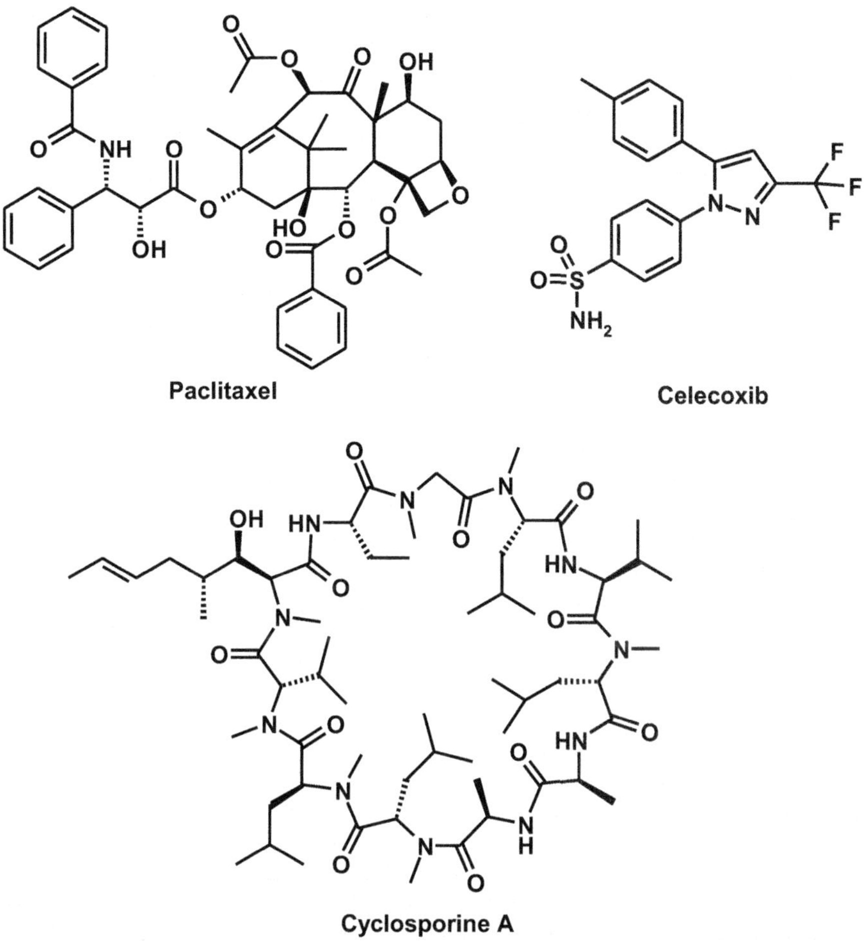

Fig. 4.20 Structure of antineoplastic agents used for cyclodextrin complexation

the drug by ten and eighty times (Malaekeh-Nikouei et al. 2007). The formation of an inclusion complex between cyclosporine A and cyclodextrins was supported by differential scanning calorimetric analysis.

Anti-inflammatory and anti-rheumatic drugs generally have simple structures (Fig. 4.21). They belong to various chemical classes such as butylpyrazolidines, oxicams, and propionic acid derivatives. Oxyphenbutazone is a non-steroidal anti-inflammatory drug with higher activity. A binary host-guest supramolecular system based on oxyphenbutazone, β- or γ-cyclodextrin have been obtained by kneading method (Veiga and Merino 2002). Cyclodextrin/oxyphenbutazone complexes were analyzed by various thermal methods such as differential scanning calorimetry, thermogravimetry and hot stage microscopy. It was observed by differential scanning calorimetric analysis that only γ-cyclodextrin forms an inclusion complex with

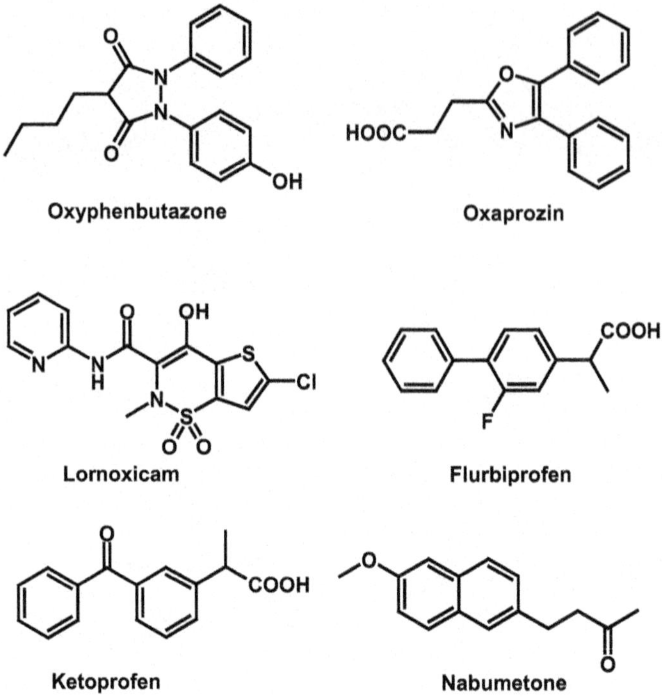

Fig. 4.21 Anti-inflammatory and anti-rheumatic structures used for cyclodextrin complexation

oxyphenbutazone if kneading method was used. Oxaprozin have similar diphenyl-heterocyclic moiety but low water solubility. An extensive study on the complexation of oxaprozin with natural and chemically modified cyclodextrins have been performed (Maestrelli et al. 2009). Complexes of oxaprozin with α-, β- and γ-cyclodextrin as well as with heptakis(2,6-*O*-dimethyl)-β-cyclodextrin, randomly methylated β-cyclodextrin and low methylated β-cyclodextrin were obtained by blending, kneading, co-grinding, sealed-heating, co-evaporation and co-lyophillization techniques at 1:1 host:guest molar ratios. They were analyzed by X-ray diffractometry, Fourier transform-infrared spectroscopy, scanning electron microscopy and differential scanning calorimetry. The fusion endotherm peak of oxaprozin from 161.3 °C is significantly reduced after complexation. Oxicams also belong to non-steroidal anti-inflammatory drug class and lornoxicam was studied for the possibility to enhance its therapeutic efficiency by encapsulation with β-cyclodextrin and 2-hydroxypropyl-β-cyclodextrin at host:guest molar ratios in the range of 1:1 to 4:1. Differential scanning calorimetric analysis of complexes revealed that the lornoxicam related endothermal peak is significantly reduced. Flurbiprofen and nabumetone have hydrophobic biphenyl or naphthyl moieties that can properly be encapsulated in cyclodextrin cavities (Todorova and Schwarz 2007). There are some studies related to flurbiprofen and nabumetone complexation by β- and γ-cyclodextrin as well as by randomly methylated β-cyclodextrin and analyzed

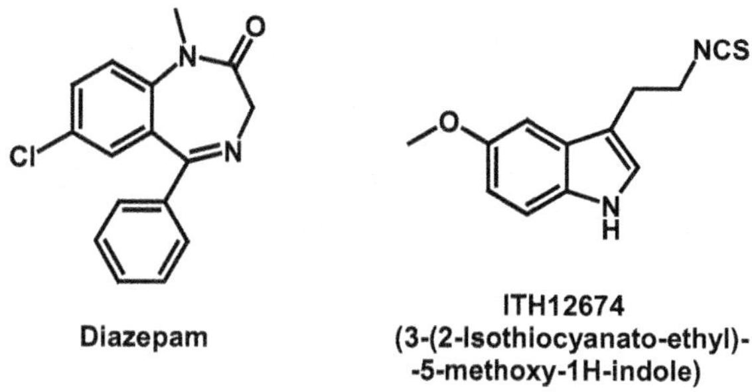

Fig. 4.22 Structures of psycholeptic drugs used for cyclodextrin complexation

by thermal methods. Isothermal titration calorimetric method was used for evaluating the natural β- and γ-cyclodextrin/flurbiprofen or nabumethone interactions. It was revealed that the binding constant of flurbiprofen with β-cyclodextrin is about 9520 M^{-1}, according to isothermal titration calorimetric results (Todorova and Schwarz 2007). Flurbiprofen was also evaluated for the capability to be included in alkylcarbonate derivatives of γ-cyclodextrin and complexes were analyzed by differential scanning calorimetry and X-ray diffractometry (Cavalli et al. 2007). The disappearance of the melting peak of flurbiprofen after cyclodextrin complexation was observed. Recently, flurbiprofen was complexed with randomly methylated β-cyclodextrin in order to enhance the apparent water solubility and dissolution properties. A single-step, organic solvent-free supercritical complexation process have been used. Moreover, the formation of the inclusion complex was evaluated by X-ray diffractometry and differential scanning calorimetry. Finally, ketoprofen and 2-hydroxypropyl-β-cyclodextrin were complexed by lyophilization in the presence of *tert*butyl alcohol (Wang et al. 2007). The porous powder complex was analyzed by Fourier transform-infrared spectroscopy, X-ray diffractometry, scanning electron microscopy and differential scanning calorimetry, showing that the material appears in an amorphous state.

Acetyl salicylic acid is a well-known analgesic and antipyretic having side-effects due to its acidic properties. Complexation with β-cyclodextrin and thermogravimetry/differential thermal analysis reveals the shifting of the temperature peaks corresponding to water release (Szafranek and Szafranek 1993).

Psycholeptic compounds are important drugs from the bioavailability point of view, because it is necessary to have constant blood concentration. Diazepam (Fig. 4.22) was studied for the possibility to enhance the bioavailability and consequently the water solubility of diazepam by molecular encapsulation in γ-cyclodextrin and its alkylcarbonate derivatives (Cavalli et al. 2007). Differential scanning calorimetry and X-ray diffractometry were the main techniques for proving the formation of the host-guest inclusion complexes. Recently, a multitarget drug designed to treatment the brain ischemia, coded as ITH12674 (Fig. 4.22), was used for obtaining complexes with 2-hydroxypropyl-β-cyclodextrin in order to increases its water solubility (Michalska et al. 2017). The formation of the inclusion complex was confirmed by various methods, including differential scanning calorimetry. Similar studies were performed for α- and β-cyclodextrin/tryptophan

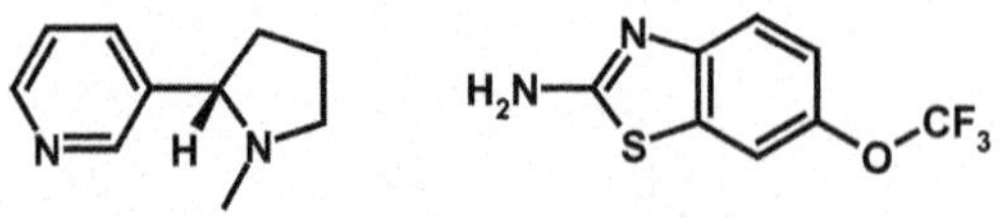

Nicotine Riluzole

Fig. 4.23 Structures of two drugs used for the treatment of nervous system disorders. They were complexed by β-cyclodextrin and its 2-hydroxypropyl derivative

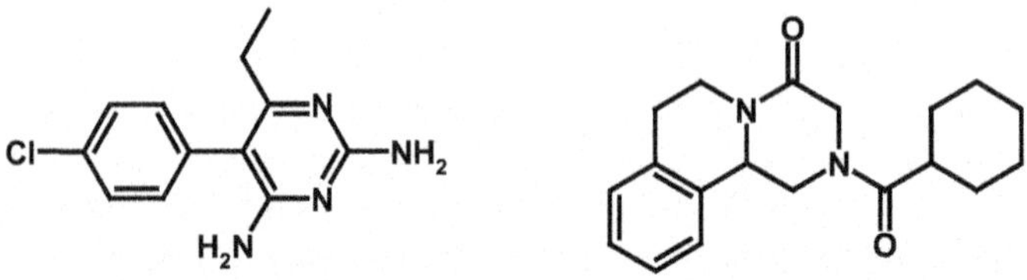

Pyrimethamine Praziquantel

Fig. 4.24 Structures of antiprotozoal and anthelmintic drugs used for cyclodextrin complexation

complexes, which were analyzed by thermogravimetry and differential thermal analysis (Song et al. 2006).

Drugs used in the treatment of some nervous system disorders such as nicotine, which is used in nicotine dependence, were studied for cyclodextrin complexation capability. Nicotine (Fig. 4.23) is the main drug-related compound used for smoking cessation. The protection of nicotine-containing formulations against degradation and controlled release properties were evaluated by β-cyclodextrin complexation. The water content and inclusion behavior of β-cyclodextrin/nicotine supramolecular system were evaluated by thermogravimetry, Karl Fischer titration and molecular docking experiments (Hădărugă et al. 2010). On the other hand, riluzole was mainly used for the treatment of amyotrophic lateral sclerosis. It has low water solubility and the 2-hydroxypropyl-β-cyclodextrin complexation was applied to enhance its bioavailability (Wang et al. 2016). Both differential scanning calorimetry and thermogravimetry support the formation of an inclusion complex between 2-hydroxypropyl-β-cyclodextrin and riluzole.

Antiprotozoal and anthelmintic drugs were studied for cyclodextrin complexation possibility. Pyrimethamine is active against protozoal parasites but is has poor water solubility (Fig. 4.24). The 2-hydroxypropyl-β-cyclodextrin/pyrimethamine complexes were obtained by kneading, co-evaporation and freeze-drying methods (Onyeji et al. 2009). They were analyzed by classical Fourier transform-infrared spectroscopy, X-ray diffractometry and differential scanning calorimetry. The amorphization of the drug after cyclodextrin complexation was emphasized. Another hydrophobic drug having anti-schistosomiasis activity is praziquantel (Fig. 4.24). Its low water solubility and bioavailability were enhanced by encapsulation using methylated β-cyclodextrin and the formation of the host-guest inclusion complex was demonstrated by differential scanning calorimetric analysis (by means of the disappearance of the praziquantel transition temperature).

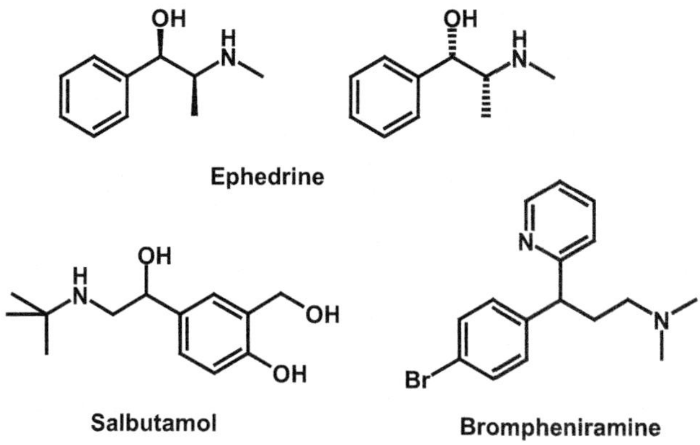

Fig. 4.25 Structures of ephedrine, salbutamol and brompheniramine used for cyclodextrin complexation

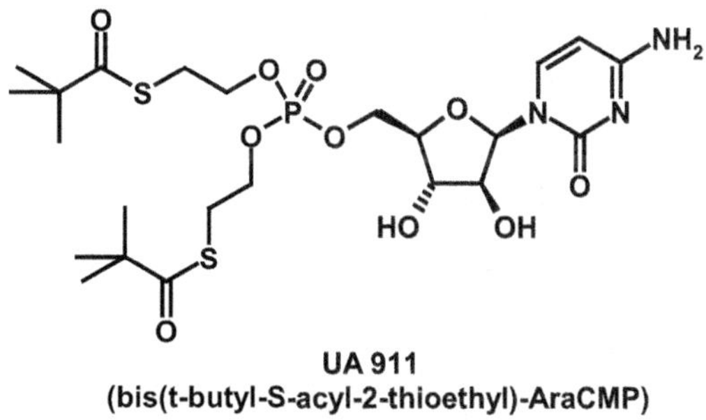

Fig. 4.26 The structure of the prodrug UA911 used for the 2-hydroxypropyl-β-cyclodextrin complexation

Ephedrine (Fig. 4.25) have sympathomimetic activity, while brompheniramine have antihistaminic activity. They were used as guest molecule for obtaining β-cyclodextrin/drug supramolecular systems (Wszelaka-Rylik 2017). Isothermal titration calorimetry was used for evaluating the formation of α- and β-cyclodextrin/drug inclusion complexes at 1:1 molar ratio. Salbutamol (Fig. 4.25) is a drug designed for systemic use, having adrenergic activity. Natural and semi-synthetically modified cyclodextrins such as α-, β-, γ-cyclodextrin and heptakis(2,6-*O*-dimethyl)-β-cyclodextrin were used as host molecules for nanoencapsulating salbutamol by freeze-drying method (Cabral Marques et al. 1990). The formation of the host-guest inclusion complex was proved by differential scanning calorimetric analysis by means of shifting or disappearing of the peaks corresponding to the melting or decomposition processes.

Prodrugs were also studied for their capabilities of encapsulation in cyclodextrins. Bis(*t*-butyl-*S*-acyl-2-thioethyl)-AraCMP, coded as UA911 (Fig. 4.26), is a prodrug that revealed anti-tumor activities by releasing intracellularly the monophosphorylated deoxynucleoside analogue. Its water solubility was significantly enhanced by 2-hydroxypropyl-β-cyclodextrin complexation at a host:guest molar ratio of 3:1. Complexes were analyzed by ^{1}H–nuclear magnetic resonance and differential scanning calorimetry (Jordheim et al. 2009).

4.3.2 Cyclodextrin/Odorant Compound or Extract Complexes

Odorant compounds and essential oil based compounds are generally small molecules with high hydrophobicity. They especially belong to terpenoid and phenolic compounds classes. Odorant compounds or odorant-containing extracts were nanoencapsulated in cyclodextrins in order to protect them against degradation under oxygen/air, light, and other environmental or matrix-related factors. Another advantages are the reducing of their volatility and the smell intensity, obtaining of solid pharmaceutical, food or cosmetic formulations containing odorant compounds, and revealing controlled release properties (Astray et al. 2009; Cabral Marques 2010; Duchêne and Bochot 2016; Kurkov and Loftsson 2013; Lima et al. 2016). A synopsis of odorants used for cyclodextrin encapsulation that were analyzed by thermal techniques is presented in Table 4.2.

Cyclodextrin complexes containing odorant compounds and extracts have been investigated by thermal analyses, especially by thermogravimetry and differential scanning calorimetry, in order to evaluate the water/moisture content as well as the dissociation process of the host-guest supramolecular system.

2-Nonanone has a pleasant fruity odor, but it has a high volatility and thermo-lability. Consequently, it was nanoencapsulated in β-cyclodextrin using co-precipitation method at three molar ratios of 1:0.5, 1:1 and 1:2 (Abarca et al. 2016). Thermal characterization of the solid complex was performed by differential scanning calorimetry and thermogravimetry, but X-ray diffractometry and Fourier transform-infrared spectroscopy have also been used. The differential scanning calorimetry endothermic peaks from the temperature range of 80–150 °C were attributed to the complexation process. Citral is a generic name for a mixture of two acyclic monoterpenoid aldehydes, neral or citral *a*, the *cis* isomer, and geranial or citral *b*, which is the *trans* isomer (Fig. 4.27). They occur in many plants such as lemongrass (*Cymbopogon flexuosus*) and have a strong lemon-like odor. Unfortunately, they are chemically unstable. The main degradation processes are polymerization, condensation and oxidation. Consequently, they were used as guest compounds for nanoencapsulation in monochlorotriazinyl-β-cyclodextrin. Cyclodextrin derivative was used in fabric finishing with antibacterial and UV-protecting properties (Zhu et al. 2015). Complexes were obtained by freeze-drying and analyzed by electron microscopies and Fourier transform-infrared spectroscopy, as well as by thermal methods, especially thermogravimetry-differential thermogravimetry, for estimating the citral-loaded efficiency. Unfortunately, no further specific water analysis of complexes have been performed in order to correctly estimate the volatile citral and water/solvent as thermogravimetric mass loss. Citronellal (Fig. 4.27) is another acyclic monoterpenoid aldehyde that was extensively studied for the possibility to obtain complexes with natural β-cyclodextrin

Table 4.2 Cyclodextrin/odorant, essential oil, vegetable extract complexes analyzed by thermal methods (type of cyclodextrins used for obtaining complexes and the corresponding thermal methods used for characterization were included in the table)

N°	Guest compound	Host compound (cyclodextrin)[a]	Thermal analyses[b]	References
	Odorants			
1	2-Nonanone	β-CD	DSC, TG	Abarca et al. 2016
2	Citral	MCT-β-CD	TG	Zhu et al. 2015
3	Citronellal	β-CD	DSC	Songkro et al. 2012
4	Citronellal	β-CD	DSC, TG	Santos et al. 2016
5	Citronellol	α-CD, β-CD, γ-CD	EGD, TG-MS	Novák et al. 2006
6	Citronellol	β-CD	DSC	Songkro et al. 2012
7	Citronellyl acetate	α-CD, β-CD, γ-CD	EGD, TG-MS	Novák et al. 2006
8	Geraniol	β-CD	DSC, TG	Menezes et al. 2012
9	Geraniol	β-CD	DSC	Mourtzinos et al. 2008
10	Linalool	β-CD	DSC, TG	Bonetti et al. 2016
11	Linalool	HP-β-CD	DSC	Kfoury et al. 2014
12	Linalool	β-CD	DSC, TG	Menezes et al. 2014
13	α-Terpineol	β-CD, HP-β-CD	DSC	dos Santos et al. 2011
14	α-Terpineol	β-CD, HP-β-CD	DSC	dos Santos et al. 2012
15	α-Terpineol	β-CD	DSC	Mazzobre et al. 2011
16	α-Terpineol	β-CD	TG	Oliveira et al. 2016
17	β-Pinene	HP-β-CD	DSC	Kfoury et al. 2014
18	Limonene	α-CD, β-CD	TG	dos Passos Menezes et al. 2017
19	β-Caryophyllene	β-CD	DTA	Liu et al. 2013a, b, c
20	Nootkatone	β-CD, HP-β-CD	TG, DTA, ITC	Kfoury et al. 2017
21	Cinnamaldehyde	β-CD	DSC, TG	Li et al. 1997
22	Cinnamaldehyde	β-CD	DSC	Ponce Cevallos et al. 2010
23	Cinnamyl alcohol	β-CD	DSC, TG	Yu et al. 1997
24	*p*-cymene	β-CD	DSC, TG	Serafini et al. 2012
25	*trans*-Cinnamaldehyde	β-CD	DSC	Hill et al. 2013
26	Carvacrol	β-CD	DSC	Guimarães et al. 2015
27	Carvacrol	β-CD	DSC	Santos et al. 2015
28	Estragole	β-CD, HP-β-CD	DSC	Kfoury et al. 2015
29	Eugenol	β-CD	DSC	Hill et al. 2013

(continued)

Table 4.2 (continued)

N°	Guest compound	Host compound (cyclodextrin)[a]	Thermal analyses[b]	References
30	Eugenol	α-CD, β-CD, γ-CD, HP-β-CD	DSC, TG	Nuchuchua et al. 2009
31	Eugenol	β-CD	TG	Wang et al. 2011a, b
32	Methyl salicylate	β-CD	TG, DTA	Szafranek and Szafranek 1993
33	Thymol	β-CD	DSC	Tao et al. 2014
34	Thymol	HP-β-CD	DSC	Kfoury et al. 2014
35	Thymol	β-CD	DSC	Mourtzinos et al. 2008
36	Thymol	β-CD	DSC	Ponce Cevallos et al. 2010
37	*trans*-Anethole	β-CD	DSC, TG	Zhang et al. 2015
38	Vanillin	β-CD	DSC	Karathanos et al. 2007
39	Oleanolic acid	RAMEB	DSC	Ren et al. 2016
	Essential oils and extracts			
40	Cinnamon bark oil (*Cinnamomum* species)	β-CD	DSC	Hill et al. 2013
41	Garlic oil (*Allium sativum* L.)	β-CD	DSC, TG	Hădărugă et al. 2012b
42	Garlic oil (*Allium sativum* L.)	β-CD	DSC	Wang et al. 2011a
43	Citronella oil (*Cymbopogon* species)	β-CD	DSC	Songkro et al. 2012
44	Clove bud oil (*Syzygium aromaticum*)	β-CD	DSC	Hill et al. 2013
45	Clove oil (*Syzygium aromaticum*)	β-CD	TG, GC-TOF-MS	Xu et al. 2008
46	Sweet orange oil (*Citrus sinensis* L.)	β-CD	DSC, TG	Galvão et al. 2015
47	Dill oil (*Anethum graveolens* L.)	β-CD	DSC, TG	Hădărugă et al. 2012b
48	Caraway oil (*Carum carvi* L.)	β-CD	DSC	Partanen et al. 2002
49	Caraway oil (*Carum carvi* L.)	β-CD	DSC, TG	Hădărugă et al. 2012b
50	Fennel oil (*Foeniculum vulgare* L.)	β-CD	DSC, TG	Hădărugă et al. 2012b
51	Coriander oil (*Coriandrum sativum* L.)	β-CD	DSC, TG	Hădărugă et al. 2012b
52	"Cidreira do campo" oil (*Hyptis martiusii* Benth)	β-CD	DSC, TG	Andrade et al. 2017
53	"Small leaved mint" oil (*Mentha x villosa* Hudson)	β-CD	TG, DTA	Martins et al. 2007

(continued)

Table 4.2 (continued)

N°	Guest compound	Host compound (cyclodextrin)[a]	Thermal analyses[b]	References
54	Basil oil (*Ocimum basilicum* L.)	β-CD	DSC, TG	Menezes et al. 2016
55	Thyme oil (*Thymus vulgaris* L.)	β-CD	DSC	Tao et al. 2014
56	"Alecrim-da-chapada" oil (*Lippia gracilis* H.B.K.)	HP-β-CD	TG	Marreto et al. 2008
57	Juniper oil (*Juniperus communis* L.)	β-CD	DSC, TG	Hădărugă et al. 2012b
58	Red pepper alcoholic extract (*Capsicum annuum* L.)	β-CD	DSC, TG	Hegheş et al. 2015
59	Red pepper alcoholic extract (*Capsicum annuum* L.)	HP-β-CD	DSC	de Lima Petito et al. 2016
60	Red pepper alcoholic extract (*Capsicum annuum* L.)	β-CD	DSC	Martins Mendes Gomes et al. 2014

[a] *α-CD* α-cyclodextrin; *β-CD* β-cyclodextrin; *γ-CD* γ-cyclodextrin; *HP-β-CD* 2-hydroxypropyl-β-cyclodextrin; *RAMEB* randomly methylated β-cyclodextrin; *MCT-β-CD* monochlorotriazinyl-β-cyclodextrin

[b] *TG* thermogravimetry; *DSC* differential scanning calorimetry; *ITC* isothermal titration calorimetry; *DTA* differential thermal analysis; *EGD* evolved gas detection; *TG-MS* thermogravimetry-mass spectrometry; *GC-TOF-MS* gas chromatography-time-of-flight-mass spectrometry

(Santos et al. 2016; Songkro et al. 2012). Citronellal occurs in various plant species that belong to *Corymbia* and *Cymbopogon* genders. It has specific citronella odor but it also has antifungal and insect repellent properties (Santos et al. 2016). Complexes with β-cyclodextrin were obtained by co-evaporation and kneading methods at 1:1 and 2:1 molar ratios. Both thermogravimetry-differential thermogravimetry and differential scanning calorimetric analyses support the formation of the β-cyclodextrin/citronellal inclusion complex, especially by disappearing the calorimetric peak of 144 °C corresponding to the odorant compound in the complex (Santos et al. 2016; Songkro et al. 2012). The same behavior was observed for citronellol (Fig. 4.27), the corresponding alcohol, which have a characteristic calorimetric peak at 151 °C (Songkro et al. 2012). Moreover, similar situation was discussed for the β-cyclodextrin/citronella oil complex at 121 °C (Songkro et al. 2012). Citronellol and its acetate (Fig. 4.27) were complexed with all main natural cyclodextrins by co-precipitation method at 1:1 molar ratio (Novák et al. 2006). Evolved gas detection and thermogravimetry-mass spectrometry coupling was used for the analysis of the complexes in order to evaluate the monoterpenoid stability. It was observed that α- and γ-cyclodextrin complexes were the most stable during the dynamic heating (Novák et al. 2006).

Geraniol and linalool (Fig. 4.27) are two widely used acyclic alcoholic monoterpenoids in food, cosmetic and pharmaceutical fields. They were nanoencapsulated in β-cyclodextrin or 2-hydroxypropyl-β-cyclodextrin by various techniques such as co-precipitation, slurry, kneading, spray-drying and freeze-drying methods (Bonetti

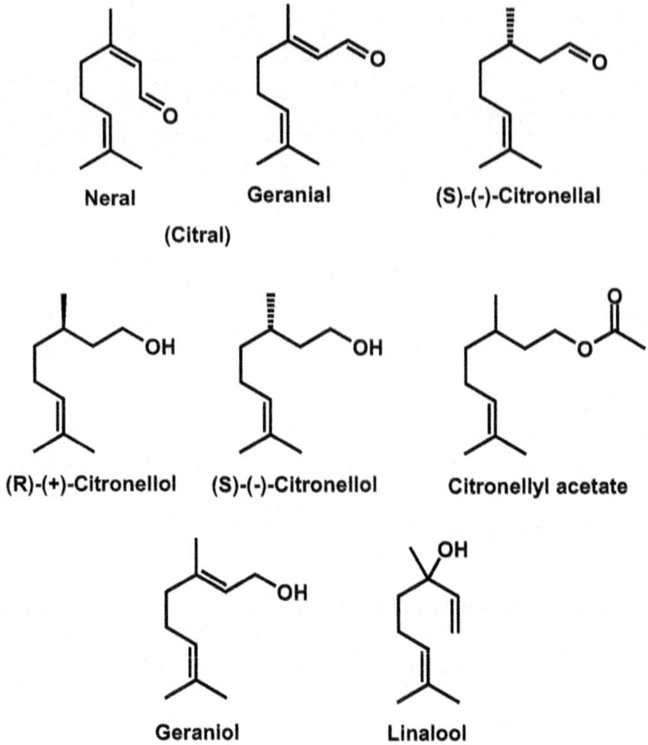

Fig. 4.27 Structures of acyclic monoterpenoids used for cyclodextrin complexation

et al. 2016; Kfoury et al. 2014; Menezes et al. 2014; 2012; Mourtzinos et al. 2008). The main thermal methods used for confirmation the formation of the inclusion complex were thermogravimetry and differential scanning calorimetry, especially through the disappearance of the endothermic peak corresponding to the volatilization of geraniol or linalool at 198–199 °C (Menezes et al. 2014; Menezes et al. 2012). Moreover, oxidative differential scanning calorimetry demonstrates the protection capacity of β-cyclodextrin against oxidation of monoterpene alcohol (Mourtzinos et al. 2008). Linalool, geraniol and other monoterpenoids such as α- and β-pinene, limonene, eucalyptol and pulegone, were complexed with 2-hydroxypropyl-β-cyclodextrin using freeze-drying method (Kfoury et al. 2014). Differential scanning calorimetric analysis of 2-hydroxypropyl-β-cyclodextrin complexes with linalool and β-pinene also confirm the formation of the host-guest inclusion complex due to the disappearing of the peak corresponding to vaporization of the guest.

Cyclic mono- and sesquiterpenoids (Fig. 4.28) were studied for their capability to interact with natural and semi-synthetically modified cyclodextrins. The main thermal analyses of these complexes were differential scanning calorimetry and thermogravimetry. α-Terpineol was used as guest compound for obtaining supramolecular systems with β-cyclodextrin or 2-hydroxypropyl-β-cyclodextrin. Different

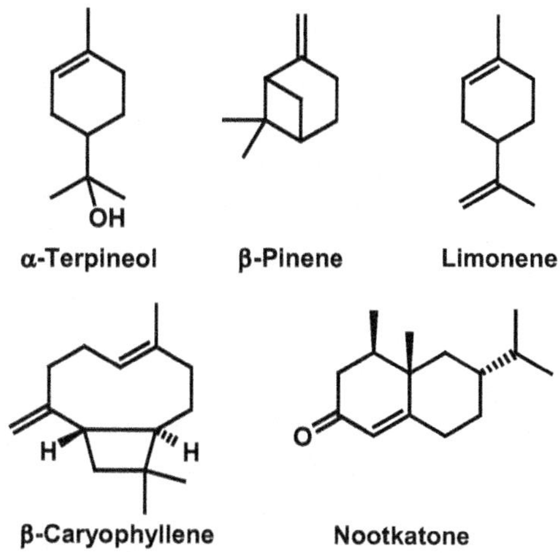

Fig. 4.28 Structures of cyclic mono- and sesquiterpenoids used for cyclodextrin complexation

complexation techniques such as co-precipitation, slurry or freeze-drying at different molar ratios of 1:1 and 3:1 have been used (dos Santos et al. 2012; Mazzobre et al. 2011; Oliveira et al. 2016). Detailed discussion on the results of thermal analysis of cyclodextrin/α-terpineol complexes was provided by dos Santos and collaborators (dos Santos et al. 2011). Moreover, ternary complexes of α-terpineol with β-cyclodextrin and 2-hydroxypropyl-β-cyclodextrin in the presence of myristic acid were investigated by differential scanning calorimetry (dos Santos et al. 2012). Limonene is one of the most common monoterpene that occurs in essential oils. It has a middle citric odor, but it has very low oxidative stability. Its hydrophobic character and relatively rigid structure make this odorant compound an "ideal" guest for cyclodextrin molecular encapsulation. Complexes with α- and β-cyclodextrin were obtained by kneading and slurry techniques at 1:1 molar ratio and analyzed by X-ray diffractometry, Fourier transform-infrared spectroscopy, ^{1}H–nuclear magnetic resonance and thermogravimetry-differential thermogravimetry, which confirm the formation of the inclusion complex (dos Passos Menezes et al. 2017). β-Caryophyllene belongs to sesquiterpene class and has a bicyclic structure. This provides a relatively rigid structure that can fit into the β-cyclodextrin cavity. Co-precipitation method was used for obtaining β-cyclodextrin/β-caryophyllene complex. Differential thermal analysis was selected from the thermal methods for proving the formation of the inclusion complex. The disappearance of the temperature peak corresponding to the boiling point of β-caryophyllene was superficially discussed (Liu et al. 2013c). Nootkatone belongs to the same sesquiterpenoid class having low water solubility and stability. Beside its specific odor, it has important insect repellent properties. Consequently, it was used as guest molecule for obtaining complexes with natural β-cyclodextrin and its semi-synthetically modified 2-hydroxypropyl-β-cyclodextrin (Kfoury et al. 2017). Thermal properties of cyclodextrin/nootkatone complexes were investigated by thermogravimetry, differential thermal analysis as well as isothermal titration calorimetry, revealing the enhancing of thermal stability of encapsulated sesquiterpene ketone.

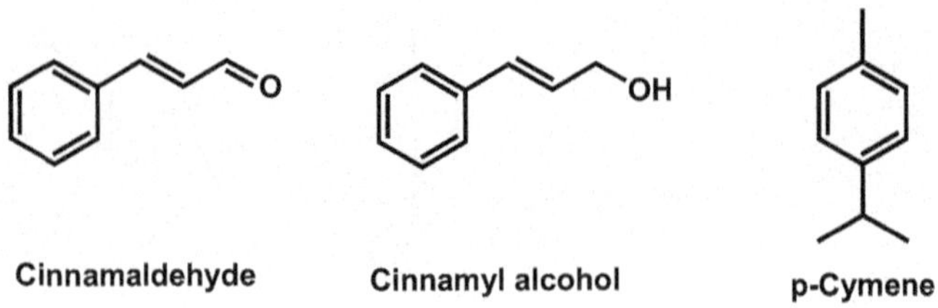

Fig. 4.29 Structure of non-phenolic aromatic compounds used for cyclodextrin complexation

Aromatic odorants are appropriate molecules for cyclodextrin nanoencapsulation due to the presence of the phenyl moiety, a rigid and highly hydrophobic ring. They belong to aromatic hydrocarbons such as *p*-cymene, non-phenolic or phenolic aromatic compounds. It is the case of compounds like cinnamic aldehyde and cinnamic alcohol for non-phenolic compounds and carvacrol, estragole, eugenol, thymol, anethole or vanillin for phenolic aromatic compounds. However, the last class was extensively studied for cyclodextrin encapsulation ability and thermal analyses techniques (Table 4.2).

Cinnamaldehyde and the corresponding alcohol (Fig. 4.29) were complexed by β-cyclodextrin and characterized by thermogravimetry and differential scanning calorimetry (Hill et al. 2013; Li et al. 1997; Ponce Cevallos et al. 2010; Yu et al. 1997). For example, freeze-drying or co-precipitation allows the interaction between cinnamaldehyde and β-cyclodextrin cavity, as was revealed by differential scanning calorimetry (Li et al. 1997; Ponce Cevallos et al. 2010). *p*-Cymene (Fig. 4.29) was used as highly hydrophobic guest compound for β-cyclodextrin complexation using slurry and kneading methods (Serafini et al. 2012). Differential scanning calorimetry and thermogravimetry-differential thermogravimetry analyses showed a high content of *p*-cymene that was encapsulated by β-cyclodextrin for both methods of preparation.

Carvacrol, thymol, estragole, eugenol, anethole, vanillin and methyl salicylate (Fig. 4.30) were complexed with various cyclodextrins and analyzed especially by differential scanning calorimetry, thermogravimetry and differential thermal analysis (Guimarães et al. 2015; Hill et al. 2013; Karathanos et al. 2007; Kfoury et al. 2014; Mourtzinos et al. 2008; Nuchuchua et al. 2009; Ponce Cevallos et al. 2010; Santos et al. 2015; Szafranek and Szafranek 1993; Wang et al. 2011b; Zhang et al. 2015). β-Cyclodextrin/carvacrol complexes were obtained by slurry, kneading and freeze-drying methods (Guimarães et al. 2015; Santos et al. 2015). Differential scanning calorimetry indicated a completely different behavior for complexes in comparison with the starting materials or their physical mixtures. Moreover, the endothermic peak corresponding to the boiling point of carvacrol at ~239 °C was completely disappeared in the complex. Similar situations were observed for eugenol having a boiling point of ~241 °C. Its α-, β-, γ-cyclodextrin and 2-hydroxypropyl-β-cyclodextrin complexes did not reveal this calorimetric peak. The same behavior was observed for thymol and its β-cyclodextrin and 2-hydroxypropyl-β-cyclodextrin complexes at the melting point of ~49 °C, for anethole and its β-cyclodextrin com-

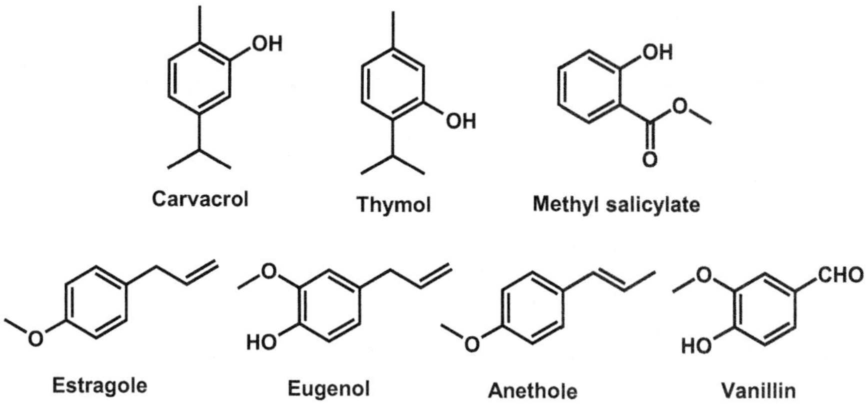

Fig. 4.30 Structures of phenolic compounds used for cyclodextrin complexation

Fig. 4.31 Structure of triterpenoid oleanolic acid for obtaining complexes with randomly methylated β-cyclodextrin

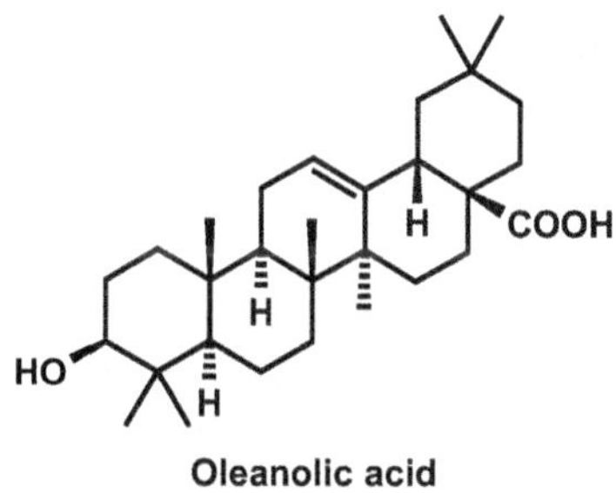

plex, when the disappearance of the peak corresponding to the melting point at ~21 °C can be observed. Vanillin has a melting point at 81–83 °C and another endothermic peak at 210 °C appear, which is probably due to the oxidation of vanillin. These calorimetric effects did not appear in the β-cyclodextrin/vanillin complex (Karathanos et al. 2007; Mourtzinos et al. 2008; Nuchuchua et al. 2009; Ponce Cevallos et al. 2010; Tao et al. 2014; Wang et al. 2011b; Zhang et al. 2015).

Attempts to obtain supramolecular structures based on cyclodextrins were also performed with triterpenoids such as oleanolic acid (Fig. 4.31). Complexes with randomly methylated β-cyclodextrin had a 1:1 inclusion molarity that was revealed by ^{1}H–nuclear magnetic resonance, X-ray diffractometry and differential scanning calorimetry (Ren et al. 2016).

Important studies on the complexation of essential oils or vegetable extracts with natural and semi-synthetically modified cyclodextrins have been performed (Table 4.2). They generally contain odorant compounds presented above, at different concentrations. However, the relative composition of such essential oils can be modified after nanoencapsulation.

Essential oils used for cyclodextrin complexation belong to Magnoliids, Monocots, Commelinids, Malvids (Rosids) and Asterids clades. Thus, cinnamon bark oil (*Cinnamomum* species) from the first clade was complexed by β-cyclodextrin using freeze-drying method (Hill et al. 2013). Oxidative differential scanning calorimetry confirm the formation of inclusion complexes with the main components, named *trans*-cinnamaldehyde. Furthermore, thermal analysis revealed the protection capacity of β-cyclodextrin against oxidation of cinnamaldehyde. It was observed the disappearance of the corresponding peak at ~265 °C. Garlic oil or *Allium sativum* L. oil, from Monocots clade and citronella oil from *Cymbopogon* species were also studied for the possibility of encapsulation by β-cyclodextrin; differential scanning calorimetry and thermogravimetry support the formation of the inclusion complexes with the main sulfur-containing compounds and especially with diallyl di- and trisulfides (Hădărugă et al. 2012b; Songkro et al. 2012; Wang et al. 2011a). Clove or *Syzygium aromaticum* and sweet orange, having the Latin name *Citrus sinensis* L., belong to Malvids (Rosids) clade. Essential oils of clove and sweet orange peel were complexed by β-cyclodextrin and solid complexes were analyzed by the main thermal methods such as oxidative differential scanning calorimetry or thermogravimetry (Galvão et al. 2015; Hill et al. 2013). Calorimetric data for complexes obtained by co-precipitation method indicated large endothermic peaks in the range of 120–280 °C, which were attributed to the release of complexed essential oil components. Essential oils from Asterides clades were the most studied for the complexation ability, especially with β-cyclodextrin. Differential scanning calorimetry, thermogravimetry or differential thermal analysis were the main thermal analyses used for characterization of complexes. In all cases, the behavior of water release and after that the modification of the calorimetric and thermogravimetric plots up to β-cyclodextrin decomposition were evaluated. It was the case of dill - *Anethum graveolens* L., caraway - *Carum carvi* L., fennel - *Foeniculum vulgare* L., coriander - *Coriandrum sativum* L., "cidreira do campo" - *Hyptis martiusii* Benth, "small leaved mint" - *Mentha x villosa* Hudson, basil - *Ocimum basilicum* L. or thyme - *Thymus vulgaris* L. oils encapsulated in β-cyclodextrin and "alecrim-da-chapada" - *Lippia gracilis* H.B.K. oil encapsulated by 2-hydroxypropyl-β-cyclodextrin (Andrade et al. 2017; Hădărugă et al. 2012b; Martins et al. 2007; Menezes et al. 2016; Partanen et al. 2002; Tao et al. 2014). Juniper - *Juniperus communis* L. oil, from the Pinopsida clade, have been complexed by β-cyclodextrin using co-precipitation method and complexes were analyzed by differential scanning calorimetry and thermogravimetry (Hădărugă et al. 2012b). The significant reduction of the water content of complexes in comparison with β-cyclodextrin supported the formation of host-guest supramolecular systems with the juniper oil components, especially highly hydrophobic cyclic monoterpenes.

Interesting studies were performed for the nanoencapsulation of the main components from alcoholic vegetable extracts. Extensively studied were red pepper - *Capsicum annum* L. and variants alcoholic extracts, which were complexed with β-cyclodextrin and 2-hydroxypropyl-β-cyclodextrin (de Lima Petito et al. 2016; Hegheş et al. 2015; Martins Mendes Gomes et al. 2014). Alcoholic extracts especially contain capsaicinoids and carotenoids, which are geometrically compatible with β-cyclodextrin cavity. It was observed that the differential scanning calorimetric peak corresponding to possible degradation of red pepper extract components between 115–160 °C is not present in the case of β-cyclodextrin complex. On the other hand, pigments from red pepper alcoholic extracts were complexed by 2-hydroxypropyl-β-cyclodextrin at high host:guest mass ratios of 4:1 to 10:1, using lyophillization method. The water release as well as the disappearance of the calorimetric peaks corresponding to degradation of pigments after complexation were emphasized (de Lima Petito et al. 2016).

4.3.3 Cyclodextrin/Antioxidant Compound Complexes

The most known natural antioxidants are significantly different in comparison with odorant compounds from the chemical point of view. They generally belong to phenolic compounds and the main natural classes are flavonoids, flavonoid glycosides and flavonolignans, antocyanidins, anthocyanins, and other mono- or polyphenolic compounds (Table 4.3). Due to the presence of hydroxyl groups, the inclusion of these antioxidant molecules into the cyclodextrin cavity is sometimes hindered. Hydroxyl groups of antioxidants can interact with the external hydroxyl groups of cyclodextrins better than by van der Waals bonds with the cyclodextrin inner cavity.

Flavone and 3-hydroxyflavone, apigenin, baicalein, chrysin, fisetin, luteolin, morin and quercetin (Fig. 4.32) belong to flavone group and were used as guest molecules for obtaining supramolecular systems with natural α-, β- and γ-cyclodextrin as well as semi-synthetically modified cyclodextrins, such as 2-hydroxyethyl-β-cyclodextrin, 2-hydroxypropyl-β-cyclodextrin or randomly methylated β-cyclodextrin. Flavone is not really an antioxidant but a skeleton for other natural antioxidants. 3-Hydroxyflavone, baicalein and chrysin have a non-substituted phenyl group in the C^2-position of the benzopyranone moiety that allows better hydrophobic interaction with the cyclodextrin cavity (Fig. 4.33). Flavone was encapsulated in α- and β-cyclodextrin by classical and ultrasonication-assisted co-precipitation method (Hădărugă et al. 2012a). Complexes had lower content of crystallization water that suggest the replacing of the solvent molecules by the guest, as was emphasized by thermogravimetry and differential scanning calorimetry analyses. Similar results were obtained for 3-hydroxyflavone, baicalein and chrysin, bearing non-substituted phenyl group on the benzopyranone moiety (Calabrò et al. 2004; Hădărugă et al. 2012a; Zhou et al. 2013). The reduction/disappearance of the differential scanning calorimetric peak corresponding to the melting process of the guest was observed at ~170 °C for 3-hydroxyflavone, 272.5 °C for baicalein and ~284 °C for chrysin. Other flavone antioxidants used for cyclodextrin nanoencapsulation were apigenin, fisetin, luteolin, morin and quercetin (Alvarez-Parrilla et al. 2005; Calabrò et al. 2004; Dias et al. 2008; Hădărugă et al. 2012a; Huang et al. 2016; Liu et al. 2013a; Pralhad and Rajendrakumar 2004). Host molecules were especially β-cyclodextrin, but other natural of modified cyclodextrins, such as α-cyclodextrin and randomly methylated β-cyclodextrin, 2-hydroxyethyl- and 2-hydroxypropyl-β-cyclodextrin, were used. They were obtained by freeze-drying, co-precipitation, co-evaporation, kneading and supercritical antisolvent methods

Table 4.3 Cyclodextrin/antioxidant complexes analyzed by thermal methods (type of cyclodextrins used for obtaining complexes and the corresponding thermal methods used for characterization were included in the table)

N°	Guest compound	Host compound (cyclodextrin)[a]	Thermal analyses[b]	References
	Flavonoids and isoflavonoids			
1	3-Hydroxyflavone	α-CD, β-CD	DSC	(Calabrò et al. 2004)
2	Apigenin	β-CD	DSC, TG	(Hădărugă et al. 2012a)
3	Apigenin	HP-β-CD	DSC, TG	(Huang et al. 2016)
4	Baicalein	α-CD, β-CD, γ-CD	DSC	(Zhou et al. 2013)
5	Catechin	β-CD	TG, DTA	(Dias et al. 2008)
6	Chrysin	α-CD, β-CD	DSC, TG	(Hădărugă et al. 2012a)
7	Fisetin	β-CD	DSC, TG	(Hădărugă et al. 2012a)
8	Flavone	α-CD, β-CD	DSC, TG	(Hădărugă et al. 2012a)
9	Hesperetin	β-CD	DSC, TG	(Hădărugă et al. 2012a)
10	Hesperetin	HP-β-CD	DSC	(Majumdar and Srirangam 2009)
11	Luteolin	β-CD	DSC, TG	(Hădărugă et al. 2012a)
12	Luteolin	β-CD, RAMEB, HE-β-CD, HP-β-CD, Glu-β-CD	DSC	(Liu et al. 2013a)
13	Morin	α-CD, β-CD	DSC	(Calabrò et al. 2004)
14	Naringenin	β-CD	DSC, TG	(Hădărugă et al. 2012a)
15	Quercetin	β-CD	DSC	(Alvarez-Parrilla et al. 2005)
16	Quercetin	α-CD, β-CD[a]	DSC	(Calabrò et al. 2004)
17	Quercetin	β-CD	TG, DTA	(Dias et al. 2008)
18	Quercetin	β-CD, HP-β-CD	DSC	(Pralhad and Rajendrakumar 2004)
19	Daidzein, Genistein, Glycitein (isoflavone enriched fraction)	β-CD, HP-β-CD	DSC	(Yatsu et al. 2013)
	Flavonoid glycosides and flavonolignans			
20	Hesperidin	β-CD	DSC	(Corciova et al. 2015)
21	Hesperidin	HP-β-CD	DSC	(Majumdar and Srirangam 2009)
22	Naringin dihydrochalcone	HP-β-CD	DSC	(Liu et al. 2013b)

(continued)

Table 4.3 (continued)

Nº	Guest compound	Host compound (cyclodextrin)[a]	Thermal analyses[b]	References
23	Rutin	β-CD	DSC	(Alvarez-Parrilla et al. 2005)
24	Rutin	β-CD	DSC	(Haiyun et al. 2003)
25	Rutin	β-CD	DSC, TG	(Şamlı et al. 2014)
26	Scutellarin-β-CD bioconjugate	β-CD bioconjugate	TG	(Yang et al. 2013a)
27	Silybin	β-CD	DSC, TG	(Hădărugă et al. 2012a)
	Other antioxidants			
28	α-Tocopherol	β-CD	TG	(Siró et al. 2006)
29	Caffeic acid	α-CD, β-CD	DSC, TG	(Hădărugă et al. 2012a)
30	Chlorogenic acid	β-CD	DSC	(Alvarez-Parrilla et al. 2005)
31	Cinnamic acid	α-CD, β-CD	DSC, TG	(Hădărugă et al. 2012a)
32	Curcumin	HP-α-CD, HP-β-CD, HP-γ-CD	DSC	(Krishna Mohan et al. 2012)
33	Curcumin	HP-β-CD	DSC	(Popat et al. 2014)
34	Curcumin	HP-β-CD	DSC	(Serri et al. 2017)
35	Mangiferin	β-CD, HP-β-CD, SBE-β-CD, EN-β-CD	TG	(Yang et al. 2013b)
36	Mangiferin	β-CD	DSC	(da Rocha Ferreira et al. 2013)
37	Phloridzin	HP-β-CD	ITC	(Zhang et al. 2017)
38	Resveratrol	α-CD, β-CD, γ-CD, HP-β-CD, DIMEB	DSC	(Bertacche et al. 2006)
39	Resveratrol	HP-β-CD	DSC	(Sapino et al. 2009)
40	*trans*-Polydatin	β-CD, γ-CD	DSC, TG	(Zhang et al. 2016)
41	Propolis extracts	HP-β-CD	TG	(Rocha et al. 2012)
	Carotenoids			
42	Astaxanthin	HP-β-CD	TG, DTA	(Yuan et al. 2008)
43	Lycopene	β-CD	DSC	(Nerome et al. 2013)

[a]*α-CD* α-cyclodextrin; *β-CD* β-cyclodextrin; *γ-CD* γ-cyclodextrin; *HP-α-CD* 2-hydroxypropyl-α-cyclodextrin; *HP-β-CD* 2-hydroxypropyl-β-cyclodextrin; *HP-γ-CD* 2-hydroxypropyl-γ-cyclodextrin; *HE-β-CD* 2-hydroxyethyl-β-cyclodextrin; *RAMEB* randomly methylated β-cyclodextrin; *Glu-β-CD* glucosylated-β-cyclodextrin; *SBE-β-CD* β-cyclodextrin heptakis(sulfobutylether), as sodium salt; *EN-β-CD* mono(6-ethylene-diamino-6-deoxy)-β-cyclodextrin; *DIMEB* heptakis(2,6-*O*-dimethyl)-β-cyclodextrin

[b]*TG* thermogravimetry; *DSC* differential scanning calorimetry; *ITC* isothermal titration calorimetry; *DTA* differential thermal analysis

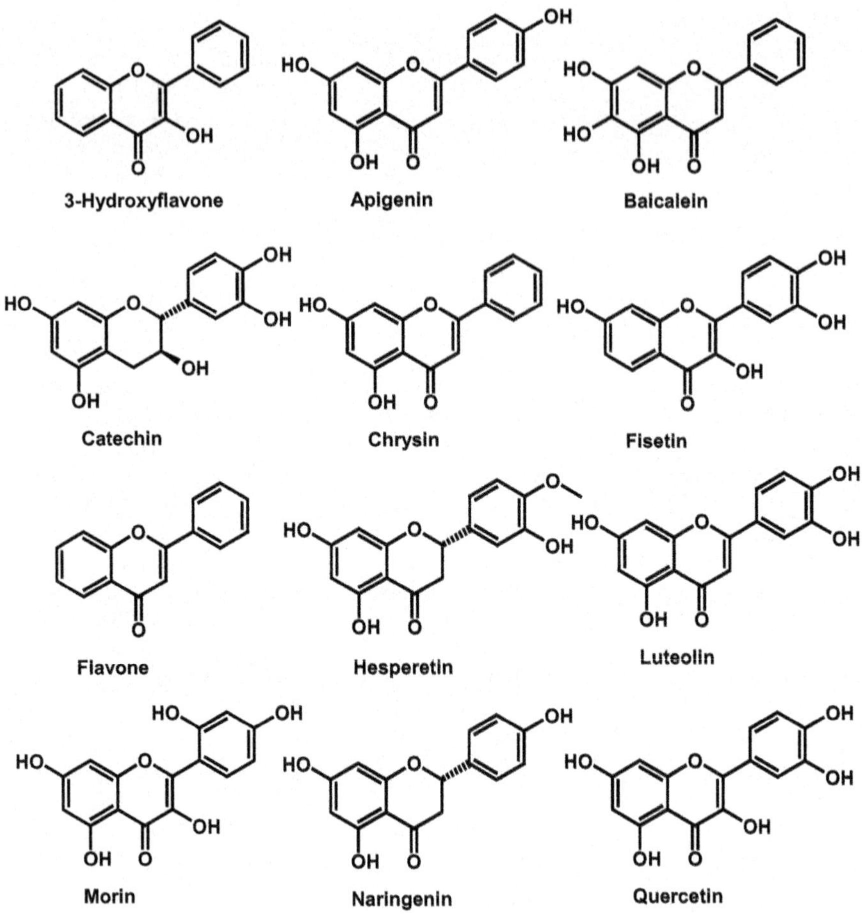

Fig. 4.32 Structures of antioxidants from flavonoid class used for cyclodextrin complexation

and analyzed by classical methods including thermal techniques. Hesperetin and naringenin are flavanone-like structures that were used for α- and β-cyclodextrin nanoencapsulation (Hădărugă et al. 2012a). As for above mentioned cyclodextrin/flavonoid complexes, differential scanning calorimetry and thermogravimetry also reveals the lowering of the crystallization water up to ~120 °C while thermograms significantly differ from the starting host and guest compounds. Soy isoflavone extracts that are enriched by daidzein, genistein and glycitein (Fig. 4.34) were used for nanoencapsulation in β-cyclodextrin and 2-hydroxypropyl-β-cyclodextrin using freeze-drying, spray-drying, kneading/microwave and co-evaporation methods. Among other characterization techniques, such as scanning electron microscopy, Fourier transform-infrared spectroscopy, X-ray diffractometry, ^{1}H- and ^{13}C–nuclear magnetic resonance, differential scanning calorimetry thermal method was used for

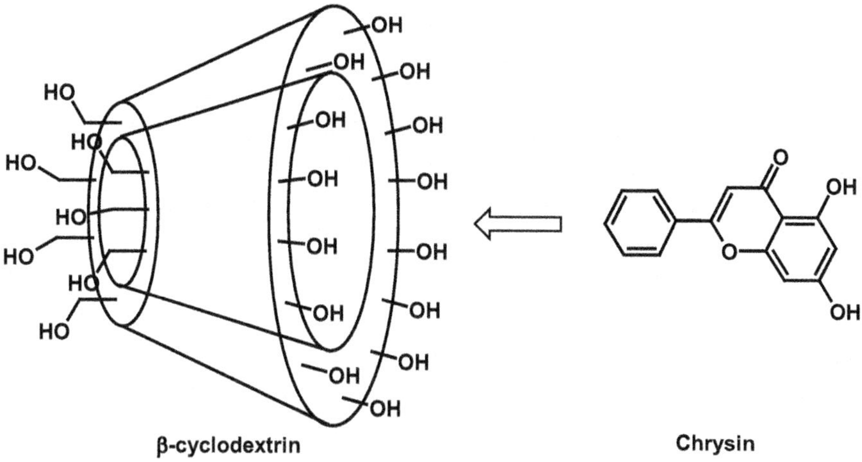

Fig. 4.33 Possible mechanism of interaction for the inclusion of chrysin into the β-cyclodextrin cavity

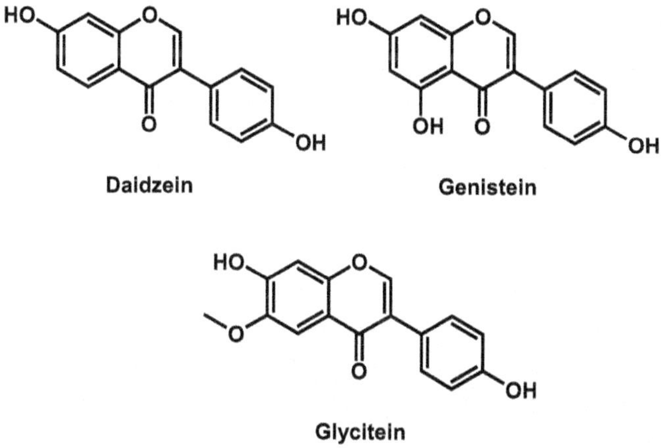

Fig. 4.34 The main isoflavone structures used for cyclodextrin complexation

revealing the significant reduction of the peak corresponding to the melting process of guest molecules after complexation (Yatsu et al. 2013).

Flavonoid glycosides have structures bearing saccharide moieties, which hinder the molecular encapsulation process. It is the case of hesperidin, rutin, and naringin dihydrochalcone (Fig. 4.35) that were studied for the possibility to nanoencapsulate by β-cyclodextrin and/or 2-hydroxypropyl-β-cyclodextrin (Alvarez-Parrilla et al. 2005; Corciova et al. 2015; Haiyun et al. 2003; Liu et al. 2013b; Majumdar and Srirangam 2009; Şamlı et al. 2014). Differential scanning calorimetric analysis of starting materials, physical mixtures and complexes reveals a shift to lower temperatures of the peaks corresponding to water release, but no clear evidence on the

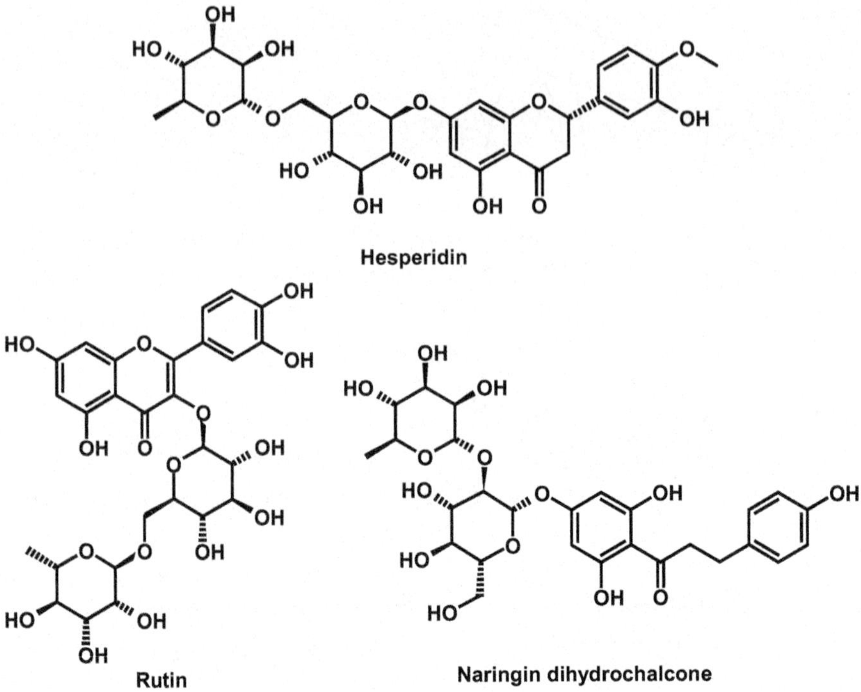

Fig. 4.35 Structures of flavonoid glycosides used for cyclodextrin complexation

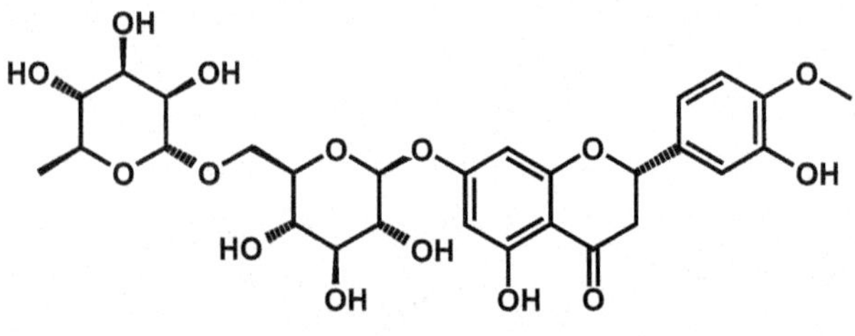

Silybin

Fig. 4.36 Structure of silybin A used for β-cyclodextrin complexation

formation of the inclusion complexes can be obtained. It was an exception for naringin dihydrochalcone (Fig. 4.35), which have a more flexible hydrophobic moiety allowing to interact with the 2-hydroxypropyl-β-cyclodextrin cavity (Liu et al. 2013b). Silybin (Fig. 4.36) is a generic name for a mixture of diastereoisomers that belong to flavonolignan class. Silybin has valuable hepatoprotective activity but it is not stable under oxidative conditions. Consequently, the cyclodextrin complexation can provide more stable materials having controlled release properties. β-Cyclodextrin/silybin complexes obtained by co-precipitation method were ana-

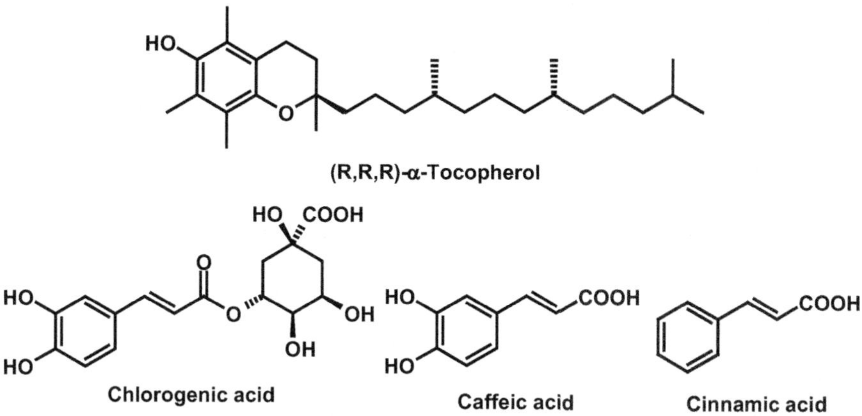

Fig. 4.37 Structures of α-tocoferol, chlorogenic, caffeic and cinnamic acids used for cyclodextrin complexation

lyzed by differential scanning calorimetry and thermogravimetry for evaluating the formation of the inclusion complex by means of replacing the crystallization of water molecules from the starting β-cyclodextrin (Hădărugă et al. 2012a).

Other natural antioxidants from phenolic/polyphenolic and carotenoid classes have been investigated for nanoencapsulation in cyclodextrins (Table 4.3). α-Tocopherol (vitamin E) is one of the most used natural antioxidants (Fig. 4.37). Its oxidative lability and very low water solubility suggest protecting this hydrophobic molecule by nanoencapsulation. Biofilms containing both α-tocopherol and β-cyclodextrin with controlled release properties were obtained. The stability of β-cyclodextrin /α-tocopherol complexes obtained by co-precipitation method was evaluated by thermogravimetry-differential thermogravimetry. It was observed that complexed α-tocopherol is stable up to 190 °C (Siró et al. 2006). Caffeic acid (Fig. 4.37) was used for obtaining complexes with α- and β-cyclodextrin by co-precipitation and ultrasonically assisted co-precipitation methods. Similar complexes were obtained with cinnamic acid that belongs to the same class, but do not reveals antioxidant activity (Fig. 4.37) (Hădărugă et al. 2012a). The disappearance of differential scanning calorimetric peaks corresponding to the melting of the guest compounds at ~133 °C for cinnamic acid and ~223 °C for caffeic acid were observed after complexation. This confirm the formation of inclusion complexes. Chlorogenic acid is an ester of caffeic acid with (−)-quinic acid (Fig. 4.37) and was used as guest for obtaining complexes with β-cyclodextrin by freeze-drying (Alvarez-Parrilla et al. 2005). Classical characterization of solid complexes confirming the host-guest interaction.

Curcumin is the main pigment from turmeric, *Curcuma longa*. It has diaryl moieties substituted with phenolic hydroxyl groups that also confer antioxidant activity (Fig. 4.38). It was complexed with 2-hydroxypropylated α-, β- and γ-cyclodextrin by freeze-drying, spray-drying and co-precipitation methods (Krishna Mohan et al. 2012; Popat et al. 2014; Serri et al. 2017). Differential scanning calorimetric analysis

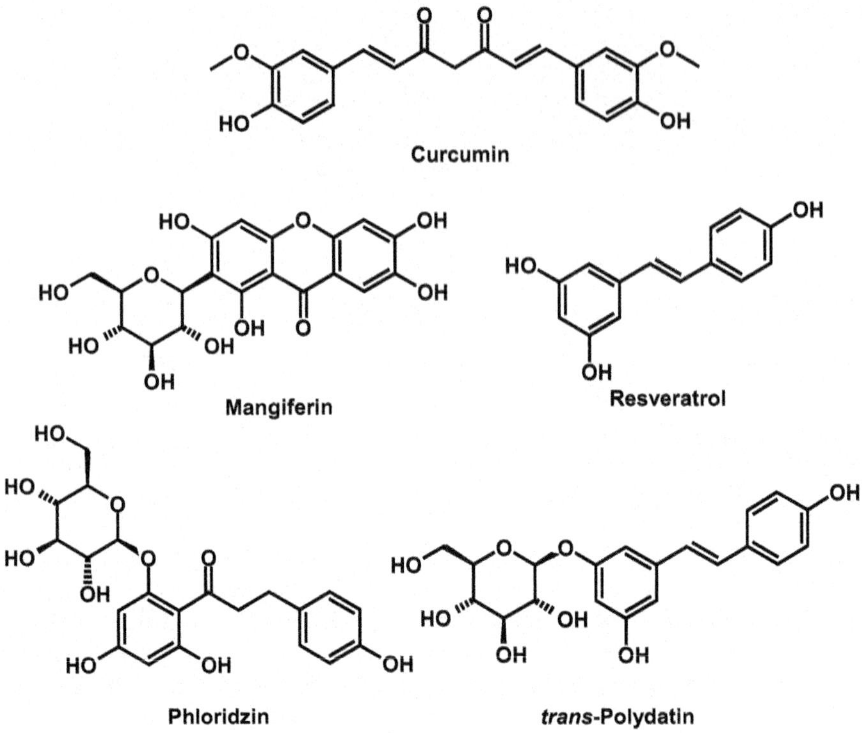

Fig. 4.38 Structures of other polyphenolic antioxidants used for cyclodextrin complexation

of complexes revealed that the endothermic peak corresponding to the melting of curcumin from 183 °C is not present in the case of complexes of curcumin with 2-hydroxypropyl-cyclodextrins. Mangiferin is a xanthonoid (Fig. 4.38) that is found in mangoes, *Mangifera indica* L. It has four phenolic hydroxyl groups which provide important antioxidant activity but the presence of a glucosyl moiety can hinder the efficiency of cyclodextrin nanoencapsulation. It has used as guest compound for obtaining complexes with natural β-cyclodextrin and its 2-hydroxypropyl, sulfobutyleter, mono(6-ethylene-diamino-6-deoxy) derivatives using co-precipitation method at 1:1 and 1:4 molar ratios (da Rocha Ferreira et al. 2013; Yang et al. 2013b). The fusion endothermic peak of mangiferin between 252–260 °C is consistently reduced after cyclodextrin complexation. This suggest the formation of the inclusion complex. The inclusion process is also proved by other analytical methods such as Fourier transform-infrared spectroscopy, X-ray diffractometry, ^{1}H- and 2D–nuclear magnetic resonance. Resveratrol is one of the most important antioxidant compound from red wine and grapes. It has a stilbene-based structure with three active hydroxyl groups (Fig. 4.38). Natural α-, β- and γ-cyclodextrin as well as semi-synthetically modified 2-hydroxypropyl-β-cyclodextrin and heptakis(2,6-*O*-dimethyl)-β-cyclodextrin were used for complexation of resveratrol

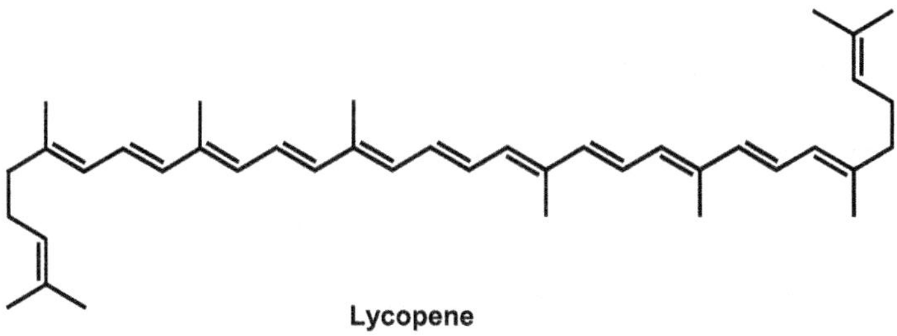

Lycopene

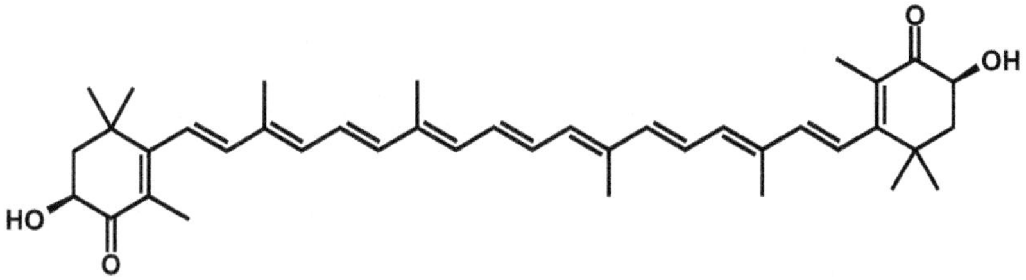

Astaxanthin

Fig. 4.39 Structures of lycopene and astaxanthin used for cyclodextrin complexation

by co-precipitation in vacuum, microwave irradiation and freeze-drying methods at 1:1 or 2:1 molar ratios (Bertacche et al. 2006; Sapino et al. 2009). The differential scanning calorimetric peak corresponding to the melting process at ~267 °C was evaluated. It was demonstrated the formation of the cyclodextrin/resveratrol inclusion complex due to the disappearance of this peak in all complexes. A resveratrol glycoside, *trans*-polydatin (Fig. 4.38), was identified in grapes and red wine. It is the main active compound from buffalo pea, *Pediomelum cuspidatum* (Zhang et al. 2016). *trans*-Polydatin was complexed with β- and γ-cyclodextrin by freeze-drying at 1:1 molar ratios and analyzed by Fourier transform-infrared spectroscopy, X-ray diffractometry, electrospray ionization-mass spectrometry and scanning electron microscopy as well as by thermal methods, named thermogravimetry and differential scanning calorimetry. The calorimetric peak corresponding to the melting of *trans*-polydatin from 228 °C disappearing after cyclodextrin complexation and confirm the formation of the inclusion complex. Furthermore, the decomposition of the guest starting at ~273 °C, according to the thermogravimetric results (Zhang et al. 2016). Another glucose-based antioxidant used for cyclodextrin complexation is phloridzin (Fig. 4.38). It has a dihydrochalcone structure with low water solubility and stability. 2-Hydroxypropyl-β-cyclodextrin was used for phloridzin nanoencapsulation and the formation of the inclusion complex was studied by isothermal titration calorimetry (Zhang et al. 2017).

Another class of antioxidants are carotenoids, some of them being encapsulated as enriched extracts or as pure compounds. It is the case of lycopene and asthaxanthin (Fig. 4.39), which were nanoencapsulated by β-cyclodextrin and 2-hydroxypropyl-β-cyclodextrin, respectively (Nerome et al. 2013; Yuan et al.

2008). Supercritical antisolvent precipitation method have been used for preparing the β-cyclodextrin/lycopene complex (Nerome et al. 2013). The differential scanning calorimetry endothermic melting peak of lycopene disappears or shifts after complexation. On the other hand, the stability of 2-hydroxypropyl-β-cyclodextrin/astaxanthin complex was investigated by thermogravimetry and differential thermal analysis (Yuan et al. 2008).

4.3.4 Cyclodextrin/Fatty Acid or Oil Complexes

Fatty acids and their glycerides are appropriate guests for obtaining cyclodextrin-based supramolecular systems. They have highly flexible hydrophobic chains that well fit into the cyclodextrin cavities. However, unsaturated fatty acids and their glycerides have low thermal and oxidative stability and the encapsulation can provide products with better stability. Obviously, oils and fats, which especially contain triglycerides of fatty acids, have not water solubility and cyclodextrin nanoencapsulation can ameliorates this property. It can be valuable for omega-3 containing oils for functional food products.

Various free fatty acids, oils and fats were used for complexation with natural or chemically modified cyclodextrins (Table 4.4). Szente and collaborators studied the complexation of different free saturated and unsaturated fatty acids with natural cyclodextrins and derivatives (Szente et al. 1993). They used lauric, myristic, palmitic and stearic acids from the saturated fatty acid class and oleic, linoleic, linolenic, eicosapentaenoic and docosahexaenoic acids from the unsaturated fatty acid class, as well as randomly methylated β-cyclodextrin and heptakis(2,6-*O*-dimethyl)-β-cyclodextrin as host compounds. The freeze-drying method at 1:1 molar ratio have been used. Thermal evolution analysis and differential scanning calorimetry have been used for evaluating the stability of eicosapentaenoic and palmitic fatty acids (Szente et al. 1993). Linoleic acid was studied from the thermal and oxidative stability point of view. Complexes with α- and β-cyclodextrin, obtained by co-precipitation, were subjected to degradation under oxidative conditions at various temperatures of 50–150 °C. The degradation compounds were evaluated by gas chromatography-mass spectrometry, as well as by thermogravimetry and differential scanning calorimetric thermal analyses (Hădărugă et al. 2006). The stability of various fatty acids, tocochromanols and carotenoids from the wheat bran, pumpkin and tomato oleoresins after α-cyclodextrin complexation was recently evaluated by differential scanning calorimetry (Durante et al. 2016). Other stability study that imply thermal analysis was performed on β-cyclodextrin and 2-hydroxypropyl-β-cyclodextrin/myristic acid complexes (dos Santos et al. 2012). The structures of the studied fatty acids are presented in Fig. 4.40.

Oils and fats are more valuable from the applicability point of view. Consequently, studies on the cyclodextrin complexation of such fatty acid containing mixtures have been performed. Fish oils are important materials for obtaining omega-3 enhanced functional food products, especially due to the presence of eicosapentaenoic

Table 4.4 Cyclodextrin/fatty acid and derivative complexes analyzed by thermal methods (type of cyclodextrins used for obtaining complexes and the corresponding thermal methods used for characterization were included in the table)

N°	Guest compound	Host compound (cyclodextrin)[a]	Thermal analyses[b]	References
	Free fatty acids			
1	EPA ((all-*Z*)-5,8,11,14,17-Eicosapentaenoic acid)	RAMEB, DIMEB	TEA, DSC	Szente et al. 1993
2	Fatty acids, Tocochromanols, carotenoids	α-CD	DSC	Durante et al. 2016
3	Linoleic acid	β-CD	DSC, TG	Hădărugă et al. 2006
4	Myristic acid	β-CD, HP-β-CD	DSC	dos Santos et al. 2012
5	Palmitic acid	RAMEB	DSC	Szente et al. 1993
	Fatty acid derivatives (e.g. triglycerides, other esters, phospholipids)			
6	Atlantic salmon oil	β-CD	DSC, TG	Hădărugă et al. 2016
7	Danube fish oils (common barbel, Pontic shad, European wels catfish, common bleak)	β-CD	DSC, TG	Hădărugă et al. 2017
8	European anchovy oil	β-CD	DSC, TG	Ünlüsayin et al. 2016
9	Cholesteryl oleate (phytosterol esters)	γ-CD	DSC	Sasako et al. 2016
10	Soybean lecithin (phospholipids)	β-CD	DSC	Wang et al. 2014

[a]*α-CD* α-cyclodextrin; *β-CD* β-cyclodextrin; *γ-CD* γ-cyclodextrin; *HP-α-CD* 2-hydroxypropyl-α-cyclodextrin; *HP-β-CD* 2-hydroxypropyl-β-cyclodextrin; *HP-γ-CD* 2-hydroxypropyl-γ-cyclodextrin; *HE-β-CD* 2-hydroxyethyl-β-cyclodextrin; *RAMEB* randomly methylated β-cyclodextrin; *DIMEB* heptakis(2,6-*O*-dimethyl)-β-cyclodextrin
[b]*TEA* thermal evolution analysis; *TG* thermogravimetry; *DSC* differential scanning calorimetry

and docosahexaenoic acids. Thus, fish oil samples obtained from various fish species such as Atlantic salmon, European anchovy, common barbel, Pontic shad, European wels catfish or common bleak, were complexed with β-cyclodextrin using kneading and co-precipitation method at 1:1, 2:1 and 3:1 molar ratios (Hădărugă et al. 2017; Hădărugă et al. 2016; Ünlüsayin et al. 2016). β-Cyclodextrin/fish oils complexes were evaluated by differential scanning calorimetry and thermogravimetric techniques. Thermal analyses were especially focused on the evaluation of water/moisture content of complexes, taking into account that the highly hydrophobic fish oil components replacing the water molecules inside the cyclodextrin cavity. The water related differential scanning calorimetry and thermogravimetric data were correlated with Karl Fischer titration results on water determination or with the fatty acid profile (Hădărugă et al. 2016; Ünlüsayin et al. 2016; Hădărugă et al. 2017). Fatty acid containing compounds such as cholesteryl oleate or phospholipids from

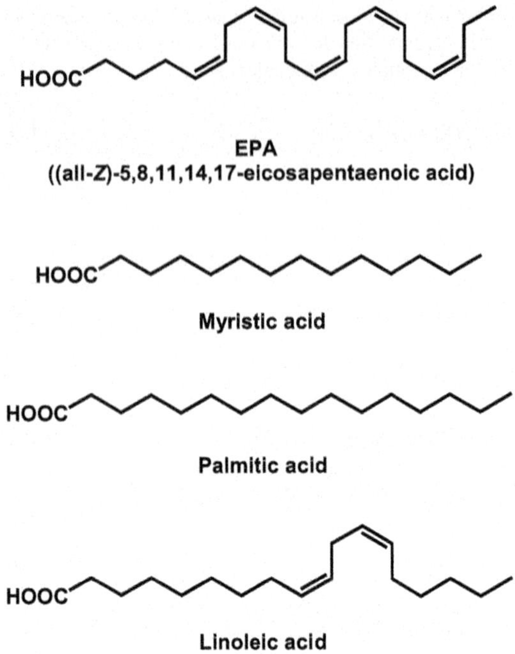

Fig. 4.40 Structures of the main free fatty acids used for cyclodextrin complexation

soybean lecithin were nanoencapsulated in β- or γ-cyclodextrin and analyzed by differential scanning calorimetry (Sasako et al. 2016; Wang et al. 2014).

4.3.5 Cyclodextrin/Other Organic and Inorganic Compound Complexes

There are various other organic, organometallic or even inorganic compounds used as guest compounds for cyclodextrin nanoencapsulation (Table 4.5). Some of these organic compounds were simple hydrophobic molecules like ethylene, benzene, ethylbenzene, toluene, *p*-xylene and styrene (Ho et al. 2011; Szafranek and Szafranek 1993; Uyar et al. 2006) or less hydrophobic organic derivatives such as phenol, aniline, alkylamines, benzoic acid, or tetrahydrofuran (Bouchemal and Mazzaferro 2012; Gatiatulin et al. 2014; Hanawa et al. 1993; Song et al. 2008; Uyar et al. 2006). Generally, thermal analysis methods such as differential scanning calorimetry, thermogravimetry and differential thermal analysis have been used for the characterization of cyclodextrin complexes. However, there were no specific applications for these cyclodextrin/organic compounds complexes, excepting for cyclodextrin/ethylene complex for fruit ripening.

There are some studies on organic compounds having more complex chemical structures that were used for nanoencapsulation by cyclodextrins. It was the case of

Table 4.5 Cyclodextrin/other organic and inorganic compound complexes analyzed by thermal methods (type of cyclodextrins used for obtaining complexes and the corresponding thermal methods used for characterization were included in the table)

N°	Guest compound	Host compound (cyclodextrin)[a]	Thermal analyses[b]	References
	Organic compounds			
1	Ethylene	α-CD	DSC, TG	Ho et al. 2011
2	Benzene	β-CD	TG, DTA	Szafranek and Szafranek 1993
3	Benzene	γ-CD	DSC, TG	Uyar et al. 2006
4	Ethylbenzene	γ-CD	DSC, TG	Uyar et al. 2006
5	*p*-Xylene	γ-CD	DSC, TG	Uyar et al. 2006
6	Styrene	γ-CD	DSC, TG	Uyar et al. 2006
7	Toluene	γ-CD	DSC, TG	Uyar et al. 2006
8	Phenol	γ-CD	DSC, TG	Uyar et al. 2006
9	Aniline	γ-CD	DSC, TG	Uyar et al. 2006
10	Diethylenetriamine	β-CD	TG	Song et al. 2008
11	Ethylenediamine	β-CD	TG	Song et al. 2008
12	Triethylamine	β-CD	TG	Song et al. 2008
13	1-Adamantylamine	β-CD	ITC	Bouchemal and Mazzaferro 2012
14	Benzoic acid	DIMEB	DSC, TG	Hanawa et al. 1993
15	2-Hydroxy-3-methoxybenzaldehyde isonicotinoylhydrazone	HP-β-CD	DSC	Ol'khovich et al. 2016
16	2-Hydroxybenzaldehyde isonicotinoylhydrazone	HP-β-CD	DSC	Ol'khovich et al. 2016
17	7-(Diethylamino)coumarin-3-carboxylic acid *N*-succinimidyl ester (in presence of urea)	β-CD, γ-CD	ITC	Maity et al. 2017
18	Alkyl-bisphosphonates	α-CD, β-CD, HP-β-CD, RAMEB	ITC	Monteil et al. 2017
19	Tetrahydrofurane	β-CD	TG-MS	Gatiatulin et al. 2014
20	Lycorine	α-CD, β-CD, γ-CD	DSC, TG	Liu et al. 2017
21	Usnic acid	β-CD, HP-β-CD	DSC	Nikolić et al. 2013
22	Ionic liquids (based on 1-dodecyl-3-methylimidazolium)	β-CD	ITC	Mofaddel et al. 2016
	Inorganic compounds or organometallic compounds			
23	Magnetite	β-CD	TG, DTA	Bocanegra-Diaz et al. 2003
24	Zinc acetate (obtaining ZnO)	β-CD	TG, DTA	Yang et al. 2003

(continued)

Table 4.5 (continued)

N°	Guest compound	Host compound (cyclodextrin)[a]	Thermal analyses[b]	References
25	Metal complexes (Co, Ni, Cu, Mn, Fe, Cr ions)	β-CD	DSC, TG	Ciesielski and Girek 2011
26	Zirconia-supported cobalt oxide catalysts	β-CD	TG-MS	Bai et al. 2013
27	Ferrocene	β-CD polymer	TG	Zhang et al. 2011
28	Phenylarsonic acid	β-CD, HP-β-CD, DIMEB	TG, GC-TOF-MS	Zhu et al. 2010

[a]*α-CD* α-cyclodextrin; *β-CD* β-cyclodextrin; *γ-CD* γ-cyclodextrin; *HP-β-CD* 2-hydroxypropyl-β-cyclodextrin; *RAMEB* randomly methylated β-cyclodextrin; *DIMEB* heptakis(2,6-*O*-dimethyl)-β-cyclodextrin
[b]*TG* thermogravimetry; *DSC* differential scanning calorimetry; *ITC* isothermal titration calorimetry; *DTA* differential thermal analysis; *TG-MS* thermogravimetry-mass spectrometry; *GC-TOF-MS* gas chromatography-time-of-flight-mass spectrometry

biologically active compounds 2-hydroxybenzaldehyde *iso*nicotinoylhydrazone and its 3-methoxylated derivative (Fig. 4.41), which were complexed with 2-hydroxypropyl-β-cyclodextrin and analyzed by differential scanning calorimetry. The disappearance of the melting peaks of the guest compounds at 248.2 and 236.7 °C after complexation have been observed (Ol'khovich et al. 2016). 7-(Diethylamino)coumarin-3-carboxylic acid *N*-succinimidyl ester (Fig. 4.41), a coumarin derivative having applications as cells labeling and to generate blue fluorescence in bioconjugates, was complexed with β- and γ-cyclodextrin in the presence of urea using co-precipitation method. Complexation process was evaluated by ^{1}H–nuclear magnetic resonance, electrospray ionization-mass spectrometry and isothermal titration calorimetry, indicating a 1:1 host:guest inclusion complex (Maity et al. 2017). Isothermal titration calorimetry was also used for evaluating the host:guest interaction of α-, β-, 2-hydroxypropyl-β-cyclodextrin and randomly methylated β-cyclodextrin with a wide range of hydrophobic alkyl substituted bisphosphonates (Fig. 4.41) (Monteil et al. 2017), as well as the interaction of some ionic liquids based on 1-dodecyl-3-methylimidazolium (Fig. 4.41) with β-cyclodextrin (Mofaddel et al. 2016). Lycorine is an alkaloid having anticancer activity (Fig. 4.41). Its relatively high rigidity and hydrophobicity make this structure a good candidate for a guest compound in natural cyclodextrin complexation. Complexes were obtained by co-precipitation method at a molar ratio of 1:2 and analyzed by differential scanning calorimetry and thermogravimetry (Liu et al. 2017). The disappearance of the peaks corresponding to pure lycorine were investigated in order to confirm the formation of the inclusion complexes. Usnic acid (Fig. 4.41) is the active compound from *Usnea* species (lichens), which have antibiotic activities. This compound was nanoencapsulated in β-cyclodextrin and 2-hydroxypropyl-β-cyclodextrin by co-precipitation method at equimolar ratios and analyzed by Fourier transform-infrared spectroscopy, X-ray diffractometry, ^{1}H–nuclear magnetic resonance and differential scanning calorimetry (Nikolić et al.

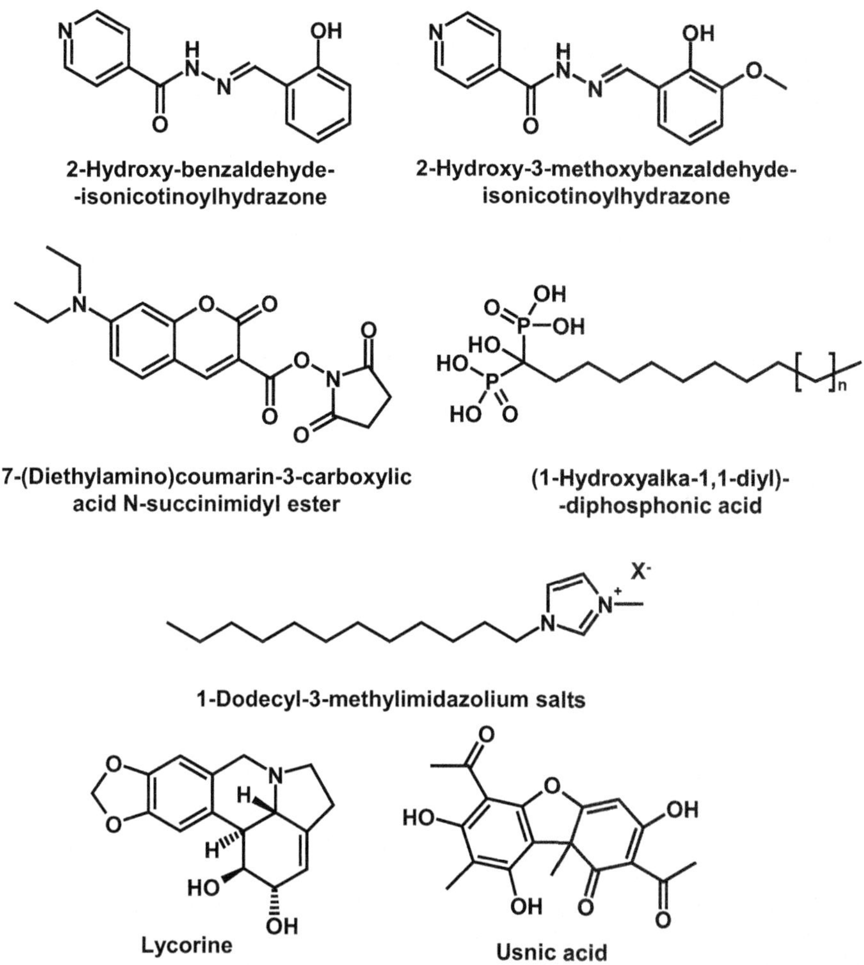

Fig. 4.41 Structures of some organic compounds used for cyclodextrin complexation

2013). It was observed that the differential scanning calorimetric melting peak disappearing after cyclodextrin complexations, which support the formation of the inclusion complexes.

There are some organometallic compounds that were complexed with β-cyclodextrin, its derivatives or polymers. Attempts to obtain β-cyclodextrin complexes using zinc acetate (for obtaining zinc oxide), ferrocene or phenylarsonic acid have been performed (Fig. 4.42). Co-precipitation method have been used in all cases (Yang et al. 2003; Zhang et al. 2011; Zhu et al. 2010). Various thermal techniques such as thermogravimetry, differential thermal analysis and gas chromatography-time-of-flight-mass spectrometry have been performed for the characterization of complexes. Some studies tried to discuss the possible complexation of inorganic

Fig. 4.42 Structures of some organometallic compounds used for cyclodextrin complexation

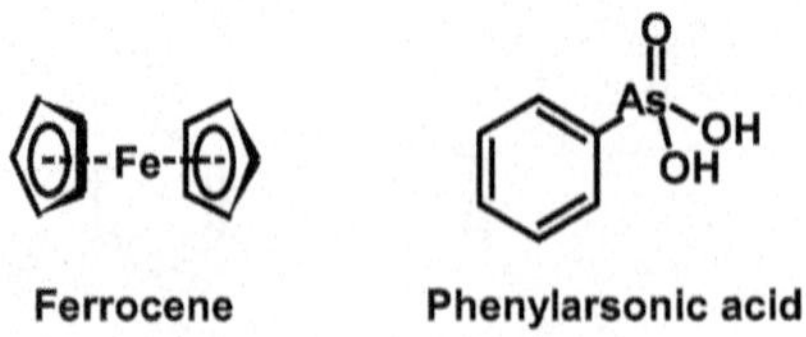

compounds such as magnetite or metal salts. Complexes were characterize by various methods, including thermogravimetric, differential thermal analysis and differential scanning calorimetric thermal techniques (Bocanegra-Diaz et al. 2003; Ciesielski and Girek 2011). However, there was no proof of the formation of an inclusion complex between cyclodextrin cavity and studied inorganic compounds.

4.4 Conclusion

Cyclodextrins remains some of the top bio-carriers due to their special structural architectures and specific properties. The main cyclodextrins, named α-, β- and γ-cyclodextrin, are natural compounds that are approved by Food and Drug Administration and Generally Recognized as Safe. However, they are already used in many fields that are related to bioactive compounds such as pharmaceutical, food and cosmetic fields. Particular attention is related to the advantages and disadvantages of the interaction of cyclodextrins with lipids influencing the human health. It is the case of fatty acids and their glycerides, cholesterol, as well as phospholipids.

The importance of thermal analyses for evaluating the cyclodextrin inclusion complexes was reviewed with emphasis to recent studies. The host-guest supramolecular structures were classified according to the type of biological activity incidence related to applications for cyclodextrin complexation. The most studied bioactive guests for cyclodextrin complexation were drugs, which are grouped by Anatomical Therapeutic Chemical classification, followed by odorants, essential oils and extracts, natural antioxidants, especially flavonoids and isoflavonoids, flavonoid glycosides, flavonolignans, other polyphenols, carotenoids, fatty acids, oils and fatty acid based derivatives, as well as other organic, organometallic and inorganic compounds, which have no specific applications. By far, differential scanning calorimetry was the most used thermal method for the characterization of cyclodextrin/bioactive compound complexes. Studies were focused on the proving of the formation of cyclodextrin inclusion complexes by disappearance of the thermal characteristic peaks of the guest compound after nanoencapsulation, such as melting of boiling points. Another aspect studied by differential scanning calorimetry, thermogravimetry-differential thermogravimetry and differential thermal analysis was related to the behavior of hydration water molecules of complexes. The stability of the guest compound during heating up to the decomposition of cyclodextrin molecule was investigated using above mentioned methods, as well as combined techniques such as thermogravimetry-mass spectrometry or gas chromatography-time-of-flight-mass spectrometry. Thermal methods are some of the most valuable techniques for analyzing cyclodextrin complexes and together with other methods can complete the information related to the host-guest molecular inclusion process.

References

Abarca RL, Rodríguez FJ, Guarda A, Galotto MJ, Bruna JE (2016) Characterization of beta-cyclodextrin inclusion complexes containing an essential oil component. Food Chem 196:968–975. https://doi.org/10.1016/j.foodchem.2015.10.023

Alvarez-Parrilla E, de la Rosa LA, Torres-Rivas F, Rodrigo-Garcia J, González-Aguilar GA (2005) Complexation of apple antioxidants: chlorogenic acid, quercetin and rutin by β-cyclodextrin (β-CD). J Incl Phenom Macrocycl Chem 53:121–129. https://doi.org/10.1007/s10847-005-1620-z

Ammar HO, Ghorab M, Mahmoud AA, Makram TS, Noshi SH (2012) Topical liquid crystalline gel containing lornoxicam/cyclodextrin complex. J Incl Phenom Macrocycl Chem 73:161–175. https://doi.org/10.1007/s10847-011-0039-y

Andrade TA, Freitas TS, Araújo FO, Menezes PP, Dória GAA, Rabelo AS, Quintans-Júnior LJ, Santos MRV, Bezerra DP, Serafini MR, Menezes IRA, Nunes PS, Araújo AAS, Costa MS, Campina FF, Santos ATL, Silva ARP, Coutinho HDM (2017) Physico-chemical characterization and antibacterial activity of inclusion complexes of *Hyptis martiusii* Benth essential oil in β -cyclodextrin. Biomed Pharmacother 89:201–207. https://doi.org/10.1016/j.biopha.2017.01.158

Astray G, Gonzalez-Barreiro C, Mejuto JC, Rial-Otero R, Simal-Gándara J (2009) A review on the use of cyclodextrins in foods. Food Hydrocoll 23:1631–1640. https://doi.org/10.1016/j.foodhyd.2009.01.001

Badr-Eldin SM, Elkheshen SA, Ghorab MM (2008) Inclusion complexes of tadalafil with natural and chemically modified β-cyclodextrins. I: Preparation and in-vitro evaluation. Eur J Pharm Biopharm 70:819–827. https://doi.org/10.1016/j.ejpb.2008.06.024

Bai L, Wyrwalski F, Lamonier J-F, Khodakov AY, Monflier E, Ponchel A (2013) Effects of β-cyclodextrin introduction to zirconia supported-cobalt oxide catalysts: from molecule-ion associations to complete oxidation of formaldehyde. Appl Catal B Environ 138-139:381–390. https://doi.org/10.1016/j.apcatb.2013.03.015

Bertacche V, Lorenzi N, Nava D, Pini E, Sinico C (2006) Host–guest interaction study of resveratrol with natural and modified cyclodextrins. J Incl Phenom Macrocycl Chem 55:279–287. https://doi.org/10.1007/s10847-006-9047-8

Bettinetti GP, Sorrenti M, Rossi S, Ferrari F, Mura P, Faucci MT (2002) Assessment of solid-state interactions of naproxen with amorphous cyclodextrin derivatives by DSC. J Pharm Biomed Anal 30:1173–1179

Biliaderis CG (1983) Differential scanning calorimetry in food research. A review. Food Chem 10:239–265

Bocanegra-Diaz A, Mohallem NDS, Sinisterra RD (2003) Preparation of a ferrofluid using cyclodextrin and magnetite. J Braz Chem Soc 14(6):936–941

Boldescu V, Bratu I, Borodi G, Kacso I, Bende A, Duca G, Macaev F, Pogrebnoi S, Ribkovskaia Z (2012) Study of binary systems of β-cyclodextrin with a highly potential anti-mycobacterial drug. J Incl Phenom Macrocycl Chem 74:129–135. https://doi.org/10.1007/s10847-011-0091-7

Bonetti P, de Moraes FF, Zanin GM, de Cássia Bergamasco R (2016) Thermal behavior study and decomposition kinetics of linalool/β-cyclodextrin inclusion complex. Polym Bull 73:279–291. https://doi.org/10.1007/s00289-015-1486-1

Bouchemal K, Mazzaferro S (2012) How to conduct and interpret ITC experiments accurately for cyclodextrin-guest interactions. Drug Discov Today 17(11/12):623–629. https://doi.org/10.1016/j.drudis.2012.01.023

Brewster ME, Loftsson T (2007) Cyclodextrins as pharmaceutical solubilizers. Adv Drug Deliv Rev 59:645–666. https://doi.org/10.1016/j.addr.2007.05.012

Cabral Marques HM (2010) A review on cyclodextrin encapsulation of essential oils and volatiles. Flavour Fragr J 25:313–326. https://doi.org/10.1002/ffj.2019

Cabral Marques HM, Hadgraft J, Kellaway IW (1990) Studies of cyclodextrin inclusion complexes. I. The salbutamol-cyclodextrin complex as studied by phase solubility and DSC. Int J Pharm 63:259–266

Calabrò ML, Tommasini S, Donato P, Raneri D, Stancanelli R, Ficarra P, Ficarra R, Costa C, Catania S, Rustichelli C, Gamberini G (2004) Effects of α- and β-cyclodextrin complexation on the physico-chemical properties and antioxidant activity of some 3-hydroxyflavones. J Pharm Biomed Anal 35:365–377. https://doi.org/10.1016/j.jpba.2003.12.005

Calsavara LPV, Zanin GM, de Moraes FF (2012) Enrofloxacin inclusion complexes with cyclodextrins. J Incl Phenom Macrocycl Chem 73:219–224. https://doi.org/10.1007/s10847-011-0045-0

Carrier RL, Miller LA, Ahmed I (2007) The utility of cyclodextrins for enhancing oral bioavailability. J Control Release 123:78–99. https://doi.org/10.1016/j.jconrel.2007.07.018

Cavalli R, Trotta F, Trotta M, Pastero L, Aquilano D (2007) Effect of alkylcarbonates of γ-cyclodextrins with different chain lengths on drug complexation and release characteristics. Int J Pharm 339:197–204. https://doi.org/10.1016/j.ijpharm.2007.03.001

Ceborska M, Zimnicka M, Wszelaka-Rylik M, Troć A (2016) Characterization of folic acid/native cyclodextrins host-guest complexes in solution. J Mol Struct 1109:114–118. https://doi.org/10.1016/j.molstruc.2015.12.082

Cerchiara T, Luppi B, Bigucci F, Zecchi V (2003) Effect of chitosan on progesterone release from hydroxypropyl-β-cyclodextrin complexes. Int J Pharm 258:209–215. https://doi.org/10.1016/S0378-5173(03)00202-3

Ciesielski W, Girek T (2011) Study of thermal stability of β-cyclodextrin/metal complexes in the aspect of their future applications. J Incl Phenom Macrocycl Chem 69:461–467. https://doi.org/10.1007/s10847-010-9803-7

Corciova A, Ciobanu C, Poiata A, Mircea C, Nicolescu A, Drobota M, Varganici C-D, Pinteala T, Marangoci N (2015) Antibacterial and antioxidant properties of hesperidin:β -cyclodextrin complexes obtained by different techniques. J Incl Phenom Macrocycl Chem 81(1):71–84. https://doi.org/10.1007/s10847-014-0434-2

Dias K, Nikolaou S, De Giovani WF (2008) Synthesis and spectral investigation of Al(III) catechin/β -cyclodextrin and Al(III) quercetin/β -cyclodextrin inclusion compounds. Spectrochim Acta A 70(1):154–161. https://doi.org/10.1016/j.saa.2007.07.022

Duchêne D, Bochot A (2016) Thirty years with cyclodextrins. Int J Pharm 514:58–72. https://doi.org/10.1016/j.ijpharm.2016.07.030

Durante M, Lenucci MS, Marrese PP, Rizzi V, De Caroli M, Piro G, Fini P, Russo GL, Mita G (2016) α-Cyclodextrin encapsulation of supercritical CO extracted oleoresins from different plant matrices: a stability study. Food Chem 199:684–693. https://doi.org/10.1016/j.foodchem.2015.12.073

Éhen Z, Giordano F, Sztatisz J, Jicsinszky L, Novák C (2005) Thermal characterization of natural and modified cyclodextrins using TG-MS combined technique. J Therm Anal Calorim 80:419–424

Funk O, Schwabe L, Fromming K-H (1993) Composition and properties of freeze-dried products of nicotinic acid with β -cyclodextrin and heptakis (2,6-O-dimethyl)-β -cyclodextrin. J Incl Phenom Mol Recognit Chem 16:299–314

Galvão JG, Silva VF, Ferreira SG, França FRM, Santos DA, Freitas LS, Alves PB, Antunes de Souza Araújo A, Cavalcanti SCH, Nunes RS (2015) β-Cyclodextrin inclusion complexes containing *Citrus sinensis* (L.) Osbeck essential oil: an alternative to control *Aedes aegypti* larvae. Thermochim Acta 608:14–19. https://doi.org/10.1016/j.tca.2015.04.001

Gatiatulin AK, Ziganshin MA, Gorbatchuk VV (2014) Selective preparation of beta-cyclodextrin clathrates by solid-phase exchange of included tetrahydrofurane for volatile guests in absence of water. J Therm Anal Calorim 118:987–992. https://doi.org/10.1007/s10973-014-3800-9

Giordano F, Novak C, Moyano R (2001) Thermal analysis of cyclodextrins and their inclusion compounds. Thermochim Acta 380:123–151

Giron D (2002) Applications of thermal analysis and coupled techniques in pharmaceutical industry. J Therm Anal Calorim 68:335–357

Gomes Rodrigues S, de Souza Chaves I, de Melo NF-S, de Jesus MB, Fraceto LF, Fernandes SA, de Paula E, Puggina de Freitas M, de Matos Alves Pinto L (2011) Computational analysis and physico-chemical characterization of an inclusion compound between praziquantel and methyl-β-cyclodextrin for use as an alternative in the treatment of schistosomiasis. J Incl Phenom Macrocycl Chem 70:19–28. https://doi.org/10.1007/s10847-010-9852-y

Guimarães AG, Oliveira MA, dos Santos Alves R, dos Passos Menezes P, Russo Serafini M, Antunes de Souza Araújo A, Bezerra DP, Quintans Júnior LJ (2015) Encapsulation of carvacrol, a monoterpene present in the essential oil of oregano, with β-cyclodextrin, improves the pharmacological response on cancer pain experimental protocols. Chem Biol Interact 227:69–76. https://doi.org/10.1016/j.cbi.2014.12.020

Hădărugă NG, Hădărugă DI, Păunescu V, Tatu C, Ordodi VL, Bandur G, Lupea AX (2006) Thermal stability of the linoleic acid/α- and β-cyclodextrin complexes. Food Chem 99:500–508. https://doi.org/10.1016/j.foodchem.2005.08.012

Hădărugă DI, Hădărugă NG, Butnaru G, Tatu C, Gruia A (2010) Bioactive microparticles (10): thermal and oxidative stability of nicotine and its complex with β-cyclodextrin. J Incl Phenom Macrocycl Chem 68:155–164. https://doi.org/10.1007/s10847-010-9761-0

Hădărugă DI, Hădărugă NG, Bandur GN, Isengard H-D (2012a) Water content of flavonoid/cyclodextrin nanoparticles: relationship with the structural descriptors of biologically active compounds. Food Chem 132:1651–1659. https://doi.org/10.1016/j.foodchem.2011.06.004

Hădărugă NG, Hădărugă DI, Isengard H-D (2012b) Water content of natural cyclodextrins and their essential oil complexes: a comparative study between Karl Fischer titration and thermal methods. Food Chem 132:1741–1748. https://doi.org/10.1016/j.foodchem.2011.11.003

Hădărugă DI, Ünlüsayin M, Gruia AT, Birău (Mitroi) C, Rusu G, Hădărugă NG (2016) Thermal and oxidative stability of Atlantic salmon oil (*Salmo salar* L.) and complexation with β-cyclodextrin. Beilstein J Org Chem 12:179–191. https://doi.org/10.3762/bjoc.12.20

Hădărugă DI, Birău (Mitroi) CL, Gruia AT, Păunescu V, Bandur GN, Hădărugă NG (2017) Moisture evaluation of β-cyclodextrin/fish oils complexes by thermal analyses: a data review on common barbel (*Barbus barbus* L.), Pontic shad (*Alosa immaculata* Bennett), European wels catfish (*Silurus glanis* L.), and common bleak (*Alburnus alburnus* L.) living in Danube river. Food Chem 236:49–58. https://doi.org/10.1016/j.foodchem.2017.03.093

Haiyun D, Jianbin C, Guomei Z, Shaomin S, Jinhao P (2003) Preparation and spectral investigation on inclusion complex of β-cyclodextrin with rutin. Spectrochim Acta A 59:3421–3429. https://doi.org/10.1016/S1386-1425(03)00176-8

Hanawa T, Yonemochi E, Oguchi T, Nakai Y, Yamamoto K (1993) Thermal behavior of ground mixtures of heptakis (2,6-di-O-methyl)-β-cyclodextrin and benzoic acid. J Incl Phenom Mol Recognit Chem 15:91–101

He D, Deng P, Yang L, Tan Q, Liu J, Yang M, Zhang J (2013) Molecular encapsulation of rifampicin as an inclusion complex of hydroxypropyl-β-cyclodextrin: design, characterization and in vitro dissolution. Colloids Surf B Biointerfaces 103:580–585. https://doi.org/10.1016/j.colsurfb.2012.10.062

Hegheş A, Hădărugă NG, Fuliaş A-V, Bandur GN, Hădărugă DI, Dehelean C-A (2015) *Capsicum annuum* extracts/β -cyclodextrin complexes. Thermal analyses - Karl Fischer water titration correlations and antioxidant activity. J Therm Anal Calorim 120:603–615. https://doi.org/10.1007/s10973-014-4229-x

Hill LE, Gomes C, Taylor TM (2013) Characterization of beta-cyclodextrin inclusion complexes containing essential oils (*trans*-cinnamaldehyde, eugenol, cinnamon bark, and clove bud extracts) for antimicrobial delivery applications. LWT Food Sci Technol 51:86–93. https://doi.org/10.1016/j.lwt.2012.11.011

Ho BT, Joyce DC, Bhandari BR (2011) Encapsulation of ethylene gas into α-cyclodextrin and characterisation of the inclusion complexes. Food Chem 127:572–580. https://doi.org/10.1016/j.foodchem.2011.01.043

Huang Y, Zu Y, Zhao X, Wu M, Feng Z, Deng Y, Zu C, Wang L (2016) Preparation of inclusion complex of apigenin-hydroxypropyl-β-cyclodextrin by using supercritical antisolvent process for dissolution and bioavailability enhancement. Int J Pharm 511:921–930. https://doi.org/10.1016/j.ijpharm.2016.08.007

Ikeda H, Fukushige Y, Matsubara T, Inenaga M, Kawahara M, Yukawa M, Fujisawa M, Yukawa E, Aki H (2016) Improving water solubility of nateglinide by complexation of β-cyclodextrin. J Therm Anal Calorim 123:1847–1850. https://doi.org/10.1007/s10973-015-4714-x

Jadhav P, Pore Y (2016) Physicochemical, thermodynamic and analytical studies on binary and ternary inclusion complexes of bosentan with hydroxypropyl-β-cyclodextrin. Bulletin of Faculty of Pharmacy, Cairo University 55:147–154. https://doi.org/10.1016/j.bfopcu.2016.12.004

Jamrógiewicza M, Wielgomas B, Strankowskic M (2014) Evaluation of the photoprotective effect of β-cyclodextrin on the emission of volatile degradation products of ranitidine. J Pharm Biomed Anal 98:113–119. https://doi.org/10.1016/j.jpba.2014.05.014

Jordheim LP, Degobert G, Diab R, Peyrottes S, Périgaud C, Dumontet C, Fessi H (2009) Inclusion complexes of a nucleotide analogue with hydroxypropyl-beta-cyclodextrin. J Incl Phenom Macrocycl Chem 63:11–16. https://doi.org/10.1007/s10847-008-9483-8

Kacso I, Borodi G, Fărcas SI, Bratu I (2009) Inclusion compound of vitamin B13 in β-Cyclodextrin. Structural investigations. J Phys Conf Ser 182:1–5. https://doi.org/10.1088/1742-6596/182/1/012009

Karathanos VT, Mourtzinos I, Yannakopoulou K, Andrikopoulos NK (2007) Study of the solubility, antioxidant activity and structure of inclusion complex of vanillin with β-cyclodextrin. Food Chem 101:652–658. https://doi.org/10.1016/j.foodchem.2006.01.053

Kfoury M, Auezova L, Fourmentin S, Greige-Gerges H (2014) Investigation of monoterpenes complexation with hydroxypropyl-β-cyclodextrin. J Incl Phenom Macrocycl Chem 80(1):51–60. https://doi.org/10.1007/s10847-014-0385-7

Kfoury M, Auezova L, Ruellan S, Greige-Gerges H, Fourmentin S (2015) Complexation of estragole as pure compound and as main component of basil and tarragon essential oils with cyclodextrins. Carbohydr Polym 118:156–164. https://doi.org/10.1016/j.carbpol.2014.10.073

Kfoury M, Landy D, Ruellan S, Auezova L, Greige-Gerges H, Fourmentin S (2017) Nootkatone encapsulation by cyclodextrins: effect on water solubility and photostability. Food Chem 236:41–48. https://doi.org/10.1016/j.foodchem.2016.12.086

Koontz JL, Marcy JE (2003) Formation of natamycin: cyclodextrin inclusion complexes and their characterization. J Agric Food Chem 51:7106–7110. https://doi.org/10.1021/jf030332y

Kreaz RMA, Abu-Eida EY, Erős I, Kata M (1999) Freeze-dried complexes of furosemide with β-cyclodextrin derivatives. J Incl Phenom Macrocycl Chem 34:39–48

Krishna Mohan PR, Sreelakshmi G, Muraleedharan CV, Joseph R (2012) Water soluble complexes of curcumin with cyclodextrins: characterization by FT-Raman spectroscopy. Vib Spectrosc 62:77–84. https://doi.org/10.1016/j.vibspec.2012.05.002

Kurkov SV, Loftsson T (2013) Cyclodextrins. Int J Pharm 453:167–180. https://doi.org/10.1016/j.ijpharm.2012.06.055

Lelievre J, Liu H (1994) A review of thermal analysis studies of starch gelatinization. Thermochim Acta 246:309–315

Li J-H, Zhang N, Li X-T, Wang J-Y, Tian S-J (1997) Kinetic studies on the thermal dissociation of the inclusion complex of β-cyclodextrin with cinnamic aldehyde. J Therm Anal 49:1527–1533

de Lima Petito N, da Silva Dias D, Gonçalves Costa V, Quintanilha Falcão D, Gome de Lima Araujo K (2016) Increasing solubility of red bell pepper carotenoids by complexation with 2-hydroxypropyl-β-cyclodextrin. Food Chem 208:124–131. https://doi.org/10.1016/j.foodchem.2016.03.122

Lima PSS, Lucchese AM, Araújo-Filho HG, Menezes PP, Araújo AAS, Quintans-Júnior LJ, Quintans JSS (2016) Inclusion of terpenes in cyclodextrins: preparation, characterization and pharmacological approaches. Carbohydr Polym 151:965–987. https://doi.org/10.1016/j.carbpol.2016.06.040

Lin-Hui T, Zheng-Zhi P, Ying Y (1995) Inclusion complexes of α-and β-cyclodextrin with α-lipoic acid. J Incl Phenom Mol Recognit Chem 23:119–126

Liu B, Li W, Zhao J, Liu Y, Zhu X, Liang G (2013a) Physicochemical characterisation of the supramolecular structure of luteolin/cyclodextrin inclusion complex. Food Chem 141:900–906. https://doi.org/10.1016/j.foodchem.2013.03.097

Liu B, Zhu X, Zeng J, Zhao J (2013b) Preparation and physicochemical characterization of the supramolecular inclusion complex of naringin dihydrochalcone and hydroxypropyl-β-cyclodextrin. Food Res Int 54:691–696. https://doi.org/10.1016/j.foodres.2013.08.007

Liu H, Yang G, Tang Y, Cao D, Qi T, Qi Y, Fan G (2013c) Physicochemical characterization and pharmacokinetics evaluation of β-caryophyllene/β-cyclodextrin inclusion complex. Int J Pharm 450:304–310. https://doi.org/10.1016/j.ijpharm.2013.04.013

Liu D-D, Guo Y-F, Zhang J-Q, Yang Z-K, Li X, Yang B, Yang R (2017) Inclusion of lycorine with natural cyclodextrins (α-, β-and γ-CD): experimental and in vitro evaluation. J Mol Struct 1130:669–676. https://doi.org/10.1016/j.molstruc.2016.11.018

Loftsson T, Duchêne D (2007) Cyclodextrins and their pharmaceutical applications. Int J Pharm 329:1–11. https://doi.org/10.1016/j.ijpharm.2006.10.044

Loh GOK, Tan YTF, Peh K-K (2016) Enhancement of norfloxacin solubility via inclusion complexation with β-cyclodextrin and its derivative hydroxypropyl-β-cyclodextrin. Asian Journal of Pharmaceutical Sciences 11:536–546. https://doi.org/10.1016/j.ajps.2016.02.009

Maestrelli F, Cecchi M, Cirri M, Capasso G, Mennini N, Mura P (2009) Comparative study of oxaprozin complexation with natural and chemically-modified cyclodextrins in solution and in the solid state. J Incl Phenom Macrocycl Chem 63:17–25. https://doi.org/10.1007/s10847-008-9484-7

Maity B, Chatterjee A, Ahmed SA, Seth D (2017) Deciphering the perturbation effect of urea on the supramolecular host-guest interaction of biologically active hydrophobic molecule inside the nanocavity of cyclodextrins. J Lumin 183:238–250. https://doi.org/10.1016/j.jlumin.2016.11.037

Majumdar S, Srirangam R (2009) Solubility, stability, physicochemical characteristics and *in vitro* ocular tissue permeability of hesperidin: a natural bioflavonoid. Pharm Res 26(5):1217–1225. https://doi.org/10.1007/s11095-008-9729-6

Makhlof A, Miyazaki Y, Tozuka Y, Takeuchi H (2008) Cyclodextrins as stabilizers for the preparation of drug nanocrystals by the emulsion solvent diffusion method. Int J Pharm 357:280–285. https://doi.org/10.1016/j.ijpharm.2008.01.025

Malaekeh-Nikouei B, Nassirli H, Davies N (2007) Enhancement of cyclosporine aqueous solubility using α- and hydroxypropyl β-cyclodextrin mixtures. J Incl Phenom Macrocycl Chem 59:245–250. https://doi.org/10.1007/s10847-007-9321-4

Manakov AY, Rodionova TV, Aladko LS, Villevald GV, Lipkowski JS, Zelenina LN, Chusova TP, Karpova TD (2016) α-Cyclodextrin - water binary system. New data on dehydration of α-cyclodextrin hexahydrate. J Chem Thermodyn 101:251–259. https://doi.org/10.1016/j.jct.2016.06.008

Marreto RN, Almeida EECV, Alves PB, Niculau ES, Nunes RS, Matos CRS, Antunes de Souza Araújo A (2008) Thermal analysis and gas chromatography coupled mass spectrometry analyses of hydroxypropyl-β-cyclodextrin inclusion complex containing *Lippia gracilis* essential oil. Thermochim Acta 475:53–58. https://doi.org/10.1016/j.tca.2008.06.015

Martins Mendes Gomes L, Petito N, Gonçalves Costa V, Quintanilha Falcão D, de Lima Araújo KG (2014) Inclusion complexes of red bell pepper pigments with β-cyclodextrin: preparation, characterisation and application as natural colorant in yogurt. Food Chem 148:428–436. https://doi.org/10.1016/j.foodchem.2012.09.065

Martins A d P, Craveiro AA, Machado MIL, Raffin FN, Moura TF, Novák C, Éhen Z (2007) Preparation and characterization of *Mentha x villosa* Hudson oil-β-cyclodextrin complex. J Therm Anal Calorim 88:363–371

Mathot VBF, Benoist L (1994) Calorimetry and thermal analysis of polymers. Hanser Publishers, Munich

de Matos Jensen CE, Souza dos Santos RA, Denadai AML, Santos CFF, Braga ANG, Sinisterra RD (2010) Pharmaceutical composition of valsartan: β-cyclodextrin: physico-chemical characterization and anti-hypertensive evaluation. Molecules 15:4067–4084. https://doi.org/10.3390/molecules15064067

Mazzobre MF, dos Santos CI, del Pilar Buera M (2011) Solubility and stability of β-cyclodextrin-terpineol inclusion complex as affected by water. Food Biophysics 6:274–280. https://doi.org/10.1007/s11483-011-9208-1

Meier MM, Luiz MTB, Szpoganicz B, Soldi V (2001) Thermal analysis behavior of β- and γ-cyclodextrin inclusion complexes with capric and caprilic acid. Thermochim Acta 375:153–160

Meisel T (1982) Review on problems, techniques and trends in thermal analysis. Fresen J Anal Chem 312:83–95

Menczel JD, Bruce Prime R (2009) Thermal analysis of polymers. Fundamentals and applications. John Wiley & Sons, Inc, Hoboken

Menezes PP, Serafini MR, Santana BV, Nunes RS, Quintans LJ Jr, Silva GF, Medeiros IA, Marchioro M, Fraga BP, Santos MRV, Antunes de Souza Araújo A (2012) Solid-state β-cyclodextrin complexes containing geraniol. Thermochim Acta 548:45–50. https://doi.org/10.1016/j.tca.2012.08.023

Menezes PP, Serafini MR, Quintans-Júnior LJ, Silva GF, Oliveira JF, Carvalho FMS, Souza JCC, Matos JR, Alves PB, Matos IL, Hădărugă DI, Antunes de Souza Araújo A (2014) Inclusion complex of (−)-linalool and β-cyclodextrin. J Therm Anal Calorim 115(3):2429–2437. https://doi.org/10.1007/s10973-013-3367-x

Menezes PP, Serafini MR, de Carvalho YMBG, Santana DVS, Lima BS, Quintans-Júnior LJ, Marreto RN, de Aquino TM, Sabino AR, Scotti L, Scotti MT, Grangeiro-Júnior S, Araújo AAS (2016) Kinetic and physical-chemical study of the inclusion complex of β -cyclodextrin containing carvacrol. J Mol Struct 1125:323–330. https://doi.org/10.1016/j.molstruc.2016.06.062

Michalska P, Wojnicz A, Ruiz-Nuño A, Abril S, Buendia I, León R (2017) Inclusion complex of ITH12674 with 2-hydroxypropyl-β -cyclodextrin: preparation, physical characterization and pharmacological effect. Carbohydr Polym 157:94–104. https://doi.org/10.1016/j.carbpol.2016.09.072

Mofaddel N, Fourmentin S, Guillen F, Landy D, Gouhier G (2016) Ionic liquids and cyclodextrin inclusion complexes: limitation of the affinity capillary electrophoresis technique. Anal Bioanal Chem 408:8211–8220. https://doi.org/10.1007/s00216-016-9931-z

Monteil M, Lecouvey M, Landy D, Ruellan S, Mallard I (2017) Cyclodextrins: a promising drug delivery vehicle for bisphosphonate. Carbohydr Polym 156:285–293. https://doi.org/10.1016/j.carbpol.2016.09.030

Mourtzinos I, Kalogeropoulos N, Papadakis SE, Konstantinou K, Karathanos VT (2008) Encapsulation of nutraceutical monoterpenes in β-cyclodextrin and modified starch. J Food Sci S: Sensory and Food Quality 73(1):S89–S94. https://doi.org/10.1111/j.1750-3841.2007.00609.x

Muñoz-Ruiz A, Paronen P (1997) Particle and powder properties of cyclodextrins. Int J Pharm 148:33–39

Mura P (2015) Analytical techniques for characterization of cyclodextrin complexesin the solid state: a review. J Pharm Biomed Anal 113:226–238. https://doi.org/10.1016/j.jpba.2015.01.058

Mura P, Maestrelli F, Cirri M (2003) Ternary systems of naproxen with hydroxypropyl-β-cyclodextrin and aminoacids. Int J Pharm 260:293–302. https://doi.org/10.1016/S0378-5173(03)00265-5

Neoh TL, Yamauchi K, Yoshii H, Furuta T (2008) Kinetic study of thermally stimulated dissociation of inclusion complex of 1-methylcyclopropene with α-cyclodextrin by thermal analysis. J Phys Chem B 112:15914–15920. https://doi.org/10.1021/jp806233c

Nerome H, Machmudah S, Wahyudiono FR, Higashiura T, Youn Y-S, Lee Y-W, Goto M (2013) Nanoparticle formation of lycopene/β–cyclodextrin inclusion complex using supercritical antisolvent precipitation. J Supercrit Fluids 83:97–103. https://doi.org/10.1016/j.supflu.2013.08.014

Nikolić V, Stanković M, Nikolić L, Nikolić G, Ilić-Stojanović S, Popsavin M, Zlatković S, Kundaković T (2013) Inclusion complexes with cyclodextrin and usnic acid. J Incl Phenom Macrocycl Chem 76:173–182. https://doi.org/10.1007/s10847-012-0187-8

Novák C, Éhen Z, Fodor M, Jicsinszky L, Orgoványi J (2006) Application of combined thermo-analytical techniques in the investigation of cyclodextrin inclusion complexes. J Therm Anal Calorim 84:693–701. https://doi.org/10.1007/s10973-005-7605-8

Nuchuchua O, Saesoo S, Sramala I, Puttipipatkhachorn S, Soottitantawat A, Ruktanonchai U (2009) Physicochemical investigation and molecular modeling of cyclodextrin complexation mechanism with eugenol. Food Res Int 42:1178–1185. https://doi.org/10.1016/j.foodres.2009.06.006

Ol'khovich MV, Sharapova AV, Blokhina SV, Skachilov SY, Kesarev OG, Perlovich GL (2016) Physicochemical characteristics of the inclusion complexes of biologically active compounds with 2-hydroxypropyl-β-cyclodextrin. Thermochim Acta 639:1–9. https://doi.org/10.1016/j.tca.2016.07.008

Oliveira MGB, Brito RG, Santos PL, Araújo-Filho HG, Quintans JSS, Menezes PP, Serafini MR, Carvalho YMBG, Silva JC, Almeida JRGS, Scotti L, Scotti MT, Shanmugam S, Thangaraj P, Araújo AAS, Quintans-Júnior LJ (2016) α-Terpineol, a monoterpene alcohol, complexed with β-cyclodextrin exerts antihyperalgesic effect in animal model for fibromyalgia aided with docking study. Chem Biol Interact 254:54–62. https://doi.org/10.1016/j.cbi.2016.05.029

Onyeji CO, Omoruyi SI, Oladimeji FA, Soyinka JO (2009) Physicochemical characterization and dissolution properties of binary systems of pyrimethamine and 2-hydroxypropyl-β-cyclodextrin. Afr J Biotechnol 8(8):1651–1659

Ozawa T (2000) Thermal analysis - review and prospect. Thermochim Acta 355:35–42

Paczkowska M, Mizera M, Szymanowska-Powałowska D, Lewandowska K, Błaszczak W, Gościańska J, Pietrzak R, Cielecka-Piontek J (2016) β-Cyclodextrin complexation as an effective drug delivery system for meropenem. Eur J Pharm Biopharm 99:24–34. https://doi.org/10.1016/j.ejpb.2015.10.013

Partanen R, Ahro M, Hakala M, Kallio H, Forssell P (2002) Microencapsulation of caraway extract in β-cyclodextrin and modified starches. Eur Food Res Technol 214:242–247. https://doi.org/10.1007/s00217-001-0446-1

dos Passos Menezes P, Barbosa Pereira dos Santos P, Azevedo Dória GA, Hipólito de Sousa BM, Russo Serafini M, Santos Nunes P, Quintans-Júnior LJ, Lisboa de Matos I, Alves PB, Bezerra DP, Mendonça Júnior FJB, da Silva GF, de Aquino TM, de Souza Bento E, Scotti MT, Scotti L, Antunes de Souza Araújo A (2017) Molecular modeling and physicochemical properties of supramolecular complexes of limonene with α- and β-cyclodextrins. AAPS PharmSciTech 18(1):49–57. https://doi.org/10.1208/s12249-016-0516-0

Paulik F, Paulik J (1978) Simultaneous techniques in thermal analysis. Analyst 103:417–437

Pereva S, Sarafska T, Bogdanova S, Spassov T (2016) Efficiency of "cyclodextrin-ibuprofen" inclusion complex formation. Journal of Drug Delivery Science and Technology 35:34–39. https://doi.org/10.1016/j.jddst.2016.04.006

Ponce Cevallos PA, del Pilar Buera M, Elizalde BE (2010) Encapsulation of cinnamon and thyme essential oils components (cinnamaldehyde and thymol) in β-cyclodextrin: effect of interactions with water on complex stability. J Food Eng 99:70–75. https://doi.org/10.1016/j.jfoodeng.2010.01.039

Popat A, Karmakar S, Jambhrunkar S, Xu C, Yu C (2014) Curcumin-cyclodextrin encapsulated chitosan nanoconjugates with enhanced solubility and cell cytotoxicity. Colloids Surf B: Biointerfaces 117:520–527. https://doi.org/10.1016/j.colsurfb.2014.03.005

Pralhad T, Rajendrakumar K (2004) Study of freeze-dried quercetin–cyclodextrin binary systems by DSC, FT-IR, X-ray diffraction and SEM analysis. J Pharm Biomed Anal 34:333–339. https://doi.org/10.1016/S0731-7085(03)00529-6

Reddy MN, Rehana T, Ramakrishna S, Chowdary KPR, Diwan PV (2004) β-Cyclodextrin complexes of celecoxib: molecular-modeling, characterization, and dissolution studies. AAPS PharmSci 6(1):1–9. Article 7

Ren Y, Liu Y, Niu R, Liao X, Zhang J, Yang B (2016) Host-guest inclusion system of oleanolic acid with methyl-β-cyclodextrin: preparation, characterization and anticancer activity. J Mol Struct 1117:1–7. https://doi.org/10.1016/j.molstruc.2016.03.071

da Rocha Ferreira F, Valentim IB, Ramones ELC, Salles Trevisan MT, Olea-Azar C, Perez-Cruz F, de Abreu FC, Fonseca Goulart MO (2013) Antioxidant activity of the mangiferin inclusion complex with β-cyclodextrin. LWT Food Sci Technol 51:129–134. https://doi.org/10.1016/j.lwt.2012.09.032

Rocha BA, Rodrigues MR, Pires Bueno PC, de Mello Costa-Machado AR, de Oliveira Lima Leite Vaz MM, Nascimento AP, Barud HS, Berretta-Silva AA (2012) Preparation and thermal characterization of inclusion complex of Brazilian green propolis and hydroxypropyl-β-cyclodextrin. Increased water solubility of the chemical constituents and antioxidant activity. *J Therm Anal Calorim* 108:87–94. https://doi.org/10.1007/s10973-011-1713-4

Rossel CP, Carreño JS, Rodríguez-Baeza M, Alderete JB (2000) Inclusion complex of the antiviral drug acyclovir with cyclodextrin in aqueous solution and in solid phase. Quím Nova 23(6):749–752

Rudrangi SRS, Kaialy W, Ghori MU, Trivedi V, Snowden MJ, Alexander BD (2016) Solid-state flurbiprofen and methyl-β-cyclodextrin inclusion complexes prepared using a single-step, organic solvent-free supercritical fluid process. Eur J Pharm Biopharm 104:164–170. https://doi.org/10.1016/j.ejpb.2016.04.024

Şamlı M, Bayraktar O, Korel F (2014) Characterization of silk fibroin based films loaded with rutin–β-cyclodextrin inclusion complexes. J Incl Phenom Macrocycl Chem 80(1):37–49. https://doi.org/10.1007/s10847-014-0396-4

dos Santos C, del Pilar Buera M, Mazzobre MF (2011) Phase solubility studies of terpineol with β-cyclodextrins and stability of the freeze-dried inclusion complex. Procedia Food Science 1:355–362. https://doi.org/10.1016/j.profoo.2011.09.055

dos Santos C, del Pilar Buera M, Mazzobre MF (2012) Influence of ligand structure and water interactions on the physical properties of β-cyclodextrins complexes. Food Chem 132:2030–2036. https://doi.org/10.1016/j.foodchem.2011.12.044

Santos EH, Kamimura JA, Hill LE, Gomes CL (2015) Characterization of carvacrol beta-cyclodextrin inclusion complexes as delivery systems for antibacterial and antioxidant applications. LWT Food Sci Technol 60:583–592. https://doi.org/10.1016/j.lwt.2014.08.046

Santos PL, Brito RG, Oliveira MA, Quintans JSS, Guimarães AG, Santos MRV, Menezes PP, Serafini MR, Menezes IRA, Coutinho HDM, Araújo AAS, Quintans-Júnior LJ (2016) Docking, characterization and investigation of β-cyclodextrin complexed with citronellal, a monoterpene present in the essential oil of Cymbopogon species, as an anti-hyperalgesic agent in chronic muscle pain model. Phytomedicine 23:948–957. https://doi.org/10.1016/j.phymed.2016.06.007

Sapino S, Carlotti ME, Caron G, Ugazio E, Cavalli R (2009) *In silico* design, photostability and biological properties of the complex resveratrol/hydroxypropyl-β-cyclodextrin. J Incl Phenom Macrocycl Chem 63:171–180. https://doi.org/10.1007/s10847-008-9504-7

Sapte S, Pore Y (2016) Inclusion complexes of cefuroxime axetil with β-cyclodextrin: physicochemical characterization, molecular modeling and effect of L-arginine on complexation. J Pharm Anal 6:300–306. https://doi.org/10.1016/j.jpha.2016.03.004

Sasako H, Kihara F, Koyama K, Higashi K, Yamamoto K, Moribe K (2016) A novel capsule-like structure of micro-sized particles formed by phytosterol ester and γ-cyclodextrin in water. Food Chem 210:269–275. https://doi.org/10.1016/j.foodchem.2016.04.103

Sathigari S, Chadha G, Lee Y-HP, Wright N, Parsons DL, Rangari VK, Fasina O, Babu RJ (2009) Physicochemical characterization of efavirenz–cyclodextrin inclusion complexes. AAPS PharmSciTech 10(1):81–87. https://doi.org/10.1208/s12249-008-9180-3

Sbârcea L, Ledeţi I, Drăgan L, Kurunczi L, Fuliaş A, Udrescu L (2015) Fosinopril sodium–hydroxypropyl-β-cyclodextrin inclusion complex. Thermal decomposition kinetics and compatibility studies. *J Therm Anal Calorim* 120:981–990. https://doi.org/10.1007/s10973-015-4450-2

Sbârcea L, Udrescu L, Ledeţi I, Szabadai Z, Fuliaş A, Sbârcea C (2016) β-Cyclodextrin inclusion complexes of lisinopril and zofenopril. Physicochemical characterization and compatibility study of lisinopril-β-cyclodextrin with lactose. J Therm Anal Calorim 123:2377–2390. https://doi.org/10.1007/s10973-015-5045-7

Serafini MR, Menezes PP, Costa LP, Lima CM, Quintans LJ Jr, Cardoso JC, Matos JR, Soares-Sobrinho JL, Grangeiro S Jr, Nunes PS, Bonjadim LR, Antunes de Souza Araújo A (2012) Interaction of *p*-cymene with β-cyclodextrin. J Therm Anal Calorim 109:951–955. https://doi.org/10.1007/s10973-011-1736-x

Serri C, Argirò M, Piras L, Mita DG, Saija A, Mita L, Forte M, Giarra S, Biondi M, Crispi S, Mayol L (2017) Nano-precipitated curcumin loaded particles: effect of carrier size and drug complexation with (2-hydroxypropyl)-β-cyclodextrin on their biological performances. Int J Pharm 520:21–28. https://doi.org/10.1016/j.ijpharm.2017.01.049

Siró I, Fenyvesi É, Szente L, De Meulenaer B, Devlieghere F, Orgoványi J, Sényi J, Barta J (2006) Release of alpha-tocopherol from antioxidative low-density polyethylene film into fatty food simulant: influence of complexation in beta-cyclodextrin. Food Addit Contam 23(8):845–853. https://doi.org/10.1080/02652030600699064

Skiba M, Bounoure F, Barbot C, Arnaud P, Skiba M (2005) Development of cyclodextrin microspheres for pulmonary drug delivery. J Pharm Pharm Sci 8(3):409–418

Song LX, Teng CF, Yang Y (2006) Preparation and characterization of the solid inclusion compounds of α-, β-cyclodextrin with phenylalanine (D-, L- and DL-Phe) and tryptophan (D-, L- and DL-Trp). J Incl Phenom Macrocycl Chem 54:221–232. https://doi.org/10.1007/s10847-005-7970-8

Song LX, Teng CF, Xu P, Wang HM, Zhang ZQ, Liu QQ (2008) Thermal decomposition behaviors of β-cyclodextrin, its inclusion complexes of alkyl amines, and complexed β-cyclodextrin at different heating rates. J Incl Phenom Macrocycl Chem 60:223–233. https://doi.org/10.1007/s10847-007-9369-1

Songkro S, Hayook N, Jaisawang J, Maneenuan D, Chuchome T, Kaewnopparat N (2012) Investigation of inclusion complexes of citronella oil, citronellal and citronellol with β-cyclodextrin for mosquito repellent. J Incl Phenom Macrocycl Chem 72:339–355. https://doi.org/10.1007/s10847-011-9985-7

Sreenivasan K (2001) Use of differential scanning calorimetry to study the replacement of a guest molecule from cyclodextrin-guest inclusion complexes. Anal Lett 34(2):307–311

Szafranek A, Szafranek J (1993) Thermogravimetric properties of inclusion complexes of β-cyclodextrin with benzene, acetylsalicylic acid and methyl salicylate. J Incl Phenom Mol Recognit Chem 15:351–358

Szejtli J (2004) Past, present, and future of cyclodextrin research. Pure Appl Chem 76(10):1825–1845

Szente L, Szejtli J, Szemán J, Kató L (1993) Fatty acid-cyclodextrin complexes: properties and applications. J Incl Phenom Mol Recognit Chem 16:339–354

Tan J, Meng N, Fan Y, Su Y, Zhang M, Xiao Y, Zhou N (2016) Hydroxypropyl-β-cyclodextrin-graphene oxide conjugates: carriers for anti-cancer drugs. Mater Sci Eng C 61:681–687. https://doi.org/10.1016/j.msec.2015.12.098

Tao F, Hill LE, Peng Y, Gomes CL (2014) Synthesis and characterization of β-cyclodextrin inclusion complexes of thymol and thyme oil for antimicrobial delivery applications. LWT Food Sci Technol 59:247–255. https://doi.org/10.1016/j.lwt.2014.05.037

Teramoto Y (1990) Thermal analysis - as a method of material characterization. A review. Anal Sci 6:635–643

Thiry J, Krier F, Ratwatte S, Thomassin J-M, Jerome C, Evrard B (2017) Hot-melt extrusion as a continuous manufacturing process to form ternary cyclodextrin inclusion complexes. Eur J Pharm Sci 96:590–597. https://doi.org/10.1016/j.ejps.2016.09.032

Todorova NA, Schwarz FP (2007) The role of water in the thermodynamics of drug binding to cyclodextrin. J Chem Thermodyn 39:1038–1048. https://doi.org/10.1016/j.jct.2006.12.019

Ünlüsayin M, Hădărugă NG, Rusu G, Gruia AT, Păunescu V, Hădărugă DI (2016) Nano-encapsulation competitiveness of omega-3 fatty acids and correlations of thermal analysis and

Karl Fischer water titration for European anchovy (*Engraulis encrasicolus* L.) oil/β-cyclodextrin complexes. LWT Food Sci Technol 68:135–144. https://doi.org/10.1016/j.lwt.2015.12.017
Uyar T, Hunt MA, Gracz HS, Tonelli AE (2006) Crystalline cyclodextrin inclusion compounds formed with aromatic guests: guest-dependent stoichiometries and hydration-sensitive crystal structures. Cryst Growth Des 6(5):1113–1119. https://doi.org/10.1021/cg050500+
Veiga MD, Merino M (2002) Interactions of oxyphenbutazone with different cyclodextrins in aqueous medium and in the solid state. J Pharm Biomed Anal 28:973–982
Veiga MD, Merino M, Fernández D, Lozano R (2002) Characterization of some cyclodextrin derivatives by thermal analysis. J Therm Anal Calorim 68:511–516
Wang Z, Zhang X, Deng Y, Wang T (2007) Complexation of hydrophobic drugs with hydroxypropyl-β-cyclodextrin by lyophilization using a tertiary butyl alcohol system. J Incl Phenom Macrocycl Chem 57:349–354. https://doi.org/10.1007/s10847-006-9261-4
Wang J, Cao Y, Sun B, Wang C (2011a) Physicochemical and release characterisation of garlic oil-β-cyclodextrin inclusion complexes. Food Chem 127:1680–1685. https://doi.org/10.1016/j.foodchem.2011.02.036
Wang T, Li B, Si H, Lin L, Chen L (2011b) Release characteristics and antibacterial activity of solid state eugenol/β-cyclodextrin inclusion complex. J Incl Phenom Macrocycl Chem 71:207–213. https://doi.org/10.1007/s10847-011-9928-3
Wang X, Luo Z, Xiao Z (2014) Preparation, characterization, and thermal stability of β-cyclodextrin/soybean lecithin inclusion complex. Carbohydr Polym 101:1027–1032. https://doi.org/10.1016/j.carbpol.2013.10.042
Wang L, Yan J, Li Y, Xu K, Li S, Tang P, Li H (2016) The influence of hydroxypropyl-β-cyclodextrin on the solubility, dissolution, cytotoxicity, and binding of riluzole with human serum albumin. J Pharm Biomed Anal 117:453–463. https://doi.org/10.1016/j.jpba.2015.09.033
Wszelaka-Rylik M (2017) Thermodynamics of β-cyclodextrin–ephedrine inclusion complex formation and covering of nanometric calcite with these substances. J Therm Anal Calorim 127:1825–1834. https://doi.org/10.1007/s10973-016-5467-x
Wszelaka-Rylik M, Gierycz P (2013) Isothermal titration calorimetry (ITC) study of natural cyclodextrins inclusion complexes with drugs. J Therm Anal Calorim 111:2029–2035. https://doi.org/10.1007/s10973-012-2251-4
Wunderlich B (2007) Thermal analysis of macromolecules. A personal review. J Therm Anal Calorim 89(2):321–356
Xu P, Song LX, Wang HM (2008) Study on thermal decomposition behavior of survived β-cyclodextrin in its inclusion complex of clove oil by nonisothermal thermogravimetry and gas chromatography coupled to time-of-flight mass spectrometry analyses. Thermochim Acta 469:36–42. https://doi.org/10.1016/j.tca.2007.12.009
Yang Y, Li X, Chen J, Chen H, Bao X (2003) ZnO nanoparticles prepared by thermal decomposition of β-cyclodextrin coated zinc acetate. Chem Phys Lett 373:22–27. https://doi.org/10.1016/S0009-2614(03)00562-1
Yang B, Zhao Y-L, Yang X, Liao X-L, Yang J, Zhang J-H, Gao C-Z (2013a) Scutellarin-cyclodextrin conjugates: synthesis, characterization and anticancer activity. Carbohydr Polym 92:1308–1314. https://doi.org/10.1016/j.carbpol.2012.10.012
Yang X, Zhao Y, Chen Y, Liao X, Gao C, Xiao D, Qin Q, Yi D, Yang B (2013b) Host–guest inclusion system of mangiferin with β-cyclodextrin and its derivatives. Mater Sci Eng C 33:2386–2391. https://doi.org/10.1016/j.msec.2013.02.002
Yang Y, Gao J, Ma X, Huang G (2017) Inclusion complex of tamibarotene with hydroxypropyl-β -cyclodextrin: preparation, characterization, invitro and in-vivo evaluation. Asian J Pharm Sci 12:187–192. https://doi.org/10.1016/j.ajps.2016.08.009
Yatsu FKJ, Koester LS, Lula I, Passos JJ, Sinisterra R, Bassani VL (2013) Multiple complexation of cyclodextrin with soy isoflavones present in an enriched fraction. Carbohydr Polym 98:726–735. https://doi.org/10.1016/j.carbpol.2013.06.062
Yu S-Z, Li X-T, Li J-H, Wang J-Y, Tian S-J (1997) Kinetic studies on the thermal dissociation of β-cyclodextrin-cinnamyl alcohol inclusion complex. J Therm Anal 49:1517–1525

Yuan C, Jin Z, Xu X, Zhuang H, Shen W (2008) Preparation and stability of the inclusion complex of astaxanthin with hydroxypropyl-β-cyclodextrin. Food Chem 109:264–268. https://doi.org/10.1016/j.foodchem.2007.07.051

Yuan C, Liu B, Liu H (2015) Characterization of hydroxypropyl-β -cyclodextrins with different substitution patterns via FTIR, GC-MS, and TG-DTA. Carbohydr Polym 118:36–40. https://doi.org/10.1016/j.carbpol.2014.10.070

Zhang W, Chen M, Diao G (2011) Preparation and electrochemical behavior of water-soluble inclusion complex of ferrocene with β-cyclodextrin polymer. Electrochim Acta 56:5129–5136. https://doi.org/10.1016/j.electacta.2011.03.062

Zhang W, Li X, Yu T, Yuan L, Rao G, Li D, Mu C (2015) Preparation, physicochemical characterization and release behavior of the inclusion complex of *trans*-anethole and β-cyclodextrin. Food Res Int 74:55–62. https://doi.org/10.1016/j.foodres.2015.04.029

Zhang J-Q, Jiang K-M, Xie X-G, Jin Y, Lin J (2016) Water-soluble inclusion complexes of trans-polydatin by cyclodextrin complexation: preparation, characterization and bioactivity evaluation. J Mol Liq 219:592–598. https://doi.org/10.1016/j.molliq.2016.03.054

Zhang C-L, Liu J-C, Yang W-B, Chen D-L, Jiao Z-G (2017) Experimental and molecular docking investigations on the inclusion mechanism of the complex of phloridzin and hydroxypropyl-β-cyclodextrin. Food Chem 215:124–128. https://doi.org/10.1016/j.foodchem.2016.07.155

Zhou Q, Wei X, Dou W, Chou G, Wang Z (2013) Preparation and characterization of inclusion complexes formed between baicalein and cyclodextrins. Carbohydr Polym 95:733–739. https://doi.org/10.1016/j.carbpol.2013.02.038

Zhu LH, Song LX, Guo XQ, Dang Z (2010) Comparison in thermal behaviors of homologues, derivatives and adducts of β-cyclodextrin. Thermochim Acta 507-508:77–83. https://doi.org/10.1016/j.tca.2010.04.031

Zhu G, Feng N, Xiao Z, Zhou R, Niu Y (2015) Production and pyrolysis characteristics of citral-monochlorotriazinyl-β-cyclodextrin inclusion complex. J Therm Anal Calorim 120:1811–1817. https://doi.org/10.1007/s10973-015-4498-z

Chapter 5
Measuring Binding Constants of Cyclodextrin Inclusion Compounds

David Landy

Contents

Abstract The affinity of cyclodextrins for organic and even inorganic pollutants has led to the development of numerous remediation methods at the laboratory scale. Indeed, the hydrophobic cavity of cyclodextrins constitute a versatile vehicle for the efficient transfer of various pollutants from their initial environmental com-

---

D. Landy (✉)
Unité de Chimie Environnementale et Interactions sur le Vivant (UCEIV, EA 4492),
SFR Condorcet FR CNRS 3417, Université du Littoral Côte d'Opale, Dunkerque, France
e-mail: landy@univ-littoral.fr

S. Fourmentin et al. (eds.), *Cyclodextrin Fundamentals, Reactivity and Analysis*,
Environmental Chemistry for a Sustainable World,
https://doi.org/10.1007/978-3-319-76159-6_5

partment to the cyclodextrin cavity. This transfer can be applied to any environmental media such as soil, water or atmosphere, because cyclodextrins can be dissolved in water solutions or immobilized on solid supports. Both recovery or destructive processes have thus been designed on the basis of cyclodextrin affinity for the target pollutants. As a consequence, the stability of host-guest edifices is of crucial importance for the efficiency of cyclodextrin applications. Therefore, formation constants of such inclusion compounds have been thoroughly investigated, aiming at the custom design of host-guest couples for a given application. Indeed, the molecular shape of the cavity, and consequently the inclusion compound stability, can be tuned by using cyclodextrins of different size or by taking advantage of chemical modifications on the macrocycle.

Nevertheless, the rational design of the perfect cyclodextrin may be hindered by a large uncertainty on the complex stability. Indeed, large discrepancies are observed for a given complex in the cyclodextrin literature. There is a lack of a generalized scheme for the measurement of affinity. Therefore, this chapter reviews the common experimental approaches and proposes a unified framework for measuring binding constants of cyclodextrins inclusion compounds. This unified approach relies on the use of minimization algorithms and is decomposed into major associated concepts, with the description of experimental protocols, equilibriums, analytical methods and data treatments. The chapter discusses the concept of global analysis and the issues of stability accuracy, optimization of experimental conditions and evaluation of thermodynamic parameters. Future research will probably focus on the generalization of algorithmic treatments, global analysis and statistical evaluation.

5.1 Introduction

The versatile ability of cyclodextrins to form host-guest inclusion compounds constitutes the key point of their wide use in various areas. In particular, a growing interest has been devoted to the applications of cyclodextrins to environmental issues (Fig. 5.1, synthesized from Landy et al. 2012). For most of these applications, cyclodextrins efficiency is depending, at least partially, on the stability of the involved complexes. Indeed, the higher the stability of cyclodextrin-pollutant inclusion compounds, the greater the transfer of these pollutants from their initial matrix to the cyclodextrin trapping system. These transfers might be used either in recovery based processes or coupled to irreversible transformation of the pollutants, by the use of catalytic reactions or biodegradations. The relationship between complex stability and transfer efficiency applies whatever the respective initial state of the pollutants and cyclodextrins: molecules contained in water, soil or atmosphere can be extracted by cyclodextrins, which can be either dissolved in aqueous solutions or immobilized on solid supports.

The fact that the efficiency of a given remediation process may vary dramatically from one pollutant to another is thus linked to the wide range of stability observed for cyclodextrin complexes. Such disparity in host/guest affinity is illustrated in Fig. 5.2, for a set of molecules of environmental interest.

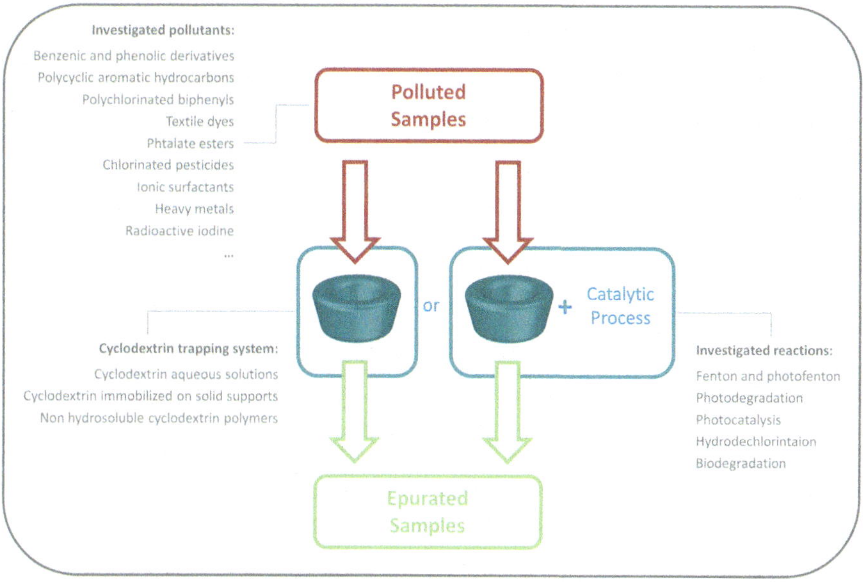

Fig. 5.1 Summary of pollutant remediation processes based on cyclodextrins, investigated at the laboratory scale. Left hand side: recovery processes; right hand side: destructive processes

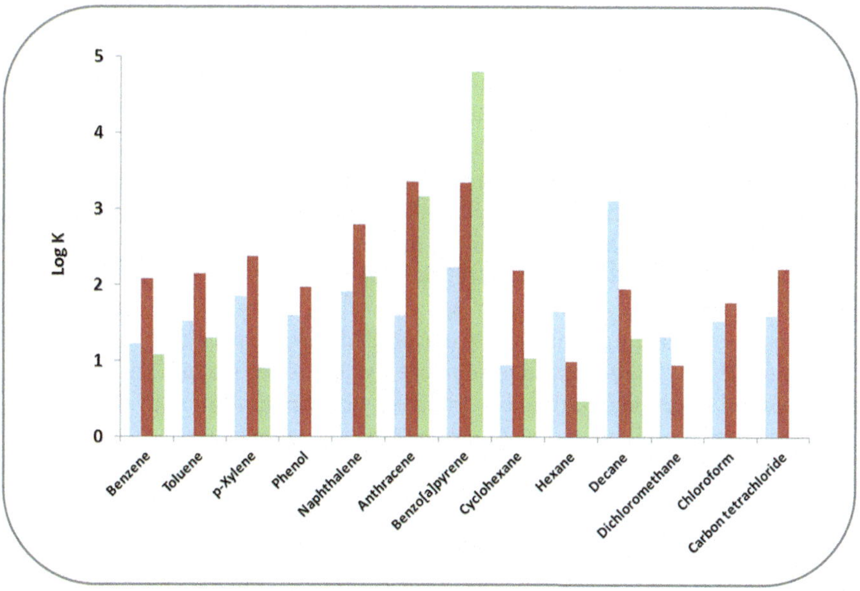

Fig. 5.2 Logarithm of the formation constants measured for α-, β- and γ-cyclodextrins (blue, red and green, respectively), with some mono- and polycyclic aromatic hydrocarbons, alkanes and chloroalkanes

Corresponding formation constants range from 10^1 to 10^5 M^{-1}, depending on the nature of both hosts and guests (Blyshak et al. 1989; Fourmentin et al. 2007; Fujiki et al. 1988; Osajima et al. 1991; Sanemasa and Akamine 1987; Sanemasa et al. 1989; 1990; Szaniszlo et al. 2005). Accordingly, the percentage of complexed pollutant evolves heavily for given experimental conditions: for instance, this percentage rises up to 90% for a formation constant equal to 10^5 M^{-1}, but it drops to 1% for a value of 10^1 M^{-1}, in the case of a 0.001 M equimolar host-guest solution. Such differences underline the crucial importance for cyclodextrin environmental applications of a fine design of host-guest couples, and as a consequence, of the fine measurement of the corresponding complex stability.

Within this scope, it has to be underlined that there is a considerable spectrum of techniques which can be employed for evaluating host/guest affinity. Indeed, as the physicochemical properties of inclusion compounds differ systematically from the properties of the free partners, any physicochemical technique may be used to study the stability of such cyclodextrin complexes. As a result, characterization of cyclodextrin-guest affinity is a recurrent topic, leading to a considerable amount of publications. Therefore, some reviews have already been devoted to the experimental techniques and methods allowing the measurement of the stability of cyclodextrin inclusion compounds (Hirayama and Uekama 1987; Szejtli 1996; Szente 1996; Connors 1996 and 1997; Singh et al. 2010; Pessine et al. 2012; Takahashi et al. 2012a, b and c).

Nevertheless, if formation constants are expected to be determined accurately, several magnitude orders discrepancies might be observed from one study to another. For instance, the estimation of the β-CD/pyrene affinity fluctuates from values as weak as 10^1 M^{-1} to intermediate ones, such as 10^3 M^{-1} (Connors 1995). Such inconsistencies may be attributed in particular to the lack of a generalized scheme for the measurement of affinity.

Accordingly, we describe in this chapter a general framework for measuring binding constants of cyclodextrins inclusion compounds. In a first section, we will present the fundamental basis of this unified approach. Then, the various experimental protocols which can be handled by this general scheme will be detailed. Next, the mathematical description of the studied equilibriums will be proposed. The formalism of the experimental responses will be generalized for a wide range of analytical technique. Finally, the universal data treatment will be described, including not only the general algorithm but also the concept of global analysis, the evaluation of uncertainty on binding parameters, the optimization of experimental conditions and the determination of thermodynamic parameters.

5.2 Unified Framework for Measuring Binding Constants

The general principle of measuring affinity is based on the modification of the physicochemical properties induced by the inclusion phenomenom. Indeed, a host-guest mixture leads to an equilibrium between free and complexed species, that is to say between the host H, guest G and inclusion compound H_mG_n (eq. (5.1)). As the physicochemical property of the inclusion compound differ from those of free molecules, the recorded signal is depending on the relative concentrations of these species. These concentrations are controlled by the formation constant defined in eq. (5.2):

$$H_m + G_n \leftrightharpoons H_m - G_n \tag{5.1}$$

$$K_{H_m - G_n} = \frac{[H_m - G_n]}{[H]^m * [G]^n} \tag{5.2}$$

It has to be mentioned that cyclodextrin complexes with a stoichiometry higher than 1:1 are generally observed in the presence of lower order complexes. As a result, for m and/or n superior to unity, it is necessary to define as many formation constants as inclusion compounds. For instance, a mixture of 1:1 and 2:1 complexes requires the definition of two formation constants:

$$K_{H-G} = \frac{[H-G]}{[H]*[G]} \tag{5.3}$$

$$K_{H_2-G} = \frac{[H_2-G]}{[H]*[H-G]} \tag{5.4}$$

As a matter of fact, these two formation constants induce the existence of a cumulative formation constant:

$$\beta_{H_2-G} = \frac{[H_2-G]}{[H]^2 * [G]} = K_{H-G} * K_{H_2-G} \tag{5.5}$$

In addition, as most complexations involving cyclodextrins are characterized by fast exchange, it is generally impossible to observe distinct experimental signals for free and complexed species, thus leading to the record of a mean signal, whatever the technique used. In addition, a full, quantitative, complexation can not be reached on a practical point of view, in such a way that the intrinsic signal of the pure complex can not be recorded. As a result, an appropriate data treatment is required to extract the concentrations of each species in the mixture, in order to evaluate the formation constant.

Accordingly, since inclusion studies are based on a succession of experimental and theoretical steps, a tremendous number of different approaches has been described in the literature, each being adapted to a particular case. Fortunately, it is possible to extract a unified framework for the evaluation of formation constants. A general scheme based on four steps might be proposed to synthesize all kind of studies (Fig. 5.3).

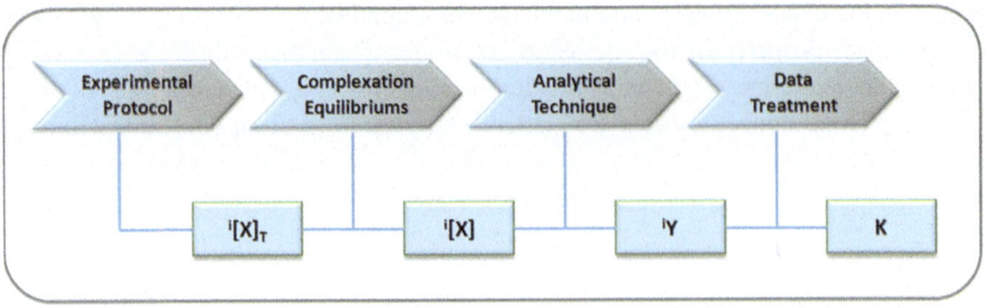

Fig. 5.3 General framework of binding studies. Total concentrations $^{i}[X]_T$, equilibrium concentrations $^{i}[X]$, analytical response (^{i}Y) and formation constants (K) are successively defined according to the four steps involved in all binding studies

In the first step, the choice of a given experimental protocol defines, for each prepared solution (denoted by superscript ^{i}), the total concentrations of each partners $^{i}[X]_T$. In the second step, the equilibriums which are taking place then control the concentration of each species ($^{i}[X]$), in free and complexed forms. In a third step, these concentrations are linked to a measurable signal (^{i}Y) by the use of an analytical technique. In the fourth step, a data treatment allows the evaluation of the formation constants, which links all $^{i}[X]_T$ values to all ^{i}Y values. Incidentally, the treatment gives also access to the intrinsic response of the pure complexes, according to the employed technique.

To be valid for all studies, this general scheme first requires that a general binding polynomial could be enounced, in order to describe most of the encountered binding equilibriums. In practice, the dimensional consistency between organic molecules and the cavity of monomeric cyclodextrins leads principally to 1:1 complexes. According to the various sizes offered by αCD, βCD, and γCD, additional equilibriums may be observed, when the cavity is large enough to accommodate two guests or if the guest is large enough to mobilize two cavities. These situations lead to 1:2 and 2:1 complexes, respectively. In addition, it may sometimes be useful, on an analytical point of view, to study the competition between two different guests. Indeed, when a given guest can not be studied directly by an experimental technique, one can investigate the perturbation induced by this guest on the cyclodextrin complexation of a reference molecule, complexation of which can be studied in a separate experiment. The observed perturbation and the stability of the cyclodextrin-reference molecule may then be combined to extract the stability of the complex of interest.

As the overwhelming majority of the cyclodextrin binding studies rely on one of the aforementioned equilibriums, Fig. 5.4 may be considered as a general framework for studying cyclodextrin complexation ability. If this representation is explicitly based on the competition of two guests (A and B) for one host (H), it has to be kept in mind that a symmetric scheme may be drawn for the competition of two hosts (A and B) for one guest (G).

Once the binding polynomial is defined, it is then necessary to convert the equilibrium concentrations into a calculated signal ^{i}Y.

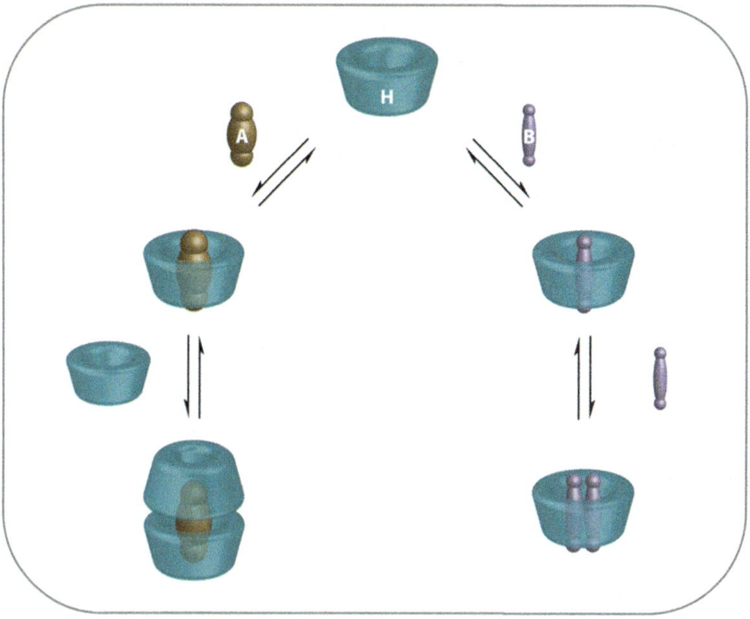

Fig. 5.4 Schematic view of the different inclusion equilibriums generally observed with cyclodextrins. H, A and B stand for host, guest A and guest B, respectively. The following complexes are considered: H-A, H-B, H_2-A and H-B_2

Finally, the general scheme presented in Fig. 5.3 requires that a single data treatment could be used whatever the studied equilibriums and whatever the employed protocols and analytical techniques. Such an assumption hinders the statement of any specific approximation during the mathematical treatment, thus enforcing the use of a numerical approach rather than a formal approach. With this respect, one can take profit of algorithmic procedures to define a universal data treatment, which is a central feature of the unified framework for measuring cyclodextrins binding constants.

Accordingly, the next sections of this chapter will describe each of the four steps of the analytical process.

5.3 Experimental Protocols

The determination of formation constants requires that total concentrations of the investigated partners should be varied. These variations are governed by different kind of experimental protocols, leading to different shapes of binding isotherms. We develop in this paragraph the various protocols that can be managed through the unified approach. For the sake of simplicity, in the following paragraphs, each protocol will be presented on the basis of a single 1:1 complex, between a host H and a guest A.

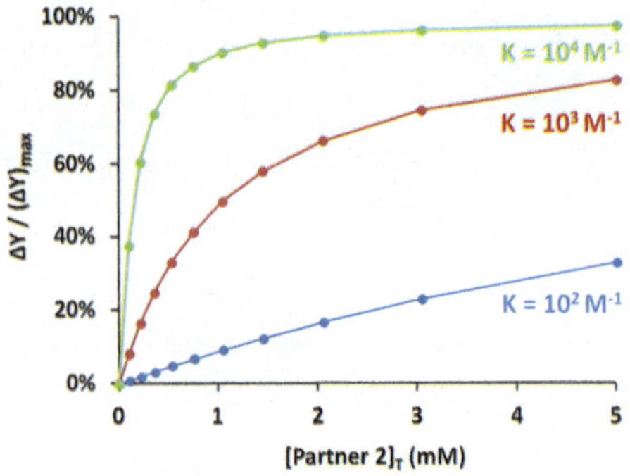

$[\text{partner 1}]_T$	Constant
$[\text{partner 2}]_T$	Variable
$\frac{[\text{partner 1}]_T}{[\text{partner 2}]_T}$	Variable
$[\text{partner 1}]_T + [\text{partner 2}]_T$	Variable

Fig. 5.5 Typical titration isotherms, for formation constants equal to 10^2 M^{-1} (blue), 10^3 M^{-1} (red) and 10^4 M^{-1} (green). Partner 1 concentration equal to 0.1 mM. Dots and lines correspond to discrete hypothetical measurements and theoretical isotherms, respectively

5.3.1 Titration

The titration protocol consists in recording an experimental response relative to partner 1, denoted as titrate, maintained at a constant concentration, as a function of the concentration of partner 2, denoted as titrant (Thordarson 2011). As a consequence, the signal evolves from the intrinsic signal of partner 1 towards the intrinsic signal of the inclusion compound. Nevertheless, this latter signal is in fact extrapolated by the data treatment, as its direct observation would require an infinite concentration of partner 2, according to the asymptotic shape of the binding curve. Corresponding isotherms are illustrated in Fig. 5.5, in the case of formation constants varying from 10^2 M^{-1} to 10^4 M^{-1}. This range of stability is typical for cyclodextrin complexes, as underlined by Connors (1997). It has to be stressed that Fig. 5.5 uses a normalized ordinate, according to relation (5.6):

$$\frac{\Delta Y}{\Delta Y_{max}} = \frac{{}^{i}Y - {}^{A}Y}{{}^{H\text{-}A}Y - {}^{A}Y} \tag{5.6}$$

With ${}^{A}Y$ and ${}^{H\text{-}A}Y$ being the experimental responses for pure species A and H-A, respectively.

As a consequence, this ordinate is no longer depending on the response factor of the various species. Such normalization allows a more straightforward comparison between the various studies, by a focus on the shape of the binding isotherm, which is completely controlled by the formation constant value. It has to be stressed that this normalization is used here only for understanding purpose. In practice, data treatment is generally applied directly on recorded signals.

This titration protocol is the most widely used, not only because it is easy to handle but also because it might allow some simplification in data treatment. Indeed, in the case of a 1:1 complex, experimental data may be linearized, and the formation constant can be extracted from the corresponding slope and intercept (Benesi and Hildebrand 1949; Scatchard 1949; Scott 1956). Nevertheless, these treatments require the use of a negligible titrate concentration if compared to titrant, which constitutes a restrictive framework. In addition, such approximation may induce significant errors on the calculated affinities (Exner 1997; Yang et al. 2000). Finally, the general equilibrium scheme presented in Fig. 5.4 does not allow the use of such graphical treatment.

5.3.2 *Perfusion*

The constancy of the titrate concentration during the titration experiment may sometimes not be perfectly respected. This is for instance the case with Isothermal Titration Calorimetry (ITC). Indeed, the delivery of the titrant is realized by injecting small volumes of a concentrated solution of partner 2 in a cell fully filled with a less concentrated solution of partner 1. As the cell volume is constant, each injection implies the simultaneous ejection of an equivalent volume, leading to a weak decrease of titrate concentration during the experiment. Such experimental conditions correspond to a perfusion model (Hansen et al. 2011; Tellinghuisen 2007).

In ITC experiments, the recorded heat is not proportional to the amount of inclusion compound which is present in solution, but to the amount of inclusion compound which have been formed during the injection. As a consequence, the shape of the isotherm, illustrated in Fig. 5.6, differs from what is observed with classical titrations.[1] As most of the cyclodextrin inclusion complexes are characterized by negative enthalpy of inclusion, recorded heat are generally represented by negative values.

5.3.3 *Dilution-Release*

Another way to vary the extent of complexation degree may be obtained by realizing multiple dilutions of a concentrated solution containing both partners (not necessarily at the same concentration). As cyclodextrins inclusion compounds are generally characterized by weak to moderate stability, each dilution induces a partial dissociation of the complexes initially formed within the concentrated solution, leading to distinct experimental responses from one dilution to another. As for the perfusion protocol, this principle, known as the release protocol, concerns mainly ITC

[1] To the contrary, similar shapes would be obtained by signal integration, i.e. if each recorded heat is added to the sum of the previous ones.

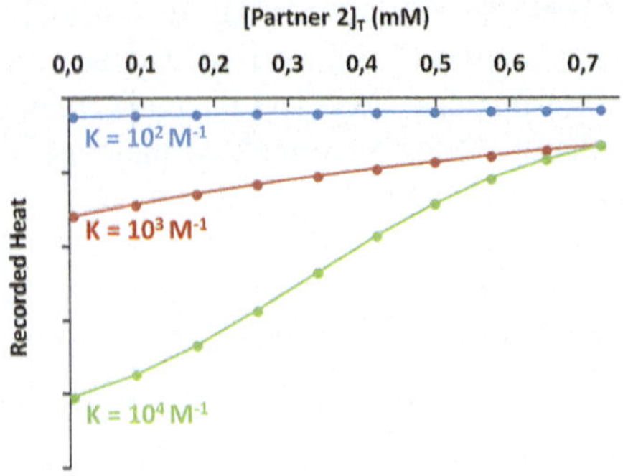

$[\text{partner 1}]_T$	Nearly Constant
$[\text{partner 2}]_T$	Variable
$\frac{[\text{partner 1}]_T}{[\text{partner 2}]_T}$	Variable
$[\text{partner 1}]_T + [\text{partner 2}]_T$	Variable

Fig. 5.6 Typical perfusion isotherms, for formation constants equal to 10^2 M^{-1} (blue), 10^3 M^{-1} (red) and 10^4 M^{-1} (green), with a negative enthalpy of inclusion. Initial concentration of Partner 1 in the cell equal to 0.5 mM. Dots and lines correspond to discrete hypothetical measurements and theoretical isotherms, respectively

(Heerklotz et al. 1999; Illapakurthy et al. 2005). Thus, the previously mentioned specificities about this technique (total concentrations controlled by the perfusion model, differential measurements) also apply here.

In particular, as the recorded heat correspond to a differential measurement, the signal is proportional to the amount of inclusion compound which is dissociated, and thus to both the amount of complexes initially formed in the syringe and to the proportion of dissociation which occurs in the cell. These two quantities evolve in reverse directions as the stability raises. Indeed, a greater formation constant is linked to a greater amount of complexes in the syringe but also to a weaker influence of dilution on the equilibrium. As a consequence, isotherms amplitudes first increase with the stability of the inclusion compound, but diminish afterwards, as can be seen in Fig. 5.7 from the variation between the curves, from K equal to 10^2 M^{-1} to K equal to 10^3 M^{-1}, and then from K equal to 10^3 M^{-1} to K equal to 10^4 M^{-1}. Such behavior implies that the release protocol is well suited to moderate formation constants, but inadequate for high stability: very weak isotherms are generally observed for formation constants superior to 10^5 M^{-1}.

5.3.4 Continuous Variations

The continuous variations principle, also known as Job Plot (Job 1928), relies on simultaneous concentration variations for both partners, but this times with a total concentration which is maintained constant (Fig. 5.8).

Whereas, in a titration experiment, the recorded signal is linked to the titrate, signals of both partner 1 and 2 may be independently used to plot the continuous variations results. Abscissa of the job plot is expressed as the concentrations ratio

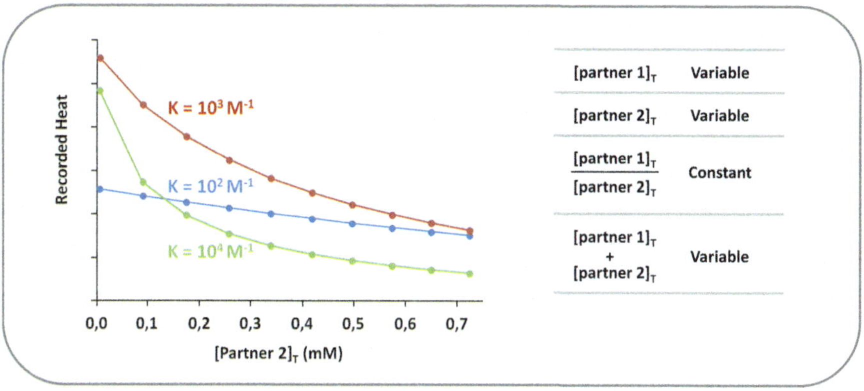

Fig. 5.7 Typical release isotherms, for formation constants equal to 10^2 M^{-1} (blue), 10^3 M^{-1} (red) and 10^4 M^{-1} (green), with a negative enthalpy of inclusion leading to a positive heat of dissociation. Concentration of partner 1 and 2 in the syringe equal to 5 mM. Dots and lines correspond to discrete hypothetical measurements and theoretical isotherms, respectively

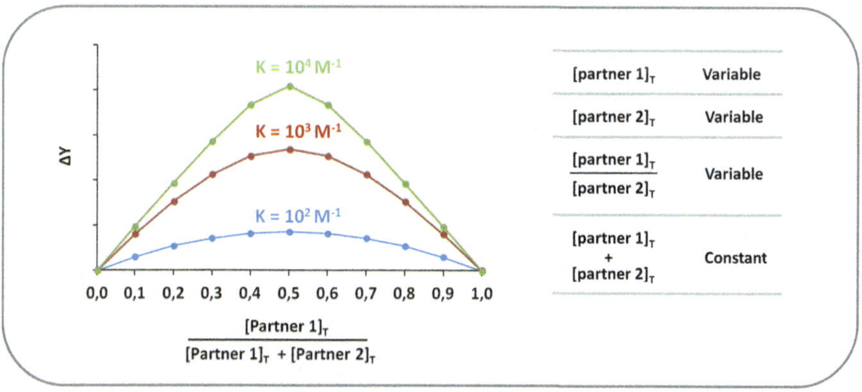

Fig. 5.8 Typical continuous variation isotherms, for formation constants equal to 10^2 M^{-1} (blue), 10^3 M^{-1} (red) and 10^4 M^{-1} (green). Sum of concentrations of partner 1 and 2 equal to 5 mM. Dots and lines correspond to discrete hypothetical measurements and theoretical isotherms, respectively

between the observed species (partner 1 for instance) and the sum of partner 1 and 2. The ordinate is equal to the difference between the recorded signal and the one that would have been obtained for the observed species at the same concentration but in the absence of the second partner. In addition, it is necessary to multiply this difference by the concentration of the studied partner, if the recorded signal is depending on species mole fraction rather than on species concentration, like in the case of NMR (Pirnau et al. 2014).

Even if such protocol could be employed for determining affinity (Smith et al. 2010; Ulatowski et al. 2016), it has been mainly used for defining the inclusion compound stoichiometry, by means of the observed ratio for the maximal ordinate of the Job Plot (R_{max}). For instance, if the highest ordinate value is obtained for a guest versus total concentration ratio equal to 0.5, then the formation of a 1:1 inclusion compound should be suspected. A 0.33 or 0.67 ratio would correspond respectively to 1:2 and 2:1 complexes. Nonetheless, it has to be underlined that such interpretation is rigorously valid only when a single complex is observed, since any additional compound adds two variables to the plot (one formation constant and one response coefficient), cancelling the direct proportionality between the ordinate and the concentration of inclusion compounds. Therefore, exact correspondence between stoichiometry and experimentally determined ratio is generally obtained only for 1:1 complexes, as cyclodextrin complexes of higher stoichiometry being often observed as a mixture of inclusion compounds. It has also to be stated that in the case of mixtures, the value of R_{max} can not be used alone in order to evaluate the respective proportion of coexisting complexes. To the contrary, simulation of the whole Job plot, according to the data treatment presented in paragraph 5.6.1., allow a full characterization of the coexisting inclusions complexes.

5.3.5 Phase Solubility

Phase solubility studies, originally developed by Higuchi and Connors (1965), consist in the measurement of the maximal partner 1 concentration which could be solubilized as a function of varying concentrations of partner 2. As cyclodextrins are generally used to increase the aqueous solubility of hydrophobic guests, a typical experiment requires to place an excess of guest in various cyclodextrins solutions, under stirring and until equilibrium is reached. After filtration, the guest concentration is then measured by any appropriate analytical technique, and the corresponding apparent solubility is plotted as a function of cyclodextrin concentration. Different kind of curves may then be observed, as illustrated in Fig. 5.9.

If the intrinsic solubility of the studied complex is at least equal to the cyclodextrin solubility, then the measured apparent solubility increases in a regular trend with the increase of the host concentration. In this case, three different patterns (namely A_L, A_P and A_N diagrams) may be observed, as a consequence of different stoichiometries. If the inclusion compound contains only one host, whatever the number of guest, then a linear increase of the apparent solubility is observed for any host concentration (A_L diagrams). At the opposite, the presence of more than one host within one of the studied complexes implies a relative increase of the inclusion compounds concentrations for high host concentrations: as a result, apparent solubility variation ceases to be linear and A_p diagrams are obtained. Finally, A_N curve type might be observed, but it is tricky to rationalize the relative decrease of the solubility at the end of the isotherm.

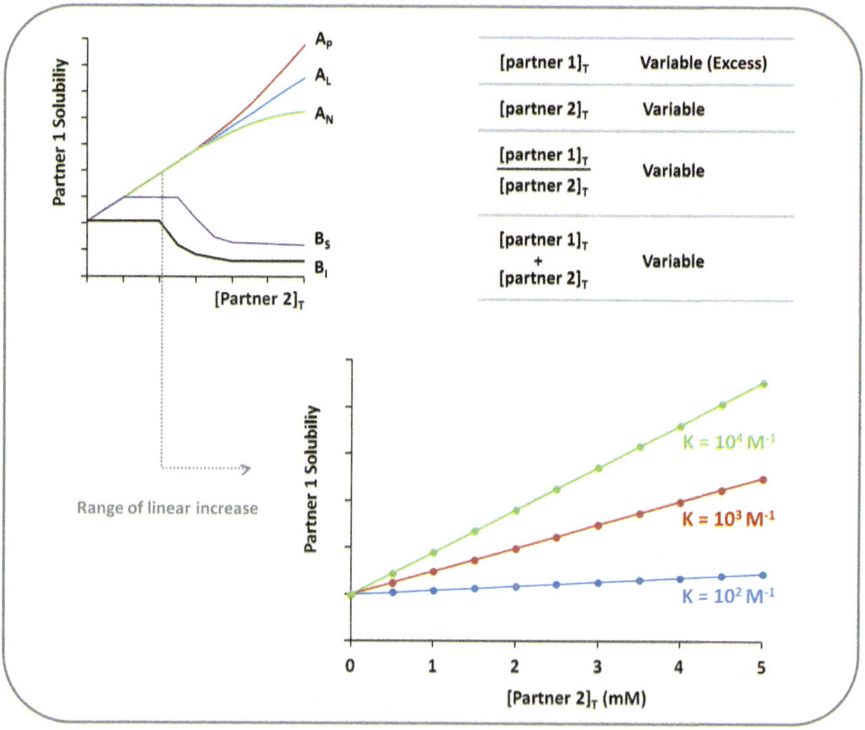

Fig. 5.9 Common types of phase solubility diagrams (upper part) and focus on the range of linear increase of the isotherms (lower part), for formation constants equal to 10^2 M^{-1} (blue), 10^3 M^{-1} (red) and 10^4 M^{-1} (green). Intrinsic solubility of partner 1 equal to 0.1 mM. Dots and lines correspond to discrete hypothetical measurements and theoretical isotherms, respectively

In contrary to the previous cases, the intrinsic solubility of the studied complex may be weaker than the cyclodextrin solubility. If the complex is more soluble than the guest, a B_S diagram is observed. If the complex is even less soluble than the guest, a B_I curve is obtained.

On the quantitative point of view, while the A_L and A_P diagrams may be simulated on the whole host concentration range, the linear part of A_N and B_S curves may also be calculated. As a consequence, all solubility diagrams, excepting B_I curves, may be exploited to extract formation constants, by the general framework exposed in this chapter.

In addition, as can be seen in Fig. 5.9, the slope of the linear part is proportional to the stability of the inclusion complex. In the case of a 1:1 complex, a simplified data treatment might be employed, since the A_L curve may be assimilated to a straight line with an intercept corresponding to the intrinsic solubility of the guest (S_0) and a slope which leads to the calculation of affinity, according to eq. (5.7):

$$K = \frac{Slope}{So(1 - Slope)} \tag{5.7}$$

According to this simplified data treatment and as hydrophobic guests are the most encountered targets in inclusion studies, a significant part of the formation constants found in the cyclodextrin literature comes from solubility measurements. Nevertheless, it has to be stressed that this method is highly sensitive to the value of the guest intrinsic solubility, but also to the presence of nanomolecular assemblies, such as a micellar behavior of the studied complexes (Loftsson et al. 2004). Formation constants may thus be biased and differ from other protocols. Finally, like in the case of the titration protocol, graphical data treatment can not describe the general equilibrium scheme of Fig. 5.4.

5.3.6 Competitive Protocols

The principle of competitive protocols relies on the determination of the stability of the studied complex by observing the perturbation induced by the latter on the experimental response linked to another complex, namely the reference system (Landy et al. 2000; Tutaj et al. 2003; Wintgens and Amiel 2004; Mohamed et al. 2009). It can be applied to study either a second guest or a second host, in addition to the reference complex.

In practice, such principle first requires the characterization of the reference system by any of the previously described direct protocols. This reference step is then followed by the competitive measurements: solutions used for the reference system are then perturbed by adding the third partner (host or guest). Since the third partner reduces the complexation of the reference system, the experimental signal is modified.

Whatever the protocol which is used, the presence of the studied complex induces a weaker concentration of the reference complex, thus leading to an experimental variation which is depending on the relative affinity of all complexes. It should be stressed that most of the articles exploiting competitive measurements has involved titration protocol as reference experiment. Such experiments are illustrated in Fig. 5.10. In addition, some examples also emerged for other methods, such as continuous variations (Landy et al. 2007), phase solubility (Ono et al. 2001), perfusion and release (Bertaut and Landy 2014). Competitive experiments may be implemented to study any host or guest, even when this studied species is transparent for the employed analytical technique. For instance, hydrocarbons with no detectable electronic absorption might be characterized by UV-Vis spectroscopy, using the phenolphtalein complex as a competitive system (Wilson et al. 1997).

Nevertheless, it should be kept in mind that the use of the competitive protocol requires the absence of interaction between competitive species but also the compatibility of the concentration/affinity characteristics of competing systems. Indeed,

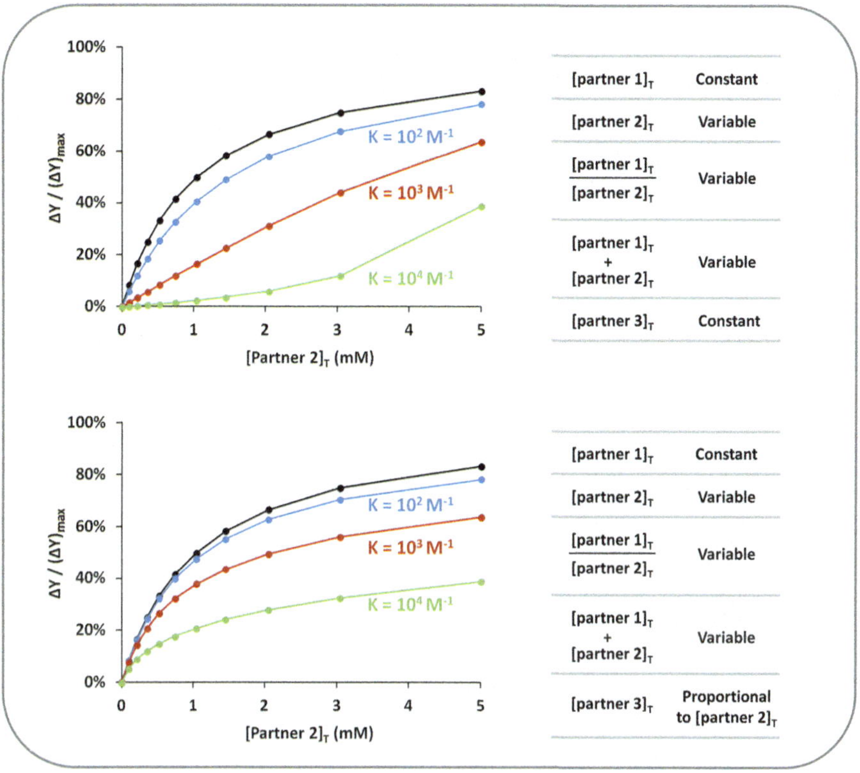

Fig. 5.10 Typical competition isotherms, concentration of partner 3 being constant (5 mM, upper part), or proportional to partner 2 (equimolar, lower part). Formation constants of the studied system equal to 10^2 M^{-1} (blue), 10^3 M^{-1} (red) and 10^4 M^{-1} (green). Titration experiments (black) used as reference experiment (formation constant of the reference system equal to 10^3 M^{-1}). Partner 1 concentration equal to 0.1 mM. Dots and lines correspond to discrete hypothetical measurements and theoretical isotherms, respectively

if the reference system is too stable or too concentrated if compared to the studied system, negligible experimental variations are obtained, leading to high uncertainty on affinity.

5.3.7 Other Protocols

All kind of concentration variations may be used to extract formation constant, even when they do not correspond to the different patterns presented in paragraph 5.3.1 to 5.3.6. For instance, non conventional ITC protocols have been proposed, either for competitive and non competitive schemes (Bertaut and Landy 2014). Their use may be complementary to classical experiments, or even more accurate for some specific cases, like the determination of weak binding constants for instance.

5.4 Binding Equilibriums

Calculation of equilibrium concentrations relies on systems of equations which are controlled by the number of partners, the number of inclusion complexes and by their stoichiometry. Solving such a system of equations consist in expressing one of the equilibrium concentration as a function of known or postulated quantities (total concentration of each partner, formation constant of each complex). Once this equilibrium concentration is obtained, the calculation of the other concentrations is straightforward.

In the case of a single 1:1 complex or of a couple of competing 1:1 complexes, an analytical form of the equilibrium concentration [H-A] can be formulated. To the contrary, the general scheme proposed in Fig. 5.4 results in the definition of a polynomial without any analytical form for its physical root. As a consequence, a numerical solving of this polynomial is required to obtain the equilibrium concentration. The following paragraphs develop the formulas to be applied for calculating equilibrium concentrations, for both simplified and general cases.

5.4.1 Simplified Case: Single 1:1 Complex

When a single cyclodextrin (Host, H) interacts with a single guest (A) under 1:1 stoichiometry, the following equations described the corresponding equilibrium:

$$H + A \leftrightharpoons H - A \tag{5.8}$$

$$K_{H-A} = \frac{[H-A]}{[H]*[A]} \tag{5.9}$$

$$[H]_T = [H] + [H-A] \tag{5.10}$$

$$[A]_T = [A] + [H-A] \tag{5.11}$$

Combining eqs. (5.9), (5.10) and (5.11) leads to eqs. (5.12) and (5.13):

$$[H]_T = \frac{[H-A]}{K_{H-A}*[A]} + [H-A] = \frac{[H-A]}{K_{H-A}*\left[[A]_T - [H-A]\right]} + [H-A] \tag{5.12}$$

$$[H]_T * K_{H-A} * \left[[A]_T - [H-A]\right] = [H-A] + K_{H-A} * \left[[A]_T - [H-A]\right] * [H-A] \tag{5.13}$$

Equation (5.13) correspond to a quadratic polynomial, physical root of which is defined by eq. (5.14):

$$[H-A]=-\frac{1}{2}\sqrt{\left(\frac{1}{K_{H-A}}+[H]_T+[A]_T\right)-4*[H]_T*[A]_T}+\frac{1}{2}\left(\frac{1}{K_{H-A}}+[H]_T+[A]_T\right) \quad (5.14)$$

At this stage, [H] and [A] equilibrium concentrations can be calculated by eqs. (5.10) and (5.11), respectively.

5.4.2 Simplified Case: Two Competing 1:1 Complexes

When a second partner is added to the previous system, leading to a second 1:1 complex, eqs. (5.15) to (5.19) appear, in addition to eq. (5.8) and (5.9):

$$H+B \leftrightharpoons H-B \quad (5.15)$$

$$K_{H-B}=\frac{[H-B]}{[H]*[B]} \quad (5.16)$$

$$[H]_T=[H]+[H-A]+[H-B] \quad (5.17)$$

$$[A]_T=[A]+[H-A] \quad (5.18)$$

$$[B]_T=[B]+[H-B] \quad (5.19)$$

The system of equations can be reformulated, by combining eqs. (5.9), (5.16), (5.17), (5.18) and (5.19):

$$[A]=\frac{[A]_T}{[1+K_{H-A}*[H]]} \quad (5.20)$$

$$[B]=\frac{[B]_T}{[1+K_{H-B}*[H]]} \quad (5.21)$$

$$[H]_T=[H]+K_{H-A}*[H]*\frac{[A]_T}{[1+K_{H-A}*[H]]}+K_{H-B}*[H]*\frac{[B]_T}{[1+K_{H-B}*[H]]} \quad (5.22)$$

Equation (5.22) correspond to a polynomial with a degree equal to 3. In this specific case, an analytical form of the physical root has been proposed (Wang 1995; Sigurskjold 2000). This analytical form allows to calculate the equilibrium concentration [H-A] by means of eq. (5.23):

$$[\mathrm{H-A}] = -\frac{[\mathrm{A}]_{\mathrm{T}}\left\{2\sqrt{(a^2-3b)}\cos\left(\frac{\theta}{3}\right)-a\right\}}{3\mathrm{K}_{\mathrm{H-A}}+\left\{2\sqrt{(a^2-3b)}\cos\left(\frac{\theta}{3}\right)-a\right\}} \tag{5.23}$$

with:

$$\mathrm{a} = \mathrm{K}_{\mathrm{H-B}} + \mathrm{K}_{\mathrm{H-B}} + [\mathrm{B}]_{\mathrm{T}} + [\mathrm{A}]_{\mathrm{T}} - [\mathrm{H}]_{\mathrm{T}} \tag{5.24}$$

$$b = \mathrm{K}_{\mathrm{H-A}} + ([\mathrm{B}]_{\mathrm{T}} - [\mathrm{H}]_{\mathrm{T}}) + \mathrm{K}_{\mathrm{H-B}} * ([\mathrm{A}]_{\mathrm{T}} - [\mathrm{H}]_{\mathrm{T}}) + \mathrm{K}_{\mathrm{H-B}} * \mathrm{K}_{\mathrm{H-A}} \tag{5.25}$$

$$c = -\,\mathrm{K}_{\mathrm{H-B}} * \mathrm{K}_{\mathrm{H-A}} * [\mathrm{H}]_{\mathrm{T}} \tag{5.26}$$

$$\theta = \arccos\frac{-2a^3+9ab-27c}{2\sqrt{(a^2-3b)^3}} \tag{5.27}$$

Finally, [A], [H], [H-B] and [B] equilibrium concentrations can be calculated by eqs. (5.18), (5.9), (5.17) and (5.19), respectively.

5.4.3 *General Case*

According to Fig. 5.4, if H_2-A and H-B_2 complexes are formed in addition to the competing 1:1 complexes formed with partner A and B, equilibriums are described by eqs. (5.8), (5.9), (5.15), (5.16) and (5.28) to (5.34):

$$\mathrm{H} + \mathrm{H-A} \leftrightharpoons \mathrm{H_2-A} \tag{5.28}$$

$$\mathrm{K}_{\mathrm{H_2-A}} = \frac{[\mathrm{H_2-A}]}{[\mathrm{H-A}]*[\mathrm{H}]} \tag{5.29}$$

$$\mathrm{H-B} + \mathrm{B} \leftrightharpoons \mathrm{H-B_2} \tag{5.30}$$

$$\mathrm{K}_{\mathrm{H-B_2}} = \frac{[\mathrm{H-B_2}]}{[\mathrm{H-B}]*[\mathrm{B}]} \tag{5.31}$$

$$[\mathrm{A}]_{\mathrm{T}} = [\mathrm{A}] + [\mathrm{H-A}] + [\mathrm{H_2-A}] \tag{5.32}$$

$$[\mathrm{B}]_{\mathrm{T}} = [\mathrm{B}] + [\mathrm{H-B}] + 2*[\mathrm{H-B_2}] \tag{5.33}$$

$$[H]_T = [H] + [H-A] + 2*[H_2-A] + [H-B] + [H-B_2] \tag{5.34}$$

$[A]_T$ may be defined as a function of [H], by combining eqs. (5.32) with (5.9) and (5.29):

$$[A]_T = [A]*\left[1 + K_{H-A}*[H] + K_{H_2-A}*K_{H-A}*[H]^2\right] \tag{5.35}$$

which leads to:

$$[A] = \frac{[A]_T}{\left[1 + K_{H-A}*[H] + K_{H_2-A}*K_{H-A}*[H]^2\right]} \tag{5.36}$$

$[B]_T$ may also be defined as a function of [H], by combining eqs. (5.33) with (5.16) and (5.31):

$$[B]_T = [B] + K_{H-B}*[H]*[B] + 2*K_{H_2-B}*K_{H-B}*[H]*[B]^2 \tag{5.37}$$

The physical root of this quadratic polynomial is then defined by eq. (5.38):

$$[B] = \sqrt{\frac{[B]_T}{2*K_{H-B_2}*K_{H-B}*[H]} + \left[\frac{(1+K_{H-B}*[H])}{4*K_{H-B_2}*K_{H-B}*[H]}\right]^2} - \frac{(1+K_{H-B}*[H])}{4*K_{H-B_2}*K_{H-B}*[H]} \tag{5.38}$$

Defining $[H]_T$ as a function of [H] leads to eq. (5.39)

$$\begin{aligned}[H]_T = [H] &+ K_{H-A}*[H]*[A] + 2*K_{H_2-A}*K_{H-A}*[H]^2*[A] \\ &+ K_{H-B}*[H]*[B] + K_{H_2-B}*K_{H-B}*[H]*[B]^2\end{aligned} \tag{5.39}$$

As [A] and [B] are complex functions of [H], eq. (5.39) corresponds to a binding polynomial of degree 4. As a consequence, there is no analytical form for the corresponding physical root, which has to be obtained by a numerical approach, such as secant or Newton's method.

Finally, [H-A], [H-B], $[H_2\text{-}A]$, and $[H\text{-}B_2]$ equilibrium concentrations can be calculated by the use of eqs. (5.9), (5.16), (5.29) and (5.31), respectively.

In the specific case of phase solubility studies, it has to be mentioned that the total guest concentration in the liquid phase is not known, in such way that eq. (5.36) and/or (5.38) can not be applied. Nevertheless, in this case, the free guest concentration corresponds to the intrinsic solubility of this guest, whatever the cyclodextrin concentration. As a consequence, in eq. (5.39), one can substitute [A] and/or [B] by the corresponding aqueous solubility S_0^A and/or S_0^B, if A and/or B are used at their solubility limit.

5.4.4 *Other Equilibriums*

When considering modified cyclodextrins, other equilibriums than the ones described in Fig. 5.4 might occur. This is especially the case for cyclodextrin dimers, since a cooperative 1:1 complex may exist for both partner A and B, in addition to non cooperative inclusion compounds. In these more complicated cases, the general solution proposed by the root of eq. (5.39) does not hold anymore. As a consequence, it is necessary to rewrite the binding polynomial according to the expected equilibriums, and solved it numerically, in a similar way to paragraph 5.4.3.

5.5 Typology of Experimental Responses

The fact that any physicochemical property might be used to evaluate the stability of inclusion compounds has been thoroughly exploited in the literature, with a considerable variety of analytical techniques (Connors 1996; Valente and Söderman 2014). Nevertheless, the response of most of these techniques may be described by a unique formalism.

The recorded analytical signal (${}^{i}Y$) correspond to the sum of the contribution (${}^{i}Y_X$) of all species present when the equilibrium is reached, for each measurement. In the case of the equilibriums presented in Fig. 5.4, ${}^{i}Y$ may be described by eq. (5.40):

$$ {}^{i}Y = {}^{i}Y_H + {}^{i}Y_A + {}^{i}Y_B + {}^{i}Y_{H-A} + {}^{i}Y_{H-B} + {}^{i}Y_{H_2-A} + {}^{i}Y_{H-B_2} \tag{5.40} $$

It is obvious that eq. (5.40) may be simplified in the case of non competitive studies, in the absence of 2:1 or 1:2 stoichiometry, or if some species does not contribute to the observed signal. Nonetheless, it offers a general framework, which can be used with most of the encountered equilibriums for cyclodextrin complexes.

In addition, whatever the experimental technique which is used, the analytical signal of each species correspond to the response factor (f_X) multiplied by a function of the concentration of species X (g([X])):

$$ {}^{i}Y = f_X * g([X]) \tag{5.41} $$

Four formalisms exist for g([X]), depending on the analytical technique to be employed. It may correspond to the species concentration itself ([X]), to the mole fraction of this species ($[X]/[X]_T$), to the logarithm of the concentration (log [X]) or to the number of moles of complexes (Δn) which are formed upon addition of some partners on the others.[2] The various forms of g([X]), associated to the corresponding techniques and response factors f_X, are summarized in Table 5.1, which also includes related references.

[2]The use of Δn is specific to ITC measurements. The exact equation defining Δn can be found in Bertaut and Landy 2014.

Table 5.1 Typology of experimental responses, corresponding to the most widely used analytical techniques for measuring cyclodextrins binding constants

g([X])	Analytical Techniques	f_X	References
[X]	Optical spectroscopy (UV-Vis or NIR, fluorescence, circular dichroism,)	Molar extinction coefficient (ε), Proportionality constant (k),Molar ellipticity (Δε)	Anand et al. 2012; Dahab and El 2012; De Feyter et al. 1996; Fakayode et al. 2009; Granadero et al. 2010; Hamai and Koshiyama 1999; Landy et al. 2000; Lu et al. 2008; Maskevich et al. 2013; Monti et al. 2003; Tran and De Paoli Lacerda 2002
	Polarimetry	Molar optical activity (Θ)	Lo meo et al. 2006
	Headspace-GC	Proportionality constant (k)	Fourmentin et al. 2013
	Solubility measurements	1 or 0 (if X contributes or not to the measured solubility)	Ono et al. 2001; Zughul and Badwan 1998
$[X]/[X]_T$	Nuclear magnetic resonance (NMR)	Chemical shift (δ), Diffusion coefficient (D)	Dodziuk et al. 2003; Fielding 2000; Pessine et al. 2012; Simova and Berger 2005
	Capillary electrophoresis	Electrophoretic mobility (μ)	Fang and Chen 2005; Rundlett and Armstrong 1997; Wahl et al. 2017
	High performance liquid chromatography (HPLC)	Capacity factor (k)	López-Nicolás et al. 2006
	Surface Plasmon resonance (SPR)	Response units (R)	Abe et al. 2011; Brown et al. 2003; Kobayashi et al. 2011
Log [X]	Potentiometry	Slope of the Nernstian equation	Funasaki et al. 2005
Δn	Isothermal titration calorimetry	Inclusion enthapy (ΔH)	Benk and Király 2012; Bertaut and Landy 2014; Cooper et al. 2005; Freiburger et al. 2012; Illapakurthy et al. 2005; Li et al. 2011

5.6 Data Treatment

The unified framework synthesized in this chapter is mainly based on the ability of a single scheme of data treatment to analyze any experiment, whatever the employed protocols or analytical techniques. This scheme not only allows to determine the stability of the studied complexes on the basis of one binding experiment, but it can also be used to handle multiple experiments simultaneously, to evaluate the uncertainty on the binding parameters, to optimize the experimental conditions (as it can simulate any isotherm) and to derive all thermodynamic parameters.

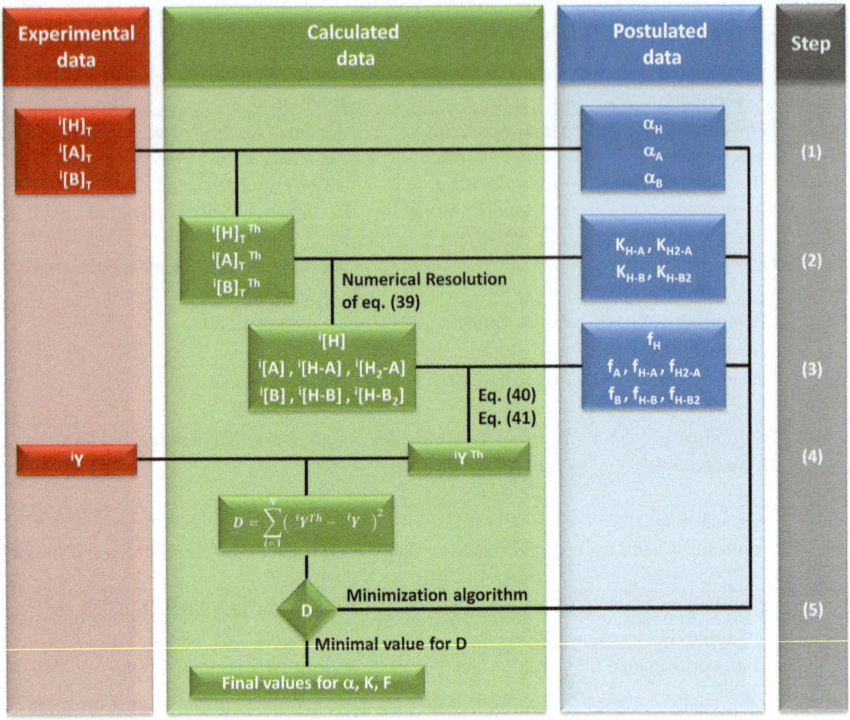

Fig. 5.11 General scheme of the algorithmic data treatment. Total concentrations, equilibrium concentrations, theoretical responses and total difference between experimental and theoretical signal are calculated in step 1 to 4, respectively. Step 5 corresponds to the minimization of this difference by varying the postulated parameters

5.6.1 Algorithmic Data Treatment

A binding study is totally defined by taking into account the experimental protocols, the postulated equilibriums and the employed analytical techniques, respectively by means of the total concentrations, equilibrium concentrations (based on the resolution of eq. (5.39)) and analytical signals (eqs. (5.40) and (5.41)). Accordingly, the general scheme that can be employed for estimating formation constants is presented in Fig. 5.11. Data which are on the same stage are combined through given equations in order to calculate the data on the following stage. Superscipt ^{i} is linked to each solution which has been analyzed. Superscipt ^{Exp} and Th refers to experimental and theoretical data, respectively.

Step 1 consist in multiplying the total concentration of each partner by a corrective factor (α_H, α_A, α_B respectively for host and guests), in order to obtain a theoretical total concentration $[X]_T{}^{Th}$, for each partner. These factors primarily accounts for any deviation between the expected concentration and the real concentration. For instance, a α value equal to 0.9 will be obtained by the numerical treatment if the real concentration is 10% lower than expected (according to any error in masses, molecular weights, volumes, water contents of the initial powder...). In addition, in the case of polymeric cyclodextrins, α factors also account for the number of available cavities per polymer molecule. For instance, a α_H factor equal to ten means that polymers contain on average ten available cyclodextrin cavities. In addition, the symetric formalism which takes into account two hosts and one guest also allows to define the number of each kind of cavity for an heterocyclodextrin polymer, by means of the corresponding α values.

Step 2 couples the postulated formation constants to the corrected total concentrations, by means of eq. (5.39), in order to calculate equilibrium concentrations.

Step 3 uses the previously calculated concentrations to simulate the analytical signals by means of eq. (5.40) and (5.41).

Step 4 calculates the difference (D) between experimental and theoretical signal. As the initial values of α, K and F are not the true ones, calculated total concentrations, equilibrium concentrations and thus analytical signals differ from the real experimental ones, in such a way that the D value is initially high.

At step 5, an algorithm is needed to minimize this D difference by varying the postulated values (α, K, f). When a minimal value is reached for D (according to a convergence criterion), the last values employed for α, K and f correspond to the ones which best reproduce the experimental isotherm. Any usual minimization algorithm might be used (Newton-Raphson, Levenberg-Marquardt, simplex, conjugate gradient...). The quality of fit between experimental and theoretical isotherms may be used to evaluate the relevance of the complexation model and of the calculated formation constants.

With this scheme, each stoichiometric hypothesis included in Fig. 5.4 may be independently tested, by setting non operating formation constants close to 0. For instance, the fit between experimental data and the single 1:1 model require that $K_{H\text{-}B}$, $K_{H2\text{-}A}$, $K_{H\text{-}B2}$, remain fixed to a negligible value, while $K_{H\text{-}A}$ is varied by the algorithm.

5.6.2 Global Analysis

The algorithmic treatment presented in Fig. 5.11 can handle an unlimited number of experimental signals for a given protocol, but also for an unlimited number of protocols. As a consequence, it is possible to simulate various isotherms with a single set of variables (α, K, f). Such a simultaneous treatment of various protocols is known under the concept of global analysis (Schönbeck et al. 2012; Lowe et al. 2012).

The first advantage of using global analysis, rather than individual analysis, is linked to the relevance of the postulated equilibriums. As the model can not use different values for a single variable from one experiment to another in order to optimize the fit between theoretical and experimental isotherms, any irrelevant equilibriums scheme will induce evident deviations on the various fits. As the model is more constrained with a global analysis, a strong adequation observed for all isotherms may be interpreted as a validation of the postulated equilibriums.

In addition, global analysis leads to more accurate values of formation constants, because of data merging. For instance, if the signal of a given protocol is strongly controlled by the formation constant but induces an high uncertainty on the response factor of the inclusion compound, while the signal of a second protocol is strongly controlled by the response coefficient but with an high uncertainty on the affinity, individual analysis of the two experiments will fail to give an exact description of K and f. To the contrary, global analysis will use the complementarities of both protocols to extract accurate values for K and f.

5.6.3 Evaluation of Uncertainty on Binding Parameters

Uncertainty relative to formation constants represent an essential data, especially when comparing results from different studies. Nevertheless, no guideline has been established for its estimation. Generally, published uncertainties correspond in fact to the standard deviation of the formation constants calculated by individual analysis of repeated experiments. As triplicate measurements may be considered as a standard, the calculated uncertainty present a weak statistical significance.

Reliable alternatives may be implemented by evaluating the influence of the analytical signal uncertainty on the calculated formation constant (Almansa López et al. 2003; Tellinghuisen 2000 and 2004). Such approaches allow the estimation of binding uncertainty even when only one experiment has been used to estimate the formation constant.

In a first step, the uncertainty relative to experimental responses has to be known. This quantity may be obtained by two different methods. Firstly, an uncertainty typical of the experimental device may be estimated previous to all binding studies, by recording several times the signal corresponding to a given solution. The standard deviation calculated from these replicates represents an estimate of the typical uncertainty of the device. Secondly, it is also possible to estimate this uncertainty by calculating the standard deviation on the residuals (differences between theoretical and experimental isotherms) of the binding experiments. These two methods are referred, respectively, to a priori and a posteriori approaches.

In a second step, i.e. once the analytical uncertainty is known, the uncertainties on the thermodynamical parameters may be obtained by two approaches.

The first approach is based on the calculation of the variance–covariance matrix V, whose diagonal elements correspond to the variances of the binding parameters. V is given by:

$$V = \left| Z^T . W . Z \right|^{-1} \tag{5.42}$$

The matrix Z contains elements $Z_{ij} = (\partial F_i/\partial \beta_j)$, with i and j ranging from 1 to N and 1 to M respectively. F and β correspond to the simulated signal and to the adjustable parameters (α, K, f, for each species), respectively. The partial derivatives are evaluated numerically for each independent variable, on the basis of the final values of the fit parameters. W designates the weight matrix, with diagonal elements equal to the inverse data variances, $W_{ii} = \sigma_i^{-2}$, and other elements equal to 0. If a unique uncertainty is used for each signal, all σ_i correspond to the aforementioned uncertainty relative to experimental responses.

As a consequence, eq. (5.43) can be expressed in the following developed form:

$$\begin{vmatrix} (\sigma_{\beta_1})^2 & & \\ & \ddots & \\ & & (\sigma_{\beta_M})^2 \end{vmatrix} = \left(\begin{Vmatrix} \frac{\partial F_1}{\partial \beta_1} & \cdots & \frac{\partial F_1}{\partial \beta_M} \\ \vdots & \ddots & \vdots \\ \frac{\partial F_N}{\partial \beta_1} & \cdots & \frac{\partial F_N}{\partial \beta_M} \end{Vmatrix}^T \cdot \begin{vmatrix} (\sigma_1)^{-2} & 0 & 0 \\ 0 & \ddots & 0 \\ 0 & 0 & (\sigma_N)^{-2} \end{vmatrix} \cdot \begin{vmatrix} \frac{\partial F_1}{\partial \beta_1} & \cdots & \frac{\partial F_1}{\partial \beta_M} \\ \vdots & \ddots & \vdots \\ \frac{\partial F_N}{\partial \beta_1} & \cdots & \frac{\partial F_N}{\partial \beta_M} \end{vmatrix} \right)^{-1} \quad (5.43)$$

It is also possible to simulate the uncertainty on formations constants by the use of Monte Carlo trials. Each point of the isotherm is randomly modified within a window defined by the analytical uncertainty. Then, the difference between this modified experimental isotherm and the theoretical one is minimized leading to a new value of formation constant. Such a perturbation procedure is then repeated several times (typically generating from 10^3 to 10^6 isotherms). Accordingly, a reliable standard deviation can be calculated on the collected formation constants.

If both variance–covariance matrix and Monte Carlo approaches already constitute a more robust evaluation of standard deviations on affinity, it has to be underlined that the use of confidence interval rather than standard deviation has to be preferred when comparing affinities under the scope of significant difference. As a consequence, upper and lower limits of formation constants should be estimated by multiplying the standard deviations by the student t factor (1.96).

When more than one experiment has been realized, both variance–covariance matrix and Monte Carlo methods should be preferentially used under the scope of global analysis, i.e. with one set of independent variables describing all experiments. As a result, accuracy benefits from information over all experiments. In the case of competitive experiments, the accuracy on the studied complex takes into account the accuracy on the reference complex (and reciprocally).

5.6.4 *Optimization of Experimental Conditions*

Although the optimization of experimental conditions may have a significant impact on the estimated affinity, few studies address this important issue. The main difficulty relies on the specificity of each binding study, since the formation constant accuracy is depending on the affinity of each studied complex, on the response factor of each species, in addition to all total concentrations which are used. Only the isotherm simulation allows the definition of the optimal conditions for estimating the affinity. Under this scope, the previously mentioned variance–covariance matrix and Monte Carlo approaches may be useful for optimization issues (Tellinghuisen 2005 and 2012; Tellinghuisen and Chodera 2011). Indeed, the calculation of formation constant uncertainties may be realized for hypothetical systems, by the simulation of binding isotherms on the basis of postulated experimental conditions, thermodynamical parameters and theoretical form of analytical signals. The empirical modification of experimental parameters then allows the definition of the optimal conditions which minimize the affinity uncertainty. Even if each binding study would require a dedicated optimization step, some general tendencies might be extracted.

Concerning the limiting partner,[3] a basic advice would be to use the greatest achievable concentration, in order to increase the signal to noise ratio. Nonetheless, it should be kept in mind that too elevated concentrations, for either limiting or non limiting partners, may generate some aggregates, thus inducing a distortion between the postulated and the real equilibriums. In addition, a decrease of the differentiation between the theoretical isotherms may be observed for high concentrations and/or high formation constants, thus inducing a weaker accuracy on the binding constant, despite a higher intensity of the recorded signals.

In addition, the variation of the concentrations during the binding experiment should efficiently sample the isotherm. This can also be rationalized on the basis of the limiting partner, complex form of which could vary from 0 to 100% of its total concentration.

First of all, the most concentrated solution of the non limiting partner should induce the complexation of a significant part of the limiting partner. For instance, in Fig. 5.5, a titrant concentration equal to 5 mM leads, for a formation constant equal to 10^3 M^{-1}, to a complexation close to 80% at the end of the titration.

Secondly, the intermediate concentrations should lead to equidistant values of ${}^{i}Y$. For instance, the curve presented for K equal to 10^3 M^{-1} in Fig. 5.5 perfectly respect this criteria, with an increase of 8% of complexation upon each increasing titrant concentration. One can underline that it implies a non linear increase of this titrant concentration, and that it is necessary to have a crude estimate of the binding constant, in order to choose, by simulation, these intermediate concentrations.

Finally, the choice of the number of intermediate concentrations is not trivial. On one hand, using a lot of intermediate concentrations may be useful to verify that the experimental data are in agreement with the postulated equilibriums. On the other hand, and contrary to a widespread believe, the accuracy on the binding constants generally decrease with increasing number of intermediate concentrations: for a given total number of measurements, replicas of a limited number of the ideal solutions is more efficient than unique recording of a greater number of solutions.

[3] The limiting partner refers here to the partner which is used at the lower concentration during the binding experiment. For instance, in titration experiments, the limiting partner is the titrate.

Indeed, an isotherm is perfectly defined with a number of distinct signals which corresponds to the number of unknown variables. For instance, a single 1:1 complex requires two recorded signals to determine simultaneously the formation constant and response factor, if these two quantities are the only variables of the fitting procedure. Increasing the number of recorded signals lead to weaker intensity differences between each solution, and thus to a weaker signal to noise ratio per measurement. In brief, if the number of intermediate concentrations remains a compromise, one has to check that the signal difference between two successive intermediate concentrations should be significantly greater than the intrinsic noise of the employed analytical technique.

5.6.5 Determination of Thermodynamic Parameters

The formation constant, which is depending on temperature, is linked to the free energy of complexation ΔG(T):

$$\Delta G(\mathrm{T}) = -\mathrm{R} * \mathrm{T} * \ln(\mathrm{K}(\mathrm{T})) \tag{5.44}$$

ΔG(T) constitutes one of the thermodynamic parameters of binding, the other ones being the enthalpy variation ΔH(T), entropy variation ΔS(T) and heat capacity variation ΔC_p.

When used without approximation, the van't Hoff law allows to determine all thermodynamic parameters, by measuring the formation constants at different temperatures, T_0 being an arbitrary reference temperature. ΔG(T), ΔH(T), ΔS(T) and ΔC_p are then linked by eqs. (5.45) to (5.47):

$$\mathrm{K}(\mathrm{T}) = \mathrm{K}(\mathrm{T}_0) * \mathrm{e}^{\left\{\frac{\Delta H(\mathrm{T}_0)}{\mathrm{R}}\left(\frac{1}{\mathrm{T}_0}-\frac{1}{\mathrm{T}}\right)+\frac{\Delta \mathrm{C}_\mathrm{p}}{\mathrm{R}}\left(\ln\left\{\frac{\mathrm{T}}{\mathrm{T}_0}\right\}+\frac{\mathrm{T}_0}{\mathrm{T}}-1\right)\right\}} \tag{5.45}$$

$$\Delta H(\mathrm{T}) = \Delta H(\mathrm{T}_0) + \Delta \mathrm{C}_\mathrm{p} * (\mathrm{T} - \mathrm{T}_0) \tag{5.46}$$

$$\Delta G(\mathrm{T}) = \Delta H(\mathrm{T}) - \mathrm{T} * \Delta S(\mathrm{T}) \tag{5.47}$$

Unfortunately, most literature studies are based on simplified equations, by considering that ΔC_p is equal to 0, which leaves $\Delta H(T_0)$ as the only unknown variable of eq. (5.45) and allows a graphical treatment. This is a convenient but totally inadequate approximation (Mizoue and Tellinghuisen 2004; Zhukov and Karlsson 2007), since there is no objective argument stating that free and complexed species should present identical heat capacities. To the contrary, since hydrophobic forces generally constitute one of the driving forces of complexation, ΔC_p is expected to be significantly negative for most inclusion phenomena.

The use of both $\Delta H(T_0)$ and ΔC_p, as additional unknown variables in the general scheme presented in Fig. 5.11, should thus be preferred. Nonetheless, it should be mentioned that such approach often leads to an over parameterized model, and thus to significant uncertainties on both terms.

With this respect, ITC constitutes the only way to obtain true thermodynamic parameters. Indeed, as ITC measures simultaneously K(T) and ΔH(T) for any temperatures, estimation of ΔG(T) and ΔS(T) is straightforward, considering eqs. (5.44) and (5.47). As a result, global analysis of ITC experiments realized at different temperatures may be considered as the most reliable method to extract all thermodynamic information (Freiburger et al. 2012), as it contributes to a more constrained model.

Finally, in the hypothetical case of a heat capacity variation which would be depending on temperature, eq. (5.45) to (5.47) would not hold anymore. It would be then necessary to consider all ΔH(T) as unknown variables. Only one formation constant would still be needed for the fitting of all isotherms, the affinity at other temperatures being calculated by trapezoidal integration of ΔH(T) values, according to eq. (5.48).

$$\ln K^{T_j} = \ln K^{T_0} + \sum_{i=1}^{j} \frac{T_i - T_{i-1}}{2R} \left(\frac{\Delta H^{T_{i-1}}}{(T_{i-1})^2} + \frac{\Delta H^{T_i}}{(T_i)^2} \right) \quad (5.48)$$

5.7 Conclusions

Despite the tremendous diversity of binding studies on cyclodextrin complexes, a general framework can be defined for determining the corresponding formation constants, without any approximation. This framework, based on a numerical approach of the data treatment, might be applied to a wide majority of encountered equilibriums, with most of the analytical techniques and with all experimental protocols.

The systematic use of this unified framework could be of great help to improve the characterization of inclusion compounds, especially trough the global analysis of different experiments. Indeed, as all realistic experimental conditions are allowed by this approach, it also becomes possible to treat simultaneously various experiments resulting from different protocols and techniques with a single set of variables. As a result, one can use the resulting quality of fit between the experimental and simulated signals to discriminate between the different equilibrium models which are simulated within the unified framework, inducing a greater confidence on the effective stoichiometric model. In addition, as the treatment aggregates information contained in all experiments, an improved accuracy on the binding constants is expected, if compared to a classical individual analysis on separate experiments.

Moreover, this unified framework is also perfectly suited to internalize statistical evaluations on the formation constants. In particular, Monte Carlo or variance-covariance approaches not only allow a robust evaluation of the accuracy on each parameter, whatever the number of experiments, but it also permits to simulate the expected accuracy for hypothetical protocols, thus allowing the optimization of experimental conditions. Besides, it has to be underlined that the use of a rigorous and systematic estimation of confidence interval on affinity could probably eliminate one of the main causes of divergence between formation constants obtained by different techniques or authors.

Finally, it should be emphasized that global analysis of ITC experiments realized at multiple temperatures should be considered as the gold standard, since it allows a complete description of thermodynamic parameters, without any approximation, according to the direct measurement of inclusion enthalpy.

References

Abe K, Ogawa N, Nagase H, Endo T, Ueda H (2011) Evaluation of the abilities of γ-cyclodextrin to form complexes by surface plasmon resonance with a Biacore® system. J Incl Phenom Macrocycl Chem 70(3–4):385–388. https://doi.org/10.1007/s10847-010-9883-4

Almansa López E, Bosque-Sendra JM, Cuadros Rodríguez L, García Campaña AM, Aaron JJ (2003) Applying non-parametric statistical methods to the classical measurements of inclusion complex binding constants. Anal Bioanal Chem 375(3):414–423. https://doi.org/10.1007/s00216-002-1693-0

Anand R, Ottani S, Manoli F, Manet I, Monti S (2012) A close-up on doxorubicin binding to γ-cyclodextrin: an elucidating spectroscopic, photophysical and conformational study. RSC Adv 2(6):2346–2357. https://doi.org/10.1039/c2ra01221a

Benesi HA, Hildebrand JH (1949) Solubility of iodine in 1, 2-dichloroethanes, cis- and trans-dichloroethylenes and perfluoro-n-heptane. J Am Chem Soc 70(12):3978–3981. https://doi.org/10.1021/ja01192a003

Benk M, Király Z (2012) Thermodynamics of inclusion complex formation of β-cyclodextrin with a variety of surfactants differing in the nature of headgroup. J Chem Thermodyn 54:221–216. https://doi.org/10.1016/j.jct.2012.03.033

Bertaut E, Landy D (2014) Improving ITC studies of cyclodextrin inclusion compounds by global analysis of conventional and non-conventional experiments. Beilstein J Org Chem 10:2630–2641. https://doi.org/10.3762/bjoc.10.275

Blyshak LA, Dodson KY, Patonay G, Warner IM, May WE (1989) Determination of cyclodextrin formation constants using dynamic coupled-column liquid chromatography. Anal Chem 61(9):955–960. https://doi.org/10.1021/ac00184a008

Brown SE, Easton CJ, Kelly JB (2003) Surface plasmon resonance to determine apparent stability constants for the binding of cyclodextrins to small immobilized guests. J Incl Phenom Macrocycl Chem 46(3–4):167–173. https://doi.org/10.1023/A:1026311003881

Connors KA (1995) Population characteristics of cyclodextrin complex ctabilities in aqueous solution. J Pharm Sci 84(7):843–848

Connors KA (1996) Measurement of cyclodextrin complex stability constants. In: Szejtli Z, Osa T (eds) Comprehensive supramolecular chemistry, vol 3. Elsevier Science Ltd., Rugby, Netherland, pp 205–242

Connors KA (1997) The stability of cyclodextrin complexes in solution. Chem Rev 97:1325–1358. https://doi.org/10.1021/cr960371r

Cooper A, Nutley M, MacLean EJ, Cameron K, Fielding L, Mestres J, Palin R (2005) Mutual induced fit in cyclodextrin-rocuronium complexes. Org Biomol Chem 3(10):1863–1871. https://doi.org/10.1039/b415903a

Dahab AA, El DH (2012) Rapid analysis of NSAIDs binding to β-cyclodextrin using the simultaneous measurement of absorption and circular dichroism with a novel multi-cell low-volume device. Anal Bioanal Chem 404:1839–1850. https://doi.org/10.1007/s00216-012-6286-y

De Feyter S, Van Stam J, Boens N, De Schryver FC (1996) On the use of dynamic fluorescence measurements to determine equilibrium and kinetic constants. The inclusion of pyrene in β-cyclodextrin cavities. Chem Phys Lett 249(1–2):46–52. https://doi.org/10.1016/0009-2614(95)01365-2

Dodziuk H, Nowinski KS, Kozminski W, Dolgonos G (2003) On the impossibility of determination of stepwise binding constants for the 1:2 complex of (+)-camphor with α-cyclodextrin. Org Biomol Chem 1(3):581–584. https://doi.org/10.1039/b209272g

Exner O (1997) Calculating equilibrium constants from spectral data: reliability of the Benesi-Hildebrand method and its modifications. Chemom Intell Lab Syst 39(1):85–93. https://doi.org/10.1016/S0169-7439(97)00057-9

Fakayode SO, Brady PN, Pollard DA, Mohammed AK, Warner IM (2009) Multicomponent analyses of chiral samples by use of regression analysis of UV-visible spectra of cyclodextrin guest-host complexes. Anal Bioanal Chem 394(6):1645–1653. https://doi.org/10.1007/s00216-009-2853-2

Fang N, Chen DDY (2005) Enumeration algorithm for determination of binding constants in capillary electrophoresis. Anal Chem 77(8):2415–2420. https://doi.org/10.1021/ac048509q

Fielding L (2000) Determination of association constants (K(a)) from solution NMR data. Tetrahedron 56(34):6151–6170. https://doi.org/10.1016/S0040-4020(00)00492-0

Fourmentin S, Outirit M, Blach P, Landy D, Ponchel A, Monflier E, Surpateanu G (2007) Solubilisation of chlorinated solvents by cyclodextrin derivatives: a study by static headspace gas chromatography and molecular modelling. J Hazard Mater 141(1):92–97. https://doi.org/10.1016/j.jhazmat.2006.06.090

Fourmentin S, Ciobanu A, Landy D, Wenz G (2013) Space filling of β-cyclodextrin and β-cyclodextrin derivatives by volatile hydrophobic guests. Beilstein J Org Chem 9:1185–1191. https://doi.org/10.3762/bjoc.9.133

Freiburger LA, Auclair K, Mittermaier AK (2012) van 't Hoff global analyses of variable temperature isothermal titration calorimetry data. Thermochim Acta 527:148–157. https://doi.org/10.1016/j.tca.2011.10.018

Fujiki M, Deguchi T, Sanemasa I (1988) Association of naphthalene and its methyl derivatives with cyclodextrins in aqueous medium. Bull Chem Soc Jpn 61(4):1163–1167. https://doi.org/10.1246/bcsj.61.1163

Funasaki N, Nagaoka M, Hirota S (2005) Competitive potentiometric determination of binding constants between α-cyclodextrin and 1-alkanols. Anal Chim Acta 531(1):147–151. https://doi.org/10.1016/j.aca.2004.09.079

Granadero D, Bordello J, Pérez-Alvite MJ, Novo M, Al-Soufi W (2010) Host-guest complexation studied by fluorescence correlation spectroscopy: Adamantane-cyclodextrin inclusion. Int J Mol Sci 11(1):173–188. https://doi.org/10.3390/ijms11010173

Hamai S, Koshiyama T (1999) Electronic absorption, fluorescence, and circular dichroism spectroscopic studies on the inclusion complexes of tetrakis(4-sulfonatophenyl)porphyrin with cyclodextrins in basic aqueous solutions. J Photochem Photobiol A 127(1–3):135–141. https://doi.org/10.1016/S1010-6030(99)00144-6

Hansen LD, Fellingham GW, Russell DJ (2011) Simultaneous determination of equilibrium constants and enthalpy changes by titration calorimetry: methods, instruments, and uncertainties. Anal Biochem 409(2):220–229. https://doi.org/10.1016/j.ab.2010.11.002

Heerklotz HH, Binder H, Epand RM (1999) A "release" protocol for isothermal titration calorimetry. Biophys J 76(5):2606–2613. https://doi.org/10.1016/S0006-3495(99)77413-8

Higuchi T, Connors KA (1965) Phase-solubility techniques. Adv Anal Chem Instrum 4:117–212
Hirayama F, Uekama K (1987) Methods of investigating and preparing inclusion compounds. In: Duchêne D (ed) Cyclodextrins and their industrial uses. Editions de Santé, Paris, pp 131–172
Illapakurthy AC, Wyandt CM, Stodghill SP (2005) Isothermal titration calorimetry method for determination of cyclodextrin complexation thermodynamics between artemisinin and naproxen under varying environmental conditions. Eur J Pharm Biopharm 59(2):325–332. https://doi.org/10.1016/j.ejpb.2004.08.006
Job P (1928) Formation and stability of inorganic complexes in solution. Ann Chim 9:113–204
Kobayashi H, Endo T, Ogawa N, Nagase H, Iwata M, Ueda H (2011) Evaluation of the interaction between β-cyclodextrin and psychotropic drugs by surface plasmon resonance assay with a Biacore® system. J Pharm Biomed Anal 54(1):258–263. https://doi.org/10.1016/j.jpba.2010.08.012
Landy D, Fourmentin S, Salome M, Surpateanu G (2000) Analytical improvement in measuring formation constants of inclusion complexes between β-cyclodextrin and phenolic compounds. J Incl Phenom Macrocycl Chem 38(1–4):187–198. https://doi.org/10.1023/A:1008156110999
Landy D, Tetart F, Truant E, Blach P, Fourmentin S, Surpateanu G (2007) Development of a competitive continuous variation plot for the determination of inclusion compounds stoichiometry. J Incl Phenom Macrocycl Chem 57(1–4):409–413. https://doi.org/10.1007/s10847-006-9226-7
Landy D, Mallard I, Ponchel A, Monflier E, Fourmentin S (2012) Remediation technologies using cyclodextrins: an overview. Environ Chem Lett 10(3):225–237. https://doi.org/10.1007/s10311-011-0351-1
Li X, Li H, Liu M, Li G, Li L, Sun D (2011) From guest to ligand a study on the competing interactions of antitumor drug resveratrol with β-cyclodextrin and bovine serum albumin. Thermochim Acta 521(1–2):74–79. https://doi.org/10.1016/j.tca.2011.04.007
Lo Meo P, D'Anna F, Riela S, Gruttadauria M, Noto R (2006) Polarimetry as a useful tool for the determination of binding constants between cyclodextrins and organic guest molecules. Tetrahedron Lett 47(51):9099–9102. https://doi.org/10.1016/j.tetlet.2006.10.078
Loftsson T, Másson M, Brewster ME (2004) Self-association of cyclodextrins and cyclodextrin complexes. J Pharm Sci 93(5):1091–1099. https://doi.org/10.1002/jps.20047
López-Nicolás JM, Núñez-Delicado E, Pérez-López AJ, Barrachina AC, Cuadra-Crespo P (2006) Determination of stoichiometric coefficients and apparent formation constants for β-cyclodextrin complexes of trans-resveratrol using reversed-phase liquid chromatography. J Chromatogr A 1135(2):158–165. https://doi.org/10.1016/j.chroma.2006.09.013
Lowe AJ, Pfeffer FM, Thordarson P (2012) Determining binding constants from 1H NMR titration data using global and local methods: a case study using [n] polynorbornane-based anion hosts. Supramol Chem 24(8):585–594. https://doi.org/10.1080/10610278.2012.688972
Lu Z, Lu C, Meng Q (2008) An inclusion complex of β-cyclodextrin with mnt anion (mnt = maleonitriledithiolate) studied by induced circular dichroism. J Incl Phenom Macrocycl Chem 61(1–2):101–106. https://doi.org/10.1007/s10847-007-9400-6
Maskevich AA, Kurhuzenkau SA, Lickevich AY (2013) Fluorescence spectral analysis of thioflavin T-γ-cyclodextrin interaction. J Appl Spectrosc 80(1):36–42. https://doi.org/10.1007/s10812-013-9717-4
Mizoue LS, Tellinghuisen J (2004) Calorimetric vs. van't Hoff binding enthalpies from isothermal titration calorimetry: Ba2+-crown ether complexation. Biophys Chem 110(1–2):15–24. https://doi.org/10.1016/j.bpc.2003.12.011
Mohamed MH, Wilson LD, Headley JV, Peru KM (2009) A spectral displacement study of cyclodextrin/naphthenic acids inclusion complexes. Can J Chem 87(12):1747–1756. https://doi.org/10.1139/V09-140
Monti S, Marconi G, Manoli F, Bortolus P, Mayer B, Grabner G, Köhler G, Boszczyk W, Rotkiewicz KA (2003) Spectroscopic and structural characterization of the inclusion complexes of p-dimethylaminobenzonitrile with cyclodextrins. Phys Chem Chem Phys 5(6):1019–1026. https://doi.org/10.1039/b209689g

Ono N, Hirayama F, Arima H, Uekama K (2001) Analysis of the phase solubility diagram of a phenacetin/competitor/β-cyclodextrin ternary system, involving competitive inclusion complexation. Chem Pharm Bull 49(1):78–81. https://doi.org/10.1248/cpb.49.78

Osajima T, Deguchi T, Sanemasa I (1991) Association of cycloalkanes with cyclodextrins in aqueous medium. Bull Chem Soc Jpn 64(9):2705–2709. https://doi.org/10.1246/bcsj.64.2705

Pessine FBT, Calderini A, Alexandrino GL (2012) Review: cyclodextrin inclusion complexes probed by NMR techniques. In: Kim D (ed) Magnetic resonance spectroscopy, vol 12, pp 237–265. https://doi.org/10.5772/32029

Pirnau A, Floare CG, Bogdan M (2014) The complexation of flurbiprofen with β-cyclodextrin: a NMR study in aqueous solution. J Incl Phenom Macrocycl Chem 78(1–4):113–120. https://doi.org/10.1007/s10847-012-0277-7

Rundlett KL, Armstrong DW (1997) Methods for the estimation of binding constants by capillary electrophoresis. Electrophoresis 18(12–13):2194–2202. https://doi.org/10.1002/elps.1150181210

Sanemasa I, Akamine Y (1987) Association of benzene and alkylbenzenes with cyclodextrins in aqueous medium. Bull Chem Soc Jpn 60(6):2059–2066. https://doi.org/10.1246/bcsj.60.2059

Sanemasa I, Takuma T, Deguchi T (1989) Association of some polynuclear aromatic hydrocarbons with cyclodextrins in aqueous medium. Bull Chem Soc Jpn 62(10):3098–3102. https://doi.org/10.1246/bcsj.62.3098

Scatchard G (1949) The attractions of proteins for small molecules and ions. Ann NY Acad Sci 51:660–672. https://doi.org/10.1111/j.1749-6632.1949.tb27297.x

Schönbeck C, Holm R, Westh P (2012) Higher order inclusion complexes and secondary interactions studied by global analysis of calorimetric titrations. Anal Chem 84(5):2305–2312. https://doi.org/10.1021/ac202842s

Scott RL (1956) Some comments on the Benesi Hildebrand equation. Recl Trav Chim Pays-Bas 75(7):787–789. https://doi.org/10.1002/recl.19560750711

Sigurskjold BW (2000) Exact analysis of competition ligand binding by displacement isothermal titration calorimetry. Anal Biochem 277(2):260–266. https://doi.org/10.1006/abio.1999.4402

Simova S, Berger S (2005) Diffusion measurements vs. chemical shift titration for determination of association constants on the example of camphor-cyclodextrin complexes. J Incl Phenom Macrocycl Chem 53(3):163–170. https://doi.org/10.1007/s10847-005-2631-5

Singh R, Bharti N, Madan J, Hiremath SN (2010) Characterization of cyclodextrin inclusion complexes – a review. J Pharm Sci Technol 2:171–183

Smith VJ, Bogdan D, Caira MR, Bogdan M, Bourne SA, Frcaş SI (2010) Cyclodextrin inclusion of four phenylurea herbicides: determination of complex stoichiometries and stability constants using solution 1H NMR spectroscopy. Supramol Chem 22(3):172–177. https://doi.org/10.1080/10610270902980655

Szaniszlo N, Fenyves E, Balla J (2005) Structure-stability study of cyclodextrin complexes with selected volatile hydrocarbon contaminants of soils. J Incl Phenom Macrocycl Chem 53(3):241–248. https://doi.org/10.1007/s10847-005-0245-6

Szejtli J (1996) Inclusion of guest molecules, selectivity and molecular recognition by cyclodextrins. In: Szejtli Z, Osa T (eds) Comprehensive supramolecular chemistry, vol 3. Elsevier Science Ltd., Rugby, Netherland, pp 189–204

Szente L (1996) Analytical methods for cyclodextrins, cyclodextrin derivatives, and cyclodextrin complexes. In: Szejtli Z, Osa T (eds) Comprehensive supramolecular chemistry, vol 3. Elsevier Science Ltd., Rugby, Netherland, pp 253–278

Takahashi AI, Veiga FJB, Ferraz HG (2012a) A literature review of cyclodextrin inclusion complexes characterization – part I: phase solubility diagram, dissolution and scanning electron microscopy. Int J Pharm Sci Rev Res 12(1):1–6

Takahashi AI, Veiga FJB, Ferraz HG (2012b) A literature review of cyclodextrin inclusion complexes characterization – part II: X-Ray diffraction, infrared spectroscopy and nuclear magnetic resonance. Int J Pharm Sci Rev Res 12(1):8–15

Takahashi AI, Veiga FJB, Ferraz HGA (2012c) Literature review of cyclodextrin inclusion complexes characterization – part III: differential scanning calorimetry and thermogravimetry. Int J Pharm Sci Rev Res 12(1):16–20

Tellinghuisen J (2000) Monte carlo study of precision, bias, inconsistency, and non-gaussian distributions in nonlinear least squares. J Phys Chem A 104(12):2834–2844. https://doi.org/10.1021/jp993279i

Tellinghuisen J (2004) Statistical error in isothermal titration calorimetry. Methods Enzymol 383:245–282. https://doi.org/10.1016/S0076-6879(04)83011-8

Tellinghuisen J (2005) Optimizing experimental parameters in isothermal titration calorimetry. J Phys Chem B 109(42):20027–20035. https://doi.org/10.1021/jp053550y

Tellinghuisen J (2007) Calibration in isothermal titration calorimetry: heat and cell volume from heat of dilution of NaCl(aq). Anal Biochem 360(1):47–55. https://doi.org/10.1016/j.ab.2006.10.015

Tellinghuisen J (2012) Designing isothermal titration calorimetry experiments for the study of 1:1 binding: problems with the "standard protocol". Anal Biochem 424(2):211–220. https://doi.org/10.1016/j.ab.2011.12.035

Tellinghuisen J, Chodera JD (2011) Systematic errors in isothermal titration calorimetry: concentrations and baselines. Anal Biochem 414(2):297–299. https://doi.org/10.1016/j.ab.2011.03.024

Thordarson P (2011) Determining association constants from titration experiments in supramolecular chemistry. Chem Soc Rev 40(3):1305–1323. https://doi.org/10.1039/c0cs00062k

Tran CD, De Paoli Lacerda SH (2002) Determination of binding constants of cyclodextrins in room-temperature ionic liquids by near-infrared spectrometry. Anal Chem 74(20):5337–5341. https://doi.org/10.1021/ac020320w

Tutaj B, Kasprzyk A, Czapkiewicz J (2003) The spectral displacement technique for determining the binding constants of β-cyclodextrin – alkyltrimethylammonium inclusion complexes. J Incl Phenom Macrocycl Chem 47(3–4):133–136. https://doi.org/10.1023/B:JIPH.0000011783.40427.56

Ulatowski F, Dabrowa K, Bałakier T, Jurczak J (2016) Recognizing the limited applicability of job plots in studying host-guest interactions in supramolecular chemistry. J Org Chem 81(5):1746–1756. https://doi.org/10.1021/acs.joc.5b02909

Valente AJM, Söderman O (2014) The formation of host-guest complexes between surfactants and cyclodextrines. Adv Colloid Interf Sci 205:156–176. https://doi.org/10.1016/j.cis.2013.08.001

Wahl J, Furuishi T, Yonemochi E, Meinel L, Holzgrabe U (2017) Characterization of complexes between phenethylamine enantiomers and β-cyclodextrin derivatives by capillary electrophoresis-determination of binding constants and complex mobilities. Electrophoresis 38(8):1188–1200. https://doi.org/10.1002/elps.20160052

Wang ZX (1995) An exact mathematical expression for describing competitive binding of two different ligands to a protein molecule. FEBS Lett 360(2):111–114. https://doi.org/10.1016/0014-5793(95)00062-E

Wilson LD, Siddall SR, Verrall RE (1997) A spectral displacement study of the binding constants of cyclodextrin-hydrocarbon and -fluorocarbon surfactant inclusion complexes. Can J Chem 75(7):927–933. https://doi.org/10.1139/v97-111

Wintgens V, Amiel C (2004) New 4-amino-N-alkylphthalimides as fluorescence probes for β-cyclodextrin inclusion complexes and hydrophobic microdomains of amphiphilic systems. J Photochem Photobiol A 168(3):217–226. https://doi.org/10.1016/j.jphotochem.2004.06.002

Yang C, Liu L, Mu TW, Guo QX (2000) The performance of the Benesi-Hildebrand method in measuring the binding constants of the cyclodextrin complexation. Anal Sci 16(5):537–539. https://doi.org/10.2116/analsci.16.537

Zhukov A, Karlsson R (2007) Statistical aspects of van't Hoff analysis: a simulation study. J Mol Recognit 20(5):379–385. https://doi.org/10.1002/jmr.845

Zughul MB, Badwan AA (1998) SL2 type phase solubility diagrams, complex formation and chemical speciation of soluble species. J Incl Phenom Macrocycl Chem 31(3):243–264. https://doi.org/10.1023/A:1007965424219

The manufacturer's authorised representative in the EU is Springer Nature Customer Service Centre GmbH, Europaplatz 3, 69115 Heidelberg, Germany. If you have any concerns regarding our products, please contact ProductSafety@springernature.com

Printed and bound by CPI Group (UK) Ltd, Croydon, CR0 4YY
30/09/2026
02225435-0004